Speaking
Through the
Aspens

THE BASQUE SERIES

Speaking Through the Aspens

Basque Tree Carvings in California and Nevada

J. Mallea-Olaetxe

University of Nevada Press
Reno & Las Vegas

The Basque Series
Series Editor: William A. Douglass

University of Nevada Press, Reno, Nevada 89557 USA

Manufactured in the United States of America
Design by Omega Clay

Library of Congress Cataloging-in-Publication Data
Mallea-Olaetxe, J. (Joxe), 1940–
Speaking through the aspens : Basque tree carvings in California and Nevada / J. Mallea-Olaetxe.
p. cm. — (The Basque series)
Includes bibliographical references and index.
ISBN 0-87417-358-2 (hardcover)
1. Basque Americans—California—Social life and customs. 2. Basque Americans—Nevada—Social life and customs. 3. Shepherds—California—Social life and customs. 4. Shepherds—Nevada—Social life and customs. 5. Art, Basque—California. 6. Art, Basque—Nevada. 7. Wood-carving—California. 8. Wood-carving—Nevada. 9. Names carved on trees. I. Title. II. Series.
F870.B15 M35 2000
305.89′ 9920794—dc21 00-008553

The paper used in this book is a recycled stock made from 30 percent post-consumer waste materials, certified by FSC, and meets the requirements of American National Standard for Information Sciences—Permanence of Paper for Printed Library Materials, ANSI Z39.48-1992 (R 2002). Binding materials were selected for strength and durability.

University of Nevada Press Paperback Edition, 2008

This book has been reproduced as a digital reprint.

ISBN-13: 978-0-87417-762-6 (pbk.: alk. paper)

This book was funded in part by grants from the Nevada Humanities Committee, a state program of the National Endowment for the Humanities; The Program for Cultural Cooperation Between Spain's Ministry of Education and Culture and United States Universities; the Pacific Southwest Region of the USDA Forest Service; and the North American Basque Organization.

Frontispiece: Two herders' carvings in White Pine County, Nevada. *Top:* Written in Basque: "Biba Etchelar Belxa Airaco Poteros" (Hurrah for the dark one from Etchelar, Poteros from Aira), ca. late 1940s or early 1950s. *Bottom left:* The camera captured only part of the second message, written in broken Spanish: "E matau buen borrego para carne Isidro Sala 1924 y a 20 setienbre" (I have butchered a good lamb for meat, Isidro Sala 1924 on 20 September).

To the roaming ghosts
of alien sheepherders
enjoying the cool
under the aspens.

AUTHOR'S NOTE

The sheepherders' arborglyphs introduced here were carved for the exclusive consumption of fellow herders; none of the information they contain was intended to reach the wider society. It is with misgivings, and only after precautionary measures were taken to safeguard the identity of the carvers, that I have agreed to make these tree carvings public.

CONTENTS

ILLUSTRATIONS

ACKNOWLEDGMENTS

I am indebted to a host of people who in various ways have made this research possible. Many individuals—too many to be mentioned by name—called or wrote to share information about relatives who had been sheepherders and tree carvers. Others told me of aspen groves containing arborglyphs. I particularly thank every herder or ex-sheepherder who agreed to tell his story and acknowledge in particular the following individuals who shared their knowledge on the Basque sheepherding and/or tree carving experiences: Leandro Arbeloa, Al Arnold, Jess Arriaga, Santi Basterrechea, Joe Bengoechea, Pete Bengoechea, Terry Birk, Stephen Bishop, Bonnie Borda, Dana Borda, Eileen Borda, John Borda, Joe Ciscar, Buster Dufurrena, John L. Eguen, Joe Eguren, Albert Erreca, Jerry Etcheverry, Aulene Evans, Dan Foster, Fred Fulston, Albert Gallues, Hank Gallues, Dave Goicoechea, Jess Goicoechea, Mariana Goicoechea, Ana Hachquet, Mañiz Ithurralde, Erremon Jayo, Jeannette Landa, Robert Laxalt, Leon Legorburu, Jess Lopategui, Joe Mallea, Abel Mendeguia, Judy Mendeguia, Pete Mendiboure, Eugenia Mendiola, Emily Miller, Frenchie Montero, Mrs. Murphy, Andrew Nelson Jr., Candido Olano, Nikolas Olaziregi, Jesus Pedroarena, Richard Potashin, Eugenio Sarratea, Florencio Sarratea, Justo Sarria, John Skinner, William Smallwood, Charlie Teiner, Patxi Txurruka, Marcelino Ugalde, Josephine Urrutia, Jean Duque Wright, Claudio Yzaguirre, Jesusa Yzaguirre, Johnny Zabalegi, Joe Zubillaga.

I am indebted to many archaeologists and officials of the National Forest Service (NFS) of the U.S. Department of Agriculture and the Bureau of Land Management (BLM) in northern Nevada and California who provided grants, transportation, maps, film, and even horses. Their cooperation was essential to the gathering of the data.

The Nevada Division of Historic Preservation and Archaeology helped launch the Aspen Carving Project, and the late archaeologist Arnie Turner of Toiyabe National Forest in Sparks, Nevada, was its first supporter along with Ronald M. James. KNPB-TV Channel 5 of Reno and producer Curt Daniels helped spread the arborglyph news with coverage of the phenomenon. The generosity of the Nevada Centennial Commission enabled us to produce a video that is available to schools and libraries throughout the state.

However, the Basque government of Euskadi was the main supporter of the project. For several years I ate my beans thanks to Eusko Jaurlaritza's largess, and I want to say *milesker* (one thousand thanks) to them. Special thanks to Mari Karmen Garmendia—now minister of culture—for her support of the project.

The following Nevada archaeologists and historians deserve my special thanks and appreciation: Robert Neary, Regina Smith, and their coworkers of the Winnemucca BLM District; Fred Frampton of Humboldt-Toiyabe National Forest, Elko; Tim Murphy, Elko BLM District; Brian Amme, Ely BLM District; Paul DeMule, Humboldt-Toiyabe National Forest, Ely; archaeologists, rangers, and foresters of Toiyabe National Forest, Tonopah; the archaeology department of Toiyabe National Forest in Austin; Pat Barker and Dave Goicoechea, archaeologist and biologist of Reno BLM, respectively; Norman Murray and archaeologist Prill Mecham of Carson City BLM; archaeologist Terry Birk of Toiyabe National Forest, Carson City; and Susan Sturtevant, Nevada Parks Service, Carson City.

My gratitude goes far beyond the borders of Nevada. I would like to extend my thanks to neighbors as well: the people at the Tahoe National Forest in California, including archaeologists Richard Markley of Nevada City, Michael Baldrica of Sierraville, Carrie Smith and Juanita Spencer of Truckee, Hank Meals of Downieville, and all their coworkers. I appreciate archaeologists Penny Ruchs and John Maher of the Lake Tahoe Basin Management Unit for believing in the project and even occasionally getting lost in the rugged high Sierra; Wallace Woolfenden of Inyo National Forest; James Snyder, historian,

Yosemite National Park; Patricia deBunch, archaeologist, Nevada Department of Transportation; Harold Klieforth, Desert Research Institute, Reno; Cathy Sprowl, Plumas National Forest, California; John Kaiser, Frémont National Forest, Lakeview, Oregon; archaeologist Gerald Gates of Modoc National Forest, Alturas, California; Mike Butchart and Wayne Wilson, Okanagan National Forest, Washington; Pete Cenarrusa, Boise, Idaho; Susan Lindstrom, Truckee, California; David Grippo, Reno; Chris Montgomery, Grand Junction, Colorado; Gordon Ponting; Hershell Davies, Lake Tahoe Basin; Chris Boyer, Oregon; Milda Isenberg, Oregon; Isabel Espinoza, Carson City; Josie Griffith, Glendale, Arizona; Dave Thomas, Truckee, California; June Rivers, Reno; Marvin Kientz of Sonoma State University, Californa; Daniel Sendek, Board of Forestry and Fire Protection, Sacramento, California; Joe Gardella, Reno; W. L. Smallwood, Buhl, Idaho; John Lyons, Chico, California; Kirk and Annaliese Odencrantz, Reno; Catherine (Apalategui) Travaglia, Yorba Linda, California; Michael Tancheck, Society of American Foresters; Joseph Sirotnak, Great Basin National Park, Nevada; Scott Thomas, BLM, Burns, Oregon; Robert Olsen, Susanville, California; Tomas Uribetxeberria, minister of culture, Bizkaia; Shirley and Philip Altick, Reno; Anita Watson, Carson City; Cody Krenka, Elko, Nevada; Josefina Urrutia, Reno; Jan Douglass, Reno; the Txasio Family, Markina, Bizkaia; Chris Ross, Ball Canyon, California; Alan Buckley and Ann (Vergara) Prehn, Modesto, California; Felipa Errecart, California; Phil C. Weigand and Acelia Garcia de Weigand, Flagstaff, Arizona; Evelyn J. Welch, La Grange Museum, California; and Don Warrin, California State University, Hayward. And so many others, like E. Loren Kingdom of Greenville, California, who quietly and on his own went about the business of videotaping aspen carvings and then mailed the videos to me with a simple note: "I am sending you a videotape of the carvings I found in Plumas County." Thanks again, Mr. Kingdom; we need more people like you. I am grateful to the amateur archaeologists (AM-ARCS) of northern Nevada for "roughing it" in the wilderness and assisting me in the collection of the data as well as arranging it; in particular, Oyvind Frock and Dan Urriola of Reno.

I remain indebted to the Basque Studies Program of the University of Nevada at Reno, and to William A. Douglass in particular for his suggestions and revision of the manuscript. The Basque Studies Program gave support to Rafa Galdos, who helped launch the database, and to photographer Inaki Insausti. I cannot thank the friendly staff at the Basque Library enough for indulging me every time I needed to use its resources. I want to thank the Graduate School of the University of Nevada at Reno, and Kenneth W. Hunter for providing computer hardware and software. Stephen Foster of the University of Nevada Computing Services helped with the architecture of the database. Melinda Conner did a wonderful job of copyediting the manuscript.

Finally, I am grateful to my daughter, Nikane, who helped in taking pictures, and to my son, Erik, my number one assistant for years. He often gave up fishing for secretarial and photographic work, and he sometimes helped read the carvings. My heartfelt appreciation goes to my wife, Sara M. Vélez, for her many talents with computers, editorial work, and photography, but most of all for her relentless pursuit of the manuscript.

Speaking Through the Aspens

Arborglyph Areas in California and Nevada

(Map by Cameron Sutherland)

Introduction

I first saw aspen tree carvings in Elko County, Nevada, on a summer day in 1968. Jess Goicoechea, of the Goicoechea Sheep Company, had invited me to visit the company's *kanpo handia*, or main summer sheep camp, in the Jarbidge area. This was going to be my first journey into the wilderness of northern Elko County, and I eagerly accepted the invitation. I was told that we would travel about eighty miles, much of it on dirt roads, and I was advised to be ready early. I quickly learned that in sheepherder's time "early" meant 4:00 A.M.

It was still dark and rather chilly—Elko being almost a mile high in elevation—when we piled into Joe Madariaga's new Jeep Wagoneer. We drove north for an hour on the Boise highway, through sagebrush country the first fifty to sixty miles, later into lush meadows, and then climbing more sharply in tight curves flanked by aspen groves.[1] Once we were forced to stop and wait for a herd of deer to clear the roadway; we saw several groups of them.

The sun was out by the time we neared Coon Creek Summit (at 8,442 feet). Below, looking southward, we could admire a hundred miles of Nevada in all of its wild immensity. Not a single sign of human presence was visible. It certainly did not look anything like the Basque Country. Just before the summit we turned left onto a "jeep road" and drove downhill until there was no more road. The camp was there, surrounded by aspens, bushy alders, lush growth, and an array of flowers that I did not recognize. I did identify the huge nettles, however.

And I remember very well the cantankerous horse that I rode the next day. The camptender was scheduled to deliver provisions to a lone herder in a distant range, and I went along. Following the trail horizontally along a steep mountainside, I feasted my eyes on the spectacular scenery and lush greenery of the Copper Basin. It was not at all like the Nevada I knew, and the vista changed every time we rounded a fold in the terrain. Finally, after several hours of riding, we reached a remote sheep camp occupied by a lone Bizkaian sheepherder. I will never forget my first sight of the young man standing in front of his tiny triangular tent. What is a human doing here? I asked myself. He waved at us, smiling shyly. Of course he was expecting us. The dogs had warned him of our approach long before we arrived.

After the usual greetings in Basque and routine questions and answers, we unpacked and unloaded the all-important supplies. I noticed how carefully he put everything away, as if each article had its proper place on an invisible shelf inside the tent. Next came lunch, because we were rather hungry from the long trail. The herder cooked with the adroitness of a chef. The ever-present stew was already steaming, and in a just few minutes he cooked the main course: lamb chops with potatoes and green peppers fried in a pool of oil at the bottom of a huge Dutch oven. Sitting on the grassy bank we took hearty bites between swats at the pestering flies. We washed down the chunks of meat and huge slabs of fresh sheepherder's bread with wine from the *zato*, or wine bag. We even had dessert, a pie baked by Mariana (Lugea) Goicoechea, Jess's wife. It was one of the most satisfying meals I can remember eating, and I think the environment had a lot to do with it. Sitting there on that isolated hillside, I felt that all was right with the world. In fact, the world seemed light-years away.

Meanwhile, I was astounded by the attention of the dogs. While keeping the distance behooving such highly sophisticated creatures, they seemed to keep track of our conversation. Whenever we laughed a little louder, their ears perked up and they opened their jaws, letting their tongues hang extra low and breathing a little faster, as if smiling and approving of the good time we were having.

In the course of the meal I told the herder that his family in Bizkaia would not believe that he lived in such isolation, so far away from everything. "This is so remote. What if something happens to you?" I asked.

He looked at me, half smiling and half serious, and blurted out in Basque, "Remote, you say? God has yet to arrive here."

1. Goicoechea Sheep Camp at Copper Basin, Jarbidge, summer of 1968. From the left: Jess Goicoechea, the author, Juan "Antzamendi," "Gontze," Luis Jayo, Joe Madariaga, Edorta Arrate, and Juan Juaristi.

I laughed, but the two men sitting opposite me did not, so I decided that it must be an old joke among the herders. We laughed a lot that day. Jokes came easily, but time passed very quickly and we still had the long ride back to the Goicoechea camp. Just before we mounted, the herder insisted that we take one final swig of wine—that's what a good host does—so we did.

As we crossed several aspen groves on the way back, I came face-to-face with the carvings on the trees. They were impossible to miss, really, because at times the horses took us inches away from the carved trunks. At first I did not know what I was seeing, so I asked the camptender about them. He shrugged and chuckled, but I got the impression that he wasn't very interested in the carvings, or thought I wouldn't be. Perhaps it was his way of telling me that arborglyphs were not important to a sheepherder's work, but that is my reaction thirty years later.

I do remember that most of the carvings were surnames; a few were dated in the 1940s, others were from the 1950s and early 1960s. The camptender pointed out several trees carved prominently with crude human figures. I recall one particular tree that was like a billboard; it hailed the prostitutes in Elko as "the world's best." I was amused at first and then intrigued, but something did not make sense. Why would the herders carve messages—obviously meant to be read—on these trees? Who would ever come here to read them? That turned out to be an important question.

The horses continued their paced march and took us away from the aspen grove, and eventually I forgot all about the carvings.[2] I would not have forgotten them if aspen art had been a common topic of conversation among the herders, but it was and is not. I lived among Basques, many of them ex-sheepherders, and I frequently asked them questions about their lives on the range, but the tree carvings were never mentioned.

Almost twenty years went by before I rediscovered tree carvings on Peavine Mountain, north of Reno, Nevada, and decided to study them. By then I had read a number of articles about the arborglyphs, and at first I wondered if I was merely "discovering the Mediterranean." Was there really something more about arborglyphs than these short articles explained? Convinced that this sea contained uncharted waters, I proceeded. A number of archaeologists and forest managers expressed interest in the research; others were skeptical. However, the project caught the imagination of granting agencies on two continents, and from 1989 to 1997 the research was supported with more than a dozen grants. In the course of that same period I have seen approximately twenty thousand arborglyphs and have recorded and cataloged almost twelve thousand of them in a computer database. They represent a lot of hours in the mountains and even more at the office, but when I compare them with what remains to be done, I realize that I have merely scratched the surface.

The Literature on Arborglyphs

As far as I know, previous researchers who published articles on aspen carvings in America were all non-Basques attracted mainly, if not exclusively, by the carvings' artistic value. The chief reason for this, I believe, is the language barrier. The figures need no translation, of course, but the carved words are not in English. As a result, much of the large body of historical and cultural information inscribed on trees throughout the American West was unavailable to writers describing the life of the Basque sheepherders. Until now.

The first nonsheepherders to become interested in the arborglyphs were outdoorsmen and hikers. Generally their published articles are based on information gathered during weekend hikes and photographs of the figures that caught their fancy. Some were impressed by the "Picassoesque" quality of the tree carvings and published short articles in newspapers and magazines for general consumption. I have found more than forty such articles,[3] most of them published in the last ten years.

A few of the articles, however, are more substantial. In 1964 J. S. Holliday published a paper about the sheepherder's life that calls the arborglyphs "doodlings." Paul Hassel of San Francisco, California, began taking photographs of aspen carvings as early as 1950, and eleven of these are included in Holliday's article, which is also graced with several photographs of sheep and sheepherders by Ansel Adams. Hassel first thought the phenomenon was unique to the Colorado-Wyoming high country,[4] but later he learned that it was present in other parts of the West as well. In his view the arborglyphs represent lonely men's thoughts of women and whiskey.[5]

Philip I. Earl of Reno and his wife, Jean, have dedicated considerable time and effort to finding and preserving aspen art. Philip Earl has published several articles on arborglyph art, and he covers the topic in his lectures on Nevada history at Truckee Meadows Community College.[6] The Earls, as well as Frances Wallace and Hans Reiss, made numerous rubbings of the carvings—the eventual total was seventy-eight—and these have been exhibited in Reno. In 1965, Robert Laxalt, founder and director of the University of Nevada Press, acquired twenty-seven of Wallace and Reiss's alluring drawings of tree carvings.[7]

Among the earliest scientific research on aspen carvings was that conducted in Utah in 1969 by Jan Harold Brunvand and John C. Abramson.[8] The two transcribed about a dozen messages, most carved by non-Basques and nonherders. As the title of their article, "Aspen Tree Doodlings," indicates, they viewed the phenomenon as simply a folksy art and paid no attention to its historicity. Females were the most common human figures in the Wasatch Mountain arborglyphs examined by Brunvand and Abramson, and the horse was the favorite animal theme. The authors labeled their study "preliminary" research, but unfortunately they do not seem to have pursued the matter any further.[9]

Among the arborglyph themes mentioned by Brunvand and Abramson are names and dates (90 percent), love messages, hearts, obscenities, personal initials, and animal figures. The hearts and love messages were probably put there by American visitors, not by herders, but the rest of the themes the authors identified were confirmed by subsequent researchers.

In 1970 James B. DeKorne published the only book extant on arborglyphs. In this case the term *book* is deceiving; the text is only four pages long—considerably shorter than Brunvald and Abramson's article—and libraries cataloged the book under the subject "graffiti." The book includes sixty-one samples of aspen art found by DeKorne in the Carson National Forest of New Mexico in 1969. Most of the carvings are names and dates followed by figures of animals and women. Although female nudes abound in DeKorne's samples, they are mostly modern, and he says "pornographic" material is rare.[10] However, a forest archaeologist from New Mexico who has seen the groves photographed by DeKorne told me that carvings with sexual content are, in fact, common.[11]

Richard Lane spent two years researching sheepherding activities in northeastern Nevada, in the process visiting several aspen groves in three mountain ranges. Lane's research resulted in a seven-page article half filled with photographs that was published in 1971. He, too, identified the topics already mentioned, but he also found carvings that featured ethnic identity, towns, provinces, humans, buildings, stars, and whores. Lane could offer no assurance that the arborglyphs were carved by herders. He found no figures of sheep and nothing carved in French, and he noted that some of the names were Spanish. (In fact, Hegoalde [southern] Basques almost always gave their names in the official, Spanish, version. There are only a handful of exceptions.) Lane found the carvings to be "only a small part of the necessary evidence [of the sheepherder's life], but a very enjoyable part." However, in his dissertation on sheepherding he used no data from the trees.[12]

I have videotaped most of the groves Lane visited, as well as many others in northeastern Nevada, and I have found both figures of sheep and carvings in French, although not many. I think Lane was mistaken in saying that we cannot be sure that the arborglyphs are the work of herders. Basque names carved in sheep ranges must be assumed to be sheepherders' names. Could "Miguel Arizcun, 1923" or "Celestino Garamendi, 1943" have been carved by anyone but a herder? Cowboys and sheepmen did not normally share ranges, but even if they had, an American cowboy would not carve "Etcheberry" or "Zabala." Cowboys did sometimes wander in aspen areas where they could have carved their names and dates, but extremely few did.

Bureau of Land Management (BLM) and National Forest Service (NFS) archaeologists from New Mexico and Colorado to California may have gathered data in many of their districts, but little of it has been made public. There are exceptions; BLM archaeologist Roberta McGonagle of Battle Mountain, Nevada, studied the arborglyphs in the Bates Mountain area and kept a database of the recordings, and even wrote several articles, one of which appeared in the *Elko Free Press*. But more about her work later.

David Beesley and Michael Claytor have written more extensively and in greater detail on arborglyphs than anyone else. They conducted research in thirty aspen groves in Placer, Sierra, and Nevada Counties of eastern California.[13] Their articles provide few details and quotes of the inscriptions, but the authors correctly recognized six primary categories or themes, most of them already identified by others. Among the new ones they recognized were European themes, personal statements, and fantastic figures.

Beesley and Claytor seem to have been unaware of the cultural value of the messages in the arborglyphs, as the following statement suggests: "In comparison with the imaginatively carved figures the names and the statements are a lower form of art." This is the major disparity between their findings and my own. It is my view that the statements are not art at all, but history and literature. Beesley and Claytor also wrote that the "statements by carvers reflect an incredible range of emotions," but unfortunately they did not elaborate on that interesting theme.[14]

There are a number of other discrepancies—some major, most minor—between Beesley and Claytor's conclusions and mine. Certainly some of the differences are regional; Beesley and Claytor studied a relatively small area of the northern Sierra Nevada in California; I worked in a larger area in Nevada and California. Among the most important discrepancies: I found "sheepherding news" in every grove I visited, but Beesley and Claytor hardly mentioned such messages. Beesley and Claytor found bear figures in California, but I do not recall finding any in Nevada (there are few bears in the state, anyway). They also found carvings featuring sheep or lambs, while I found these to be rare in the many groves I visited. On the other hand, I found figures of fish and coyotes in the Silver State, but these did not occur in carvings on the California side of the Sierra.[15]

Another important discrepancy: Beesley and Claytor found references to France to be less frequent than references to Spain. I found the opposite to be true. In the Sierra as a whole, French Basques outnumbered Spanish Basques, and so do their carvings. The word *France* is spelled different ways, however, and is sometimes in the Basque form "Frantzia" or "Franzia," which can be mistaken for a female name. Another consideration is the overwhelming distances and the remoteness of the high Sierra. It is possible to think one has covered a huge area when in fact one has been recording arborglyphs carved by the herders employed by a single sheep company. In the late 1970s there were at least five sheep companies operating in Tahoe National Forest: Blackfoot, Carter, Mendeguia, Johnson, and McPherrin. The public land allotments of some of these sheep outfits were enormous. The center of Beesley and Claytor's recordings apparently was Tahoe National Forest, where I recorded many of the same groves from 1989 to 1994. Incidentally, I failed to find a few of their carvings, an indication that these aspens no longer stood.

In order to arrive at a balanced account of the arborglyphs, a researcher must study the carvings in mountain ranges used by at least two different sheep companies, because normally the regional background of the herders differed from outfit to outfit. One might recruit herders predominantly from the Aibar region of Nafarroa while another recruited from the Valley of Baztan or Benafarroa. Those from Aibar spoke mostly Spanish, those from Baztan spoke mostly Euskara (the Basque language), and those from Benafarroa spoke hardly anything but Euskara; logically, the legacies of these carvers could be different. Beesley and Claytor may have been aware of the limitations of their research. "What is needed," they wrote, "is a comparative study of Basque tree art from the major areas of Basque activity in the American West."[16] I believe my research makes a good beginning at filling that void.

Beesley and Claytor did not associate artifactual evidence of sheepherders such as corrals, fences, chutes, tables, campsites, and ovens in the Sierra with intensely carved areas. Further, they wrote, "the Basque language and the distinctive artistic motifs from their cultural heritage are used in many of their Sierra aspen carvings," citing swastikas, stars, and crosses.[17] Of the motifs Beesley and Claytor listed, those uniquely Basque are few; the swastikas can probably be attributed to American carvers. Unfortunately, Beesley and Claytor failed to provide hard data in their analysis. They transcribed only a few of the messages, not even the names and dates. Finally, they did not mention seeing any *harri mutil*,[18] or stone piles or markers, and I did not see any either in that area of the Sierra.

It is likely that a great many carvings have been collected or photographed by archaeologists and hikers all over the West, but this information remains in private hands and is thus inaccessible.[19] In general, it seems safe to say that previous writers on the subject of arborglyphs were seduced by the unconventional characteristics of the tree carvings—more specifically, by their appearance and by the fact that they were carved on trees. It is not my intention to underestimate their work or the artistic aspects of the carvings, but the evidence I present here will establish that arborglyphs are far more than doodlings; they are a highly visual historical resource.

Interest in Arborglyph Research

In 1989 archaeologist Arnie Turner of Toiyabe National Forest, Sparks, Nevada, became the first to fund a challenge grant to begin the systematic recording of the arborglyphs. The research was bolstered by a second grant from the Nevada Office of Historic Preservation and Archaeology in 1990.

Curt Daniels, formerly of Reno's public television station (KNPB–Channel 5), produced a program on Basque aspen carvings in February 1990 as part of the *Nevada Experience* series. It has been broadcast a good number of times since then. About the same time, the 125th Anniversary Commission of the State of Nevada sponsored the production of a longer video on arborglyphs, which was finished in 1992.

In the late 1980s and early 1990s European Basques became aware of the widespread phenomenon created by their countrymen in the United States, and before long the media became involved. Euskal Telebista (Basque television) sent a crew to document the lives of Basque Americans today, and the effort resulted in six documentaries. One of the segments included a visit to the aspen groves that was shown on Basque television in 1992.[20] The interest generated was considerable and resulted in a flurry of publications on both sides of the Pyrenees.

Euskal Telebista returned to Nevada in July 1999 on the occasion of the Renoko Aste Nagusia (Basque Cultural Week and North American Basque Organization [NABO] Convention, 19–25 July) and took the opportunity to visit another aspen grove. This time, however, two young female *bertsolariak* (improvisers of poetry), Estitxu Arozena and Oihane Enbeita, were given the opportunity to read the statements and observe the female figures carved on the aspens. It was a moving sight—a scene, almost, from another time. Could the sheepherders working there in the 1920s and 1930s have imagined that seventy years later real Basque women would visit their old camps and view the etchings of their dreams?

Arozena and Enbeita did more than admire the carvings. On the spot they improvised verses reflecting their reactions and feelings. In particular they were impressed with a woman depicted on several trees. The ideas of these two young women were as insightful and fresh as the morning dew they were walking on:

ESTITXU AROZENA:
Garai bateko artzai tokian
hasiko gara kantari
baso honetan buelta bat eman
dugu biok adi adi

Triniri buruz zuhaitz hauetan
bada marrazki ugari
irtebide bat eman nahiean
heuren bakardadeari.

In this place, which once was a sheepherder area
We will begin to sing [poetry]
We went around this grove
the two of us observing carefully
On these trees there are
a lot of carvings of Trini
They [sheepherders] were trying to deal
with their loneliness.

OIHANE ENBEITA:
Hiru esne tanta agertzen dira
bular batetik erortzen
zakil bat ere jarria dago
Triniren aurrean zuzen
prostitutekin ibiltzen ziren
bakardadea uxatzen
batzuentzako pornografia
heurentzako beharra zen.

Three drops of milk can be seen
falling from a breast
A penis, too, is placed
straight in front of Trini
The sheepherders had dealings with prostitutes
That which for some is pornography,
for them, it was a necessity.

Next the two women found an aspen on which "Gora Euzkadi" (Long live Free Basque Country) had been inscribed, and this is what they composed:

AROZENA:
Baso honetan pasatu zuten
negu luze ta gogorrik
nahiz ta etxetik urrun egon
bada oroimen xamurrik
hutsik ez dago erran beharrik
gure historia ez dela egin
Euskal Herrian bakarrik.

In this forest they endured
long and harsh winters[21]
Even though they were far from home
one finds nostalgic memories here.
Needless to say
our history has not been made
only in the Basque Country.

ENBEITA:
Zuhaitz honetan elkartu gara
denak behatzen geundela
ikusi dugu Gora Euskadi
bertan idatzi zutela
fenomenoa uler daiteke
metafora bat bezela
euskal sustraiak Euskal Herritik
aparte ere badaudela.

We have converged around this tree
All of us were looking and
we saw that they carved
Long live Free Basque Country on it
The phenomenon can be understood
like a metaphor
that Basque roots can be found
even outside the Basque Country.

The regional museum of Bizkaia in Bilbao sponsored exhibits of photographs of the tree carvings. In the town of Baigorri, Benafarroa, which sent hundreds of sheepherders to America, the Izpegi Committee displayed a photographic exhibit on arborglyphs, many carved by native sons. In addition, two real carved tree trunks were sent to the museum in Bilbao.

Archaeologists who work for the National Forest Service and the Bureau of Land Management have become increasingly aware of the cultural value of the arborglyphs under their custody. So has the Nevada Office of Historic Preservation and Archaeology (SHPO), and in January 1995, its sister agency in Sacramento and Richard Markley, Tahoe National Forest (NF) archaeologist, invited me to give a presentation on my research to state and federal archaeologists, foresters, and resource managers.

To date, the Basque government of Euskadi has contributed the most substantial monies to the arborglyph study, funding the project for three years through two of its agencies: the Amerika eta Euskaldunak Commission, which supported the drafting of the original manuscript of this book in the Basque language,[22] and the Kanporako Idazkaritza (Foreign Affairs).

In 1992 Tahoe National Forest entered into a cooperative agreement with the University of Nevada, Reno, and myself to systematically gather and interpret Basque history in the Sierra Nevada.[23] Several projects grew out of the agreement with Tahoe NF:

1. The rebuilding of the Wheeler Sheep Camp oven at Kyburz Flat, nine miles south of Sierraville, California. Archaeologist Michael Baldrica was instrumental in the publication of a brochure on the Kyburz Flat Interpretative Area in 1994.
2. The rebuilding of the Whiskey Creek Sheep Camp, of which more later.
3. Various presentations and Basque history exhibits have been mounted in the Sierra, often spiced by UNR's talented Zenbat Gara Basque Dancers, as during California's Scenic Byways Celebrations of 1993.
4. The main scope of the agreement with Tahoe NF was the recording and videotaping of the arborglyphs in the Truckee, Sierraville, and Downieville Districts. This work continued into 1994.

I enjoyed similar support from the archaeology department of Lake Tahoe Basin Management Unit of South Lake Tahoe, California, which made it possible to record thousands of arborglyphs in the mountains around Lake Tahoe. My fieldwork included videotaping the old South Camp on Genoa Peak, where the ancestors of such well-known Nevada families as the Bordas, Laxalts, Uharts, Amestoys, and others worked sheep from the early decades of the century into the 1940s.

Finally, the general public in northern Nevada has shown interest in hiking the mountains and groves in search of the arborglyphs. To satisfy their curiosity I have offered dozens of slide and video presentations, field trips through the community colleges in Carson City and in Reno, plus other private excursions to aspen groves in the company of local cultural groups.

The Basques of Europe

Who are the Basques?[24] The Basques call themselves Euskaldunak, which is usually translated as "the possessors/speakers of the language of Eusk."[25] The Basque people define themselves by the language they speak; thus they call their country Euskal Herria, "the land of the speakers of Basque." This is a cultural identification rather than the political one that for centuries has been the norm for European countries. This system of identification goes back thousands of years to Neolithic times, and perhaps even further, when language determined the divisions among peoples.

Euskal Herria (the Basque Country) is a small country—slightly larger than Clark County, Nevada—located on both sides of the western Pyrenees Mountains. It is home to 2.8 million people, of whom 91 percent live in Spain and the rest in France.

There are uncanny parallels between the Basques and Native American populations who are now striving to survive under the overwhelming presence of European "newcomers." Thousands of years ago, the Basques and neighboring peoples speaking similar tongues may have suffered the same fate at the hands of Semitic and Indo-European peoples (Phoenicians, Greeks, Celts, Romans, Barbarians, etc.) who immigrated into western Europe. Just as Indian languages in the Americas today are besieged by English, Spanish, and Portuguese, the Indo-European languages of the newcomers wiped out almost all of the earlier languages in western Europe. Euskara is the lone survivor, an isolated language that today struggles to survive in the western foothills of the Pyrenees.[26]

Many consider Euskara proof that the Basques are direct descendants of the inhabitants of primitive Europe (*before* it was called Europe), perhaps even of those responsible for the prehistoric cave paintings that abound in and around Euskal Herria.[27] An acclaimed study on human genes published in 1994 suggests that the Basques are "direct descendants of the upper Paleolithic of the Cro-Magnon types," and that the first modern humans arriving in Europe spoke "languages of this family."[28] An article published by *National Geographic* in 1995 calls the Basques "Europe's First Family."[29] Blood type offers further proof of the Basque people's uniqueness: they have the highest incidence of Rh-negative blood in the world and hardly any B-type blood.

Basques in the Americas

The Basque Country has been a land of emigrants at least since Roman times. There are two important reasons for this: overpopulation and the fact that, by tradition, the family estate is passed on to only one heir. Emigration continued throughout the Middle Ages, and many younger sons found employment in the kingdom of Castile, whether as colonists in the lands recovered from the

Moors or as mercenaries, secretaries, and pages of the kings and the aristocracy. Northern Basques were the backbone of the armies of the dukes of Aquitaine and Vasconia.

The sea was another important outlet for the excess population. The Basques were the first commercial whalers in Europe; by the early sixteenth century they were also marketing large quantities of salt-cured codfish caught on the banks of Newfoundland, a fishing ground so profitable that, according to author Mark Kurlansky, the Basques kept it a secret.[30] Fishing stimulated ship building and mariner's skills, both in great demand by the Habsburg kings as well as the French monarchy during the age of exploration and empire building in America. Sailors played fundamental roles in the expansion of the Spanish colonies, but contemporary authors and historians alike found little appeal in writing about them. Furthermore, the Basques were a minority; they never amounted to more than 4–5 percent of the colonists in Spanish America, although they constituted some 18 percent of the officials, according to one estimate.[31] The discrepancy can be explained by the fact that they all claimed to be of noble status, a prerequisite to holding public office in the Spanish Empire.

A number of influential personalities of northern Mexico and the Southwest, including Juan Bautista de Anza, who collaborated in the settling of San Francisco in 1776, were of Basque descent. Juan de Oñate, in his 1598 expedition to New Mexico, introduced 3,000 sheep as well as other livestock, seeds, and 130 families of colonists; and in 1609 he founded Santa Fe, the oldest European city in the West.[32] This early association of Basques and sheep is interesting. Under Governor Borica (1794–1800) the number of sheep in Alta California increased from 48,919 to 78,928 between 1796 and 1799.[33] Borica was one of six officials of Basque ancestry who governed Spanish Alta California; others included Jose Joaquin Arrillaga and Manuel Micheltorena, the last governor officially appointed by Mexico.[34] Several of the leading explorers of the Pacific coast and such missionaries as Fermin Lasuen, who founded nine missions in California, came from the tiny Basque homeland.[35] During the colonial period many Basques were employed as scribes and secretaries, accountants, mariners, missionaries, and explorers, a far cry from the low rank Basque sheepherders would one day hold in the U.S. West.

In the last several centuries history writing in the Western world has revolved around the dominant figure of the monarch and the notion of the state. Mainstream historians tend to aggrandize the supreme power in capital cities such as Madrid, Paris, and London and to ignore ethnic and cultural groups. Under this system, the Basques, who seldom wrote about themselves, were simply represented as Spaniards or French. This, naturally, has affected the perceptions of students of history.

The earliest modern Basque immigrants to the United States came with the hope of striking it rich quick in the California bonanza of 1848–49. Most Basques knew little about mining, although iron mines had been worked in the Basque Country since at least Roman times. Basque technology was used in the mines of Mexico in the sixteenth century, and later in the mines of California and Nevada.[36] The little we know of the early Basques in the American West comes from local newspapers; because the papers were oriented toward the English-speaking population, however, foreigners received less exposure. Initially, there were Basques working the mines in California, Nevada, Idaho, and (in the 1910s) even as far east as Pennsylvania. A few gained notoriety, among them Tomas Alcorta, discoverer of the Cordero mercury mine in McDermitt in north-central Nevada, and Pete Aguerreberry, who for years prospected for gold in the Mojave Desert.[37]

Sheepherding in California and Nevada

In the wake of the California gold rush, and for the next one hundred years, waves of young Basque males arrived in the United States to herd sheep and "to see the elephant," as the expression went among forty-niners, or would-be ones.[38] When I was growing up in Bizkaia, the names *Galipornia* (California) and *Idao* (Idaho) cropped up in daily conversation. Like many people in Bizkaia, I thought they were two countries somewhere in America, or, more accurately, that America was Idao and Galipornia, as well as Argentina.[39]

Robert Laxalt, considered by many to be the spokesman of the Basques in the United States, characterized the sheepherders as "Lonely Sentinels of the West."[40] William A. Douglass chronicled their "Lonely Lives under the Big Sky,"[41] and Michele Strutin characterized Basques as "Lords of the Range."[42]

The Basques, generally speaking, made the transition from mining to cattle and sheepherding rather quickly. Most Basques were more comfortable with sheep and cows than with a pick and a pan, and putting their expertise in animal husbandry to work, they soon began to excel in the sheep industry, dominating in southern California and then expanding their operations into the mining camps in the north. As a result, sheep production in the 1860s surpassed that of cattle.[43] According to David Beesley and Michael Claytor, sheep were raised mostly for the mining camps until the 1890s, when mining declined in the Sierra.[44]

The yearly sheep runs to the markets were a centuries-old practice borrowed from Mexico.[45] During the colonial period and even later, sheep wandered unmolested in sparsely populated northern Mexico, although traditional sheep trails existed in many places. The Santa Fe Trail, which ran east–west from New Mexico to southern California through northern Arizona, was also used by the earliest sheep runs of the gold rush era. Early on, the Los Angeles area had a number of landowning Basques with interests in dairy and cattle. In the 1880s this dynamic group supported two newspapers printed in Euskara and several hotels and restaurants.[46]

The demand for meat increased as California and the West became more populated. By all accounts, the coming of the railroad in 1869 greatly increased the profitability of sheep because wool could be shipped to Boston and other points east. In 1870 a severe drought in California forced some herders to trail their sheep into the eastern foothills of the Sierra Nevada.[47] Sketchy documentation suggests that by the 1870s a few Basques had crossed these mountains into Nevada and beyond. Among the better-known early figures in the area covered by my research were the Garats, the Indarts, and the Altube brothers.[48] By 1875 Jose Manuel Ugarriza and a few other Bizkaians had settled in northern Humboldt County.[49] The Yparraguirre brothers bought a wayside station and ranch at the foot of Nevada's Sweetwater Mountains in 1886–87.[50] The ranch house and buildings are still there, east of Bridgeport, California, on the Smith Valley highway, but the place looked abandoned when I recorded it in 1992.

Jean Baptiste Garat came to San Francisco in 1849, and Jean Pierre Indart (or Yndart) came two or three years later; both were usually called Juan.[51] They became cattlemen rather than sheepmen. Garat, Yndart, and John Arrambide were relatives and partners, and after selling their California holdings to the Miller and Lux outfit, they moved to Nevada, where in 1870 or soon afterward Garat founded the Allied Cattle and Livestock Company in northern Elko County. Their brand, YP (for Pete Yndart), is believed to be the third-oldest brand in the United States. It has been in use since 1852 and is the only brand registered in the U.S. Patent Office.[52]

2. The Sweetwater Ranch, once owned by the Yparraguirre brothers, in the late 1990s.

Pedro Altube, founder of the Spanish Ranch in Elko County, is perhaps the best-known early Basque figure of Nevada. He and his brother Bernardo hired many countrymen to help run their huge empire. Although no longer owned by the Altubes, reportedly Spanish Ranch is still the largest ranching operation in Nevada, and some of the original buildings still stand in Independence Valley, near Tuscarora.[53]

The impact of the Basques in northern Elko County, Nevada, is still attested by the old Saval Ranch in the eastern foothills of the Independence Mountains,[54] and the Holland Ranch, owned by the Goicoecheas, in the same general area. Forty miles away, ten miles south of Jarbidge, I discovered a significant carving on a downed aspen that may be related to these Basque pioneers of northern Nevada: "1901 St Ramon Garat 1901."[55]

3. Salvaging information while it can be read. This large downed tree yielded the following data: "1901 St Ramon Garat 1901" and "Estanislao Ederra año."

After the turn of the century Basques immigrated in increased numbers. Most took jobs in the sheep industry, which peaked in 1900–10. The transcontinental railroad had a decisive impact on Basque immigration in the United States. The land trip from New York to California was much shorter than the trip by boat, which involved sailing around Tierra de Fuego or crossing over via Panama.[56] There were other, less traditional routes as well, such as via Mexico or by jumping ship at New Orleans, Galveston, San Francisco, and even Port Angeles, Washington.

Not all of these Basques were herders. A small minority involved themselves in other occupations that provided the necessary social services for the Basque community. Perhaps the most important of these was the *ostatu*, a combination boardinghouse and restaurant that existed in almost every town where a number of Basques were present.[57]

Rich Public Lands

Coming as they did from a tiny, land-starved country, the Basques were quick to notice the vast expanses of unused public land in the western United States. Land in the Basque Country had been a source of wealth, but it was scarce; they could hardly believe all the public land that was available to them in the West. The high mountains in the Basque Country were and are common land open to all to exploit, and the immigrants expected to do the same in the West. Those arriving via the Argentinean Pampas, such as Altube, were especially prepared to run large herds of sheep in a seminomadic way.[58]

Although the western United States scarcely resembled the green Pyrenees, the Basques grew to love the new country and its spirit. Many herders, however, have told me that their first impression of the land was depressing. "What sort of lambs can we grow here?" one asked himself as the bus sped east from Fallon, Nevada, on U.S. 50.[59]

He and many others at first believed that the land was unsuitable to sheep raising. As it turned out, however, the Nevada grasses, even in desert areas, were even more nutritious than those in the rain-soaked fields of the Basque Country. Unflattering messages carved on trees indicate that a number of herders never overcame their negative initial view of the rangeland pastures of Nevada. But I suspect that many of these were carved in moments of depression and gloom. If Basques had not been optimistic about their prospects in America, fewer of them would have stayed or gone into the sheep business.

It is important to realize, too, that the Sierra looked different then. The extensive logging after the 1859 Comstock strike in Virginia City, Nevada, left an open range filled with grasses. Today, the reforested Sierra cannot feed the millions of sheep it once did. More important, when the sheepherders first arrived, the land was sparsely settled and the herders could lead their charges fairly unmolested wherever the feed was most promising. The Basques greedily used the bountiful range to fatten their lambs and turn a fast profit. At that time the United States was still the Promised Land.

The era of unregulated range was first affected in 1905 by the creation of the U.S. Forest Service, which took over the management of the high country. The unsupervised exploitation of the rest of public lands came to a halt in 1935 when Congress passed the Taylor Grazing Act, which, according to one observer, "was as though an atomic bomb landed on the sheep business and disintegrated it altogether."[60] These dates were momentous ones for many Basques, not only because sheepherding would never again be the same, but also because, as far as the herders were concerned, America ceased to be "Amerika," a land of quick fortunes.

The Basques and the Carving Phenomenon

> Por aqui paso el adelantado Juan de Onate . . . 16 de abril de 1605. (The adelantado Juan de Oñate went through here . . . 16 April 1605.)
>
> Soy el Capitan General . . . de buelta de los pueblos de Zuñi a los 29 de Julio de 1620. (I am the Captain General [Eulate, governor of Nuevo Mexico] . . . returning from the pueblos of Zuñi on 29 July 1620, and put them at peace . . . all of which he did, with clemency . . . as a most Christian-like [gentleman].)[61]
>
> El capitan Juan de Uribarri y Juan de La Rivas, 1709. (Captain Juan de Uribarri and Juan de La Rivas, 1709.)
>
> Dia 28 de Sep[tiembre] de 1737 llego aqui . . . Martin Elizacoechea, obispo de Durango. (On the 28th day of September . . . arrived here the Bishop of Durango, Martin Elizacoechea.)
>
> Juan Ignacio Arrasain, 1737.[62]

The above messages, recorded on sandstone rock at El Morro National Park in New Mexico, are among the earliest carved documents left by people of Basque descent in the United States.[63] Oñate, the first governor of New Mexico, was a criollo, the son of Catalina Salazar and Cristobal Oñate, conquistador and governor of Nueva Galicia in Mexico. Dr. Martin Elizacoechea was from Nafarroa and headed the diocese of Durango (Mexico), a city founded by the Basque Francisco Ibarra—a native of Durango, Bizkaia—explorer and first governor of the kingdom of Nueva Vizcaya in northern Mexico.[64]

The fact that these officials carved personal messages on rock is unusual within the context of Spanish American colonial history; explorers had scribes at their service to record their adventures and exploits on paper. What prompted such behavior? Did they intend to impress the Indians? Were they perhaps imitating the petroglyphs clearly visible on many rocks in the Southwest? Was it another way of claiming the land? This was normally done by planting the royal banner and the cross on the ground and notarizing the act—it goes without saying that the purpose of Oñate's expedition was precisely to claim the land for the Spanish king. Or did the sandstone carvings have something to do with the explorers' Basque culture? The records left to us are not very helpful in answering these questions. All that can be said is that Basques carved more than other groups.

Which Trees Were Carved?

In Nevada and the West, when we say "tree carving" we usually mean aspen carving. In fact, the herders carved quaking aspens almost exclusively.[65] Aspens have the widest range of any tree species in North America; they are found from Alaska to eastern Canada and as far south as Mexico.[66] In the West they commonly grow at elevations above five thousand feet. Idaho, Colorado, and Montana have some of the largest aspen forests; Nevada has the smallest. Aspens propagate from one root, and each tree in a grove is thus a new shoot of the original one; thus, we might say that each aspen grove is a very old "tree" indeed.

All across the United States, but especially in the West, aspen groves are dwindling at an alarming rate, crowded out by conifers and victim to wildlife, especially beavers, which find the new saplings very tasty.[67] Forest managers believe the decline has to do primarily with fire suppression. Aspen regeneration projects have been undertaken in a number of national forest districts,[68] including one in Tahoe National Forest, Sierraville District of California, specifically aimed at protecting culturally sensitive aspen groves.[69]

The life span of an aspen is quite variable, but on average the individual trees survive 60 to 80 years; the oldest on record was a thirty-nine-foot-tall specimen in California's White Mountains that was 226 years old. Ordinarily, aspens grow to between sixty and ninety feet, even taller if the conditions are right.[70] I have seen a number of giant aspens in Nevada and California that must be well over 150 years old. They usually grow in areas with adequate moisture but too dry for beavers, and protected from the wind as well as the chain saw.

Sheepherders chose smooth, mature aspens for carving.[71] Only in the absence of aspens were other types of trees, such as cottonwoods, pines, and alders, inscribed.[72] According to historian James Snyder, sheepherders and others in Yosemite National Park carved primarily lodgepole pines because there were few aspens there.[73] John L. Eguen of Reno told me that he and other herders in the California valleys carved eucalyptus trees.[74] I do not know if herders in their winter quarters in the Colorado River valley of Arizona and California (the Parker and Blyth areas, for example) carved mesquite and other local trees. I suppose cottonwoods and other varieties of trees in the American West could have been carved as well.

In treeless areas of Nevada such as Gerlach, Basques produced a few petroglyphs during the winter months. According to one source, the following rhyming message in Spanish, characteristic of Basque double-edged humor, is carved on a rock there: "Candido, contento pero jodido" (Candido, happy but fucked up).[75] I suspect that one reason so few rocks were carved was the sheepherder's obsession with keeping a sharp knife. Rock carving and sharp knives do not mix. Among the few known samples of rock carvings is a boulder in Verdi, Nevada, that I videotaped in 1989 that is chiseled with the initials FAB and dated 1905.[76]

The Canvas, the Tools, and the Technique

Aspen trees grow in well-watered meadows and canyons of the high country. Creeks with large aspen stands usually have some flow year-round, and such areas were the favorite campsites of sheepherders in the summer months. In the Sierra, aspens flourish in lush meadows, which are typically crowded by pines, but in Nevada there are few pines and the aspens tend to dominate the creek banks and high meadows. Often they can be seen from miles away. The bark of aspens, especially young ones, is smooth and inviting.

The herders preferred large trees because one could draw bigger, more detailed figures on them (a couple of Mexicans tried carving saplings, but the bark was too tender and the words quickly became unreadable). Smoothness of the trunk, however, was even more important. Two aspen varieties predominate in the geographic areas I visited: green bark and white bark. The white-bark aspens are a much better "canvas" than the green-bark ones. The former produce black scars that contrast well with the white bark, while the artistic effect is dissipated on green bark. Other differences affect the carvings as well, chiefly the several varieties of bark cankers, which thwart the overall artistic effect of the carving.[77]

Sheepherders commonly used knives to produce the carvings, but any other sharp tool—rocks and even bullets and spent shells—would do. According to Jean Lekumberry of Garderville, Nevada, a six-penny nail was the best instrument, and scratching the bark produces better results than cutting or incising it.[78] Novice artists, who wanted to see instant results, usually carved too deeply or cut too wide a groove from the bark. This is the common technique used by tourists in the forests. It not only hurts the tree, but within a decade or two the inscription also becomes unreadable and devoid of any artistic effect. Brunvand and Abramson were correct in saying that the "sheepherders are the best artists of our genre, hunters and fishermen next, and picnickers and lovers—the roughest hackers and slashers of trees—worst of all."[79]

Most herders soon learned that the best arborglyphs are produced with a single thin incision. Only a few perfected the technique, however, the reason being that the carver did not know how his work would turn out twenty years later (few stayed with the sheep that long). In a way the herder had limited control over his art; he contributed the initial incision, but the aspen and nature did the rest.

A few years after an incision is made, the tree succeeds in closing the wound with a black or dark-colored scar. On greenish bark the scar turns gray. As the tree grows, the scar widens and the thick, black letters on the white bark become more appealing. The horizontal growth of the tree trunk eventually renders the artist's single incision into a double line. The growth, which occurs between the two lines, pushes them farther and farther apart, to the point that, for example, a man will appear to have two noses, two chins, etc. This poses no problem as far as figures are concerned, but it causes havoc with writing, especially when the letters were initially carved too close together. Occasionally, the scarring turns itself into a bas-relief, a development associated with the growth of the trunk, which adds to the distinctiveness of aspen art. The tree seals the herder's masterpiece with its own brand of "art."

Trees standing on creek banks often grow faster and produce a thicker scar, thus distorting the deeply carved characters and figures more quickly. Those farther from water retain the initial arborglyphs longer and more clearly. Altitude also affects the carving. The higher the elevation, the more slowly the aspen grows, and therefore the longer the incision will remain clear.

Given the low level of education of most sheepherders, it is surprising how few errors one finds in the messages. Many carvers seem to have visualized their carvings beforehand to minimize errors, but only one carver told me that he first used a crayon to draw the figures before proceeding with the knife.[80]

The artistic presentation of a piece can greatly improve the second or third time around. Also, art that takes a long time to complete is usually perceived as more artistically valuable than art that is done quickly. Giorgio Vasari said that Leonardo da Vinci took so long to complete his *Last Supper* that his patron complained.[81] The more care and attention that are given to details, the more valuable the piece of art can be.

Sheepherders generally had no regard for such principles. Tree carving art is a relatively quick "first-try" art. It is a little like life itself: you cannot erase a carving and you cannot do it over, any more than you can live twice. Most of the carvings do not appear to be the fruit of long and arduous labor. One or two minutes would have sufficed to inscribe most names and dates, although the figures no doubt took longer. The intricacy of some designs clearly indicates that the herder spent considerable time on the operation. The work of D. Borel, for example, one of the great carvers recorded so far, is so perfect that it is hard to imagine his as "one-try" art.

Time and effort are other reasons why most herders preferred to carve aspens. Carving on trees with thicker bark requires a great deal more work because the bark has to be peeled or shaved off before a name or a figure can be inscribed. Almost none of the herders bothered to do that. Besides, a pine carving lacks the contrast of the aspen glyphs because the cut is done on dead wood and does not scarify or grow. In fact, it is hardly visible.

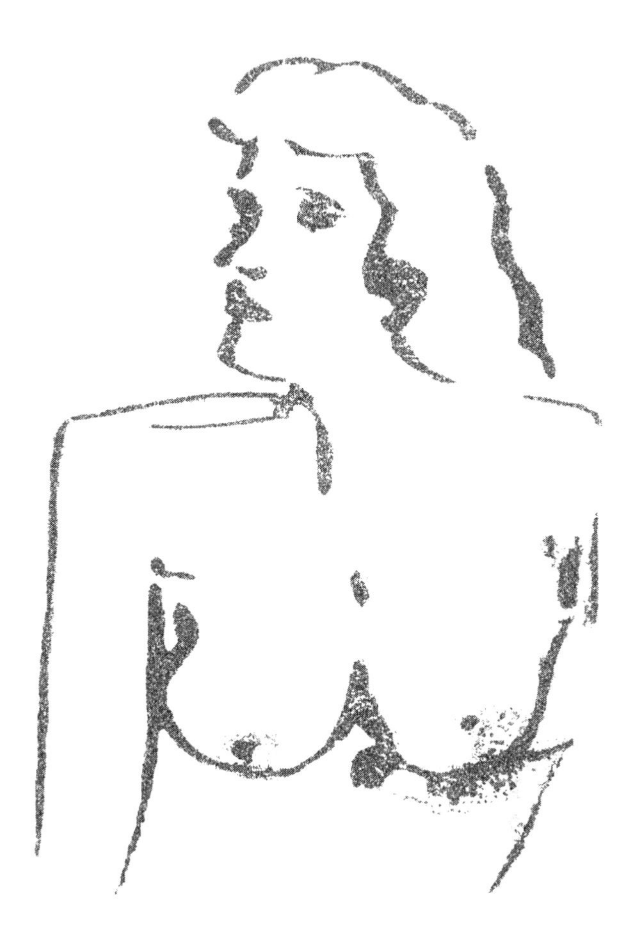

4. It is difficult to believe that this exquisite carving is seventy-five years old, as if nature defied it own laws of aging and distortion in order to protect D. Borel's art. Another herder, wanting to be as close as possible to this "angel," carved a self-portrait on the back side near her shoulder.

An Unpretentious Insider Art

During the summer months the sheepherders alone in the high country used their leisure time to carve thousands of trees. Almost all of them recorded their name and the date, but they also carved a variety of pictorial representations and statements. Most of the inscriptions are fairly primitive and crude—a characteristic of the phenomenon—but some are more elaborate, even ornate.

In time, the carving activity took on a life of its own, growing and spreading with each summer. Each year new and veteran herders arriving in the high country added their names and stories to the earlier ones. The result is an amazing data bank of news directly related to Basque immigration. This phenomenon, unique to the American West, is the legacy of hundreds of carvers and thousands of carvings, each herder contributing and adding a few pieces of information, thereby creating vignettes that illuminate the life of the Basque sheepherder.

Essentially, each herder became a part-time recorder of history, and each tree trunk a living document. Each carver provided his own interpretation, simply and candidly, in a democratic approach to history writing that may resemble the way oral traditions were formed, with the help and input of many members of particular social groups.

The researcher who finds strange messages and figures on aspens should not forget who the intended audience was. The arborglyphs were carved not for urbanites but for other herders. Those who do not understand that run the danger of asking the wrong questions. We should not expect the carvings to contain full dissertations on any topic, not even sheepherding, because the herders did not need to be lectured on their jobs.

Since the arborglyphs are never a topic of conversation among Basques, and we think that the herders were generally isolated from each other, how is it that sheepherders in Montana and ten other states all start carving trees? How did the word get around? And how did they know what to carve? Why is it that most of the themes are duplicated (as far as I know) in every state? The answer is simple: sheepherding was much the same all over the West, and therefore the comments of the workers, whether in Colorado, Montana, or Idaho, were similar. Furthermore, a number of the herders switched outfits and moved from state to state.[82]

Without a systematic search of the entire American West we can never be sure which themes are duplicated and which are not. Important topics that one might expect many sheepherders to have commented on went unremarked. This is the case for the 1936–39 Spanish Civil War, for example; only one Sierra grove has carvings relating to that.[83]

The fact that we know all the names in one grove does not give us a basis for guessing the names in the next grove. That would be like assuming that the Boise telephone directory contains the same names as the Reno one. Further, the fact that we have identified a number of themes that are repeated in fifty groves gives us no grounds to assume that the thousands of unrecorded groves in the West contain no new themes.

The Silence Broken

Arborglyphs are literally "on-site" memories, for they contain direct and reliable data on the actors of the sheep industry. And even though many messages "scream" loneliness and depression, in many other cases the sheepherders seem to have been "preaching in the desert" when they inscribed their musings on trees. My reaction was that finally, the inscrutable Euskalduna had "spoken" for history.[84] After a silence lasting thousands of years, after watching Cro-Magnon tribes, Celts, Romans, barbarians, Arabs, Napoléon, and General Franco come and go, the Basque peasants spoke. In a sense, though, the Basques kept their silence even in America. For a century they had been carving hundreds of thousands of trees, but hardly anyone knew about it. Were they trying to keep it a secret? Quite the opposite. Every sheepherder knew about them, so there was no need to discuss them.

I have asked many sheepherders if they carve or carved trees. The answer is almost always the same: "Lots of them." When I ask for the carvings' locations, they first look me up and down and say: "Oh, you cannot get there, you need a horse, you will never find them," or something of the sort. "Never mind," I answer, pretending disinterest, and then they will proceed to describe the various mountain ranges where their carvings may be found. Then, almost as an afterthought, they might add: "But do not look at them!" I do not follow their advice.

I am often uneasy in my role as writer and researcher. Walking into the old campsites in solitary and remote groves I feel the presence of bygone herders. Their names on trees watch me, their portraits, their inscrutable eyes, mildly chastise my research objectives. I am walking into their sancta sanctorum. Today the carvings are a public record, but were they public in the minds of the sheepherders who put them there in the 1920s and 1930s? Did any herder suspect that someday people back in town might read his very candid confessions? I cannot believe so.

The Basques spoke up, yes, but only because they believed their words to be safe in the vast wilderness. There in the middle of the sheepherder community, where they felt safe, they spoke up, but for each other. In front of strangers most Basques keep a stiff formality. The arborglyphs must be understood within the scope of Basque American culture, with its ethos, rules, and obligations. Although incising tree trunks would not be considered an unusual activity in Basque society, inscribing the messages and images peculiar to the herders in the United States would be. The arborglyphs reveal the Basques in their own element, among their own kind, speaking "with the utmost frankness everything that is in their minds."[85] Although I have not respected their privacy, I will guard their confidence. The names of those who revealed their innermost thoughts and secrets will remain on the trees. As any historian, I sometimes analyze, compare, and interpret data or offer my views, but for the most part, the sheepherders will speak for themselves. Rather than using the data I have gathered on the carvings to prove or refute a preconceived notion, I have been a gatherer and a compiler of information. It took some walking to get this far, and I would not want to spoil that.

A Paperless Culture

One does not need paper and pen to record the past. Most of human history has been recorded without writing, which is less than five thousand years old, as far as we know. What is five thousand years when compared to the total human record? Before writing was invented, human knowledge was consigned to memory and passed down orally.[86]

Basque peasants still follow this old method. Many famous *bertsoak* (improvised poetry) and some medieval *eresiak* (elegies) survived for decades or centuries in the collective memory of the Basque people before they were finally written down. The time to disgorge the accumulated knowledge is during the social gatherings that abound in the Basque calendar. When Basques socialize, everyone talks, everyone provides information, and together their recollections create the whole picture. If one person's memory is wrong, someone in the audience will set the record straight and consensus will be established over the accuracy of the matter. Such gatherings are inevitably accompanied by food and drink, especially during summer picnics and celebrations, when friends and relatives gather in the kitchen among heaps of bread, bowls of soup, salads, stew, meats, fish, and bottles of wine. The socializing continues until the early morning hours.

Women are more likely than men to engage in this sort of "kitchen history class," which includes the whole spectrum of the Basque community. But the preferred themes are who did what, who married or divorced, who had a baby, who went to visit the Old Country, who came, who had an accident, and so on. There are women who could easily draw a kinship chart of local Basques—entire family trees—not just for their own community, but for nearby towns and even other states. The complicated maze of *osaba* (uncles), *iloba* (nephew/niece), *biloba* (grandchild), *suhi* (son-in-law), *neba* (brother of sister), *arreba* (sister of brother), *aizpa* (sister of sister), and *amainarreba* (mother-in-law) is a living play to them. Individuals thousands of miles away or long dead are analyzed, criticized, or praised as if they were right next door. Men do engage in these conversations, but with less gusto. Today in the American West, the Basque family kitchen is the best school for one who wants to learn about the old sheepherders.

The Deep Roots of Tree Art?

According to a bumper sticker I saw on a truck in Elko, Nevada, sheepherding is the "second oldest occupation," and I can believe that it has changed little since the Neolithic. Trees, even then, were the sheepherder's ready-made canvas, and it seems logical to ask if the Neolithic sheepherders in the Pyrenees carved figures of humans and animals on trees. Unfortunately the question is impossible to answer, although I suspect that such art was not unknown to them. In fact, the prevalent theory on the origin of the Basques considers the Neolithic sheepherders to be descendants of Paleolithic Cro-Magnon people who once lived in nearby caves.[87] We can still admire the array of magnificent horses, bisons, reindeer, and other animals that these cave people painted. So the question begs itself: What happened to the Paleolithic art? Was it all forgotten at the end of the last Ice Age when people left the caves? And what about the artists? Did they also give up all artistic ambition? Or did some of these traditions survive through the millennia?[88]

The answers to these questions—if they existed—might be useful in helping us to understand the enormous carved legacy of the Basque sheepherders in the United States. As it is, I can only point out certain analogies.

1. Both cavemen and sheepherders lived very primitive lifestyles, very close to nature.
2. Both groups depicted the animals they came in contact with. For example, above Topaz Lake at the Nevada–California border I chanced upon an aspen carving of a magnificent buck deer in full flight. The figure is so lifelike and alluring that one would swear it had been chased off a cave wall of Isturitze (Benafarroa) or Lascaux (France).
3. Both groups prominently and vividly represented the objects of their desires (the cavemen yearned for meat, the herders for women).

The Roman poet Virgil (70–19 B.C.) attested to the popularity of using tree trunks for writing in ancient times. One of his characters, Mopsvs, says: "I'll try out some well-set verses which lately I carved on the green bark of a beech."[89] American pioneers frequently carved beeches. The only clue ever found to the disappearance of the Lost Colony of Roanoke Island in 1590 was reportedly the word "Croatoan" carved on a tree.[90] Henry David Thoreau was a carver in the eastern United States in 1852, as were some Mormon pioneers and frontiersmen, including Daniel Boone in 1760.[91] The problem with determining tree-carving activity is that the evidence—the trees—is not as durable as cave paintings or clay tablets.

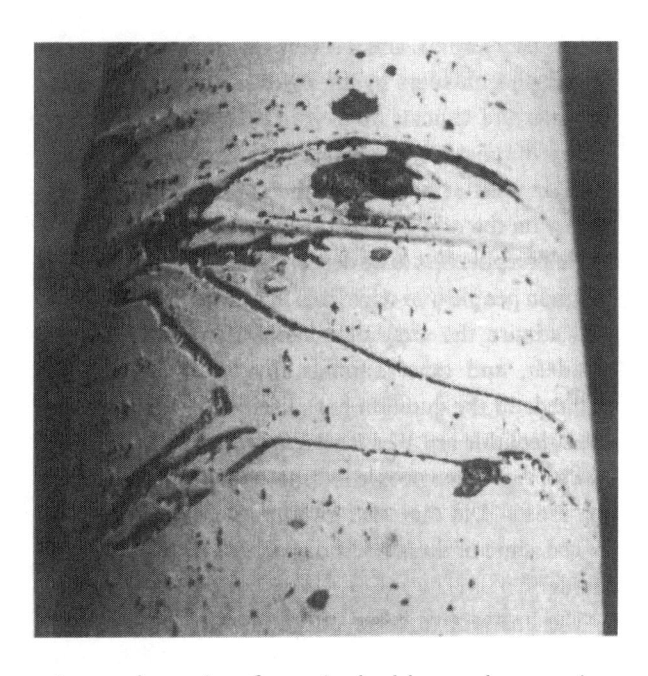

5. Spectacular carving of a running buck by an unknown artist. The deer seems to have jumped right off the wall of a Paleolithic cave in the Pyrenees. The approximate date of carving is 1925–35. The tree is in the Toiyabe National Forest on the Nevada–California border.

Around the world, forms reminiscent of tree art abound; for example, the whale-tusk carvings of the Eskimos, the birch-bark records of the Ojibway Grand Medicine Society, and Melanesian bark paintings. The Aztecs inscribed their Nahuatl language on bark and had books that folded like accordions as did the Maya. The Australian Aborigines are renowned for their intricate bark paintings, mostly figures of animals. In the absence of bark they painted rocks in caves and overhangs near water holes.[92] The men in western Finland were skillful tree carvers, and other Scandinavians also practice the craft, as do the European Basques. In the Basque Country, hikers and outdoorsmen generally carve on beech trees; their carvings are much like those found in campgrounds in the United States.[93] Richard Lane claimed that people from northern Europe, particularly the British Isles, preceded the Basques in tree carving, but he did not provide further details.[94] Carving has been practiced in Africa since time immemorial, and according to one source, a three-hundred-year-old inscription still exists on a baobab tree.[95]

Some of the earliest arborglyphs in the American West have been found near Agency Creek in the Lehmi Valley of southeastern Idaho. Local historian George E. Shoup found an arborglyph which he read thusly: "Josh Jones, 3-16-1810." Shoup interpreted Indian glyphs supposedly inscribed on the same tree to mean: "Take notice. Eight persons camped overnight at this place." If the story is true, this aspen was a very long-lived one indeed.[96]

Informants have alerted me to old carvings that they have seen here and there. For example, in Buckeye Canyon near Bridgeport, California, a forest ranger found an aspen with a carving dated 1885. It is quite legible and has a figure on it as well. In Sugar Bowl, near Donner Pass, California, I am told, there is a lodgepole pine carving dated 1855. Naturally, pine carvings last a lot longer than aspen ones. In 1999 an arborglyph was discovered in the Susanville area of California bearing the following inscription: "Peter Lassen, 1858." I am not used to seeing carvings this old, and I cannot say that it is genuine, but the lateral expansion of the letters exhibits the characteristics of the oldest carvings found. Lassen National Forest employees were planning to take a plug to determine the age of the aspen, which is alive and healthy looking. Interestingly, the tree stands very close to the Lassen Trail that early pioneers used in northern California.

A large number of old trail blazes, all on lodgepole pines, still exist in Yosemite National Park, partly because the area has been protected. The earliest of them may have been intended to provide directions for prospectors, the army, and other travelers, but other visitors to the forest, and especially sheepherders, left their marks as well. James Snyder and his team recorded more than one thousand trail blazes in the park; the majority consisted simply of initials and the date, making them difficult to interpret.[97] Snyder, a park historian, thinks that most of the Basques came just before the turn of the century or a bit later.[98] Most of the surnames on two lists of sheepherders in the area dated around 1902–5 are Basque names.[99]

How does the sheepherders' art in America compare with traditional art forms in Euskal Herria? As a rule, herders in Europe did not carve trees. The herder in Euskal Herria had two main camps, one for summer and one for winter. Every night, winter and summer, he penned his sheep in a corral. In the summer he milked them every day. The herder slept nearby in an *etxola* (*txabola* in other dialects), a stone hut or cabin, not in a tent as in America. The Pyrenean herder's hut was more than a place to sleep, however. His *gaztatei* (the area where cheese was made) was located there, and he also needed regular utensils and furniture for the cabin.

The sheepherder in the Pyrenees was not an artist as much as he was a craftsman—or a pragmatic artist, if you prefer. He used readily available materials, mostly wood, from which he built *aulki*, a special stool for milking the sheep; *ttotxu* (a type of chair); saddles; tables; and wooden containers like the *kaiku* with one handle, made from alder or birch, in which milk was boiled with hot stones. A similar but larger container, *apatza*, had two handles and was used in cheese making. In a few valleys of northeastern Euskadi the herders still make large boxwood spoons with large handles, which they decorate with engravings.

The sheep collar (*uztai*) went around the animal's neck and held the bell (*yoare*). These were made either from leather or from ash, chestnut, or walnut wood and were decorated with geometric or other drawings. The *adar ontzi* was a water container made from a cow's horn and sealed with a wooden piece on which the shepherd carved an anthropomorphic, zoomorphic, or geometric figure reminiscent of tree art in the American West. *Zimitz*, the wooden mold used for cheese making, was sometimes decorated as well. The *aska* (water trough) was made by hollowing a large tree trunk. One of the most commonly decorated instruments, one the sheepherder always kept close by, was the walking staff, or *makila*.[100] The most common designs were the sun, represented by the *lauburu* (literally "four heads"); stars; several kinds of animals; a woman; a man; a tree; flowers; and so on.[101]

6. *Ixtupa*, a sheepherder's homemade dishrag. Many herders handcrafted elaborate mops such as this one, owned by Mrs. Sario of Minden, Nevada. When cotton thread was not available, horsehair was used.

Although identical motifs can be found carved on aspens in Nevada and California, Basque sheepherders who carved trees in the West may not have been simply following the art forms developed by sheepherders in the Old Country. Many of those who came to Nevada and the West—especially in the second half of the twentieth century—had no previous experience as herders and no knowledge of the sheepherder crafts. On the other hand, most Basque peasants were familiar with wood carving.

It is very probable that tree carving became a tradition very early among the sheepherders of the West, and that the Basques who arrived after the 1890s found the art already well established. The issue is far from clear, however; there is little evidence to back up that supposition. But recently I heard an account that I think is an eye-opener. In December 1998 I learned that in the early decades of the twentieth century a sheepherder named Zabala from Munitibar, Bizkaia, from the farmstead Urkemendi, carved the following message on a beech tree on Motroilu mountain: "I am leaving for Argentina, Zabala."[102] The fact that this was carved by a sheepherder is significant. This sort of carving is fairly common on Nevada sheep ranges, and it proves that the type was not unknown in the Basque Country.[103]

Petroglyphs and Arborglyphs

The art of early herders may have been partly inspired by the Indian petroglyphs found throughout the West. There are fascinating similarities between the two phenomena, including the following:

- anthropomorphs lacking extremities;
- disproportionate, misshapen, or aberrant figures;
- a high percentage of animal representations (but fewer on trees);

art related to particular sites and their activities;

erotic and sexual images, exaggerated phalli, etc.;

poor-quality art, work of unskilled carvers; and

copies of older forms and symbols, plagiarism.[104]

It is widely believed that petroglyphs are associated with the activities that took place at the particular site where they were made.[105] Similarly, the carvings reflect the sheepherder's lifestyle in the summer months, when he set up camp in an aspen grove. Just as petroglyphs are located near Indian camps, tree carvings abound near sheep camps.

Although the similarities between these two artistic phenomena appear to be clear, they are products of peoples with very different cultures. The petroglyphs are not believed to contain "written" information. Both, however, are "outdoor arts" produced by people who lived very simple lives in close contact with nature. The animal depictions illustrate that point quite well. Ancient people carved the figures of the animals that were significant in their lives. From the Russian taiga to the caves of Euskal Herria, wildlife has been incised into stone or painted on cave walls. According to Henri Breuil and Raymond Lantier, there are 112 caves with such figures. José M. Gómez-Tabanera estimated more than 300 Quaternary caves with paintings in Europe alone.[106]

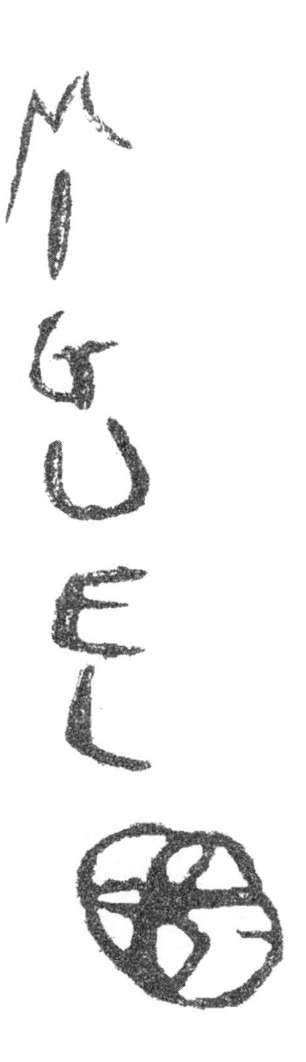

7. The *lauburu*, believed to be the Basque sun symbol, carved by a herder named Miguel.

Bear tracks are another motif found in both rock and aspen art.[107] Much more numerous are the carvings of hands, which on trees are sometimes dated and signed. The hand design is a widespread motif in rock art around the world.

Most of what scholars know about the meaning of rock art in the United States has been explained by the Indians themselves, although Campbell Grant believes that the information may not be reliable. After all, petroglyphs can be thousands of years old. How much do modern Native Americans know about their creators?

In this respect, the tree-carving researcher has a clear advantage. When I found an inscription that was not clear, I sometimes had the luxury of calling the herder who had made it and asking him what he had been thinking when he carved such-and-such image on such-and-such date. His first response was usually that he did not remember. In that case, I showed him a videotape of the tree, grove, and mountain where sixty years before he had herded sheep. Most herders could not help being touched by that, and the memories came flooding in.

Scholars believe that much of the rock art, like Paleolithic cave art, has a supernatural or religious significance. Jarl Nordbladh, for example, said that "there is general agreement that the petroglyphs are connected with religion."[108] In this aspect petroglyphs and arborglyphs may have little in common. According to the data gathered thus far, little of the aspen art deals with the supernatural, although a herder in southern Idaho reported that he saw Anbotoko Damie (the great Basque goddess or witch) flying through the skies.[109]

Symbols abound on both rocks and trees. The herders carved many stars (the most common symbol) and crosses (the next most common). Often they carved a cross first and then, below or to one side, their name and the date. In several instances, however, the crosses were

carved as a memorial for a sheepherder who had died on that particular spot in the mountain. Herders in the Pyrenees did the same thing when a comrade died.[110] A powerful concept in the Basque mentality is the *etxe* (house). The sheepherders carved this symbol often, but it is absent from petroglyphs, perhaps because the Native Americans who carved them did not have permanent abodes.

Rock art contains much information on ceremonies and rituals; for example, the many genital depictions, which denote fertility rites.[111] Relatively few petroglyphs feature figures of women, but simplistic female genitalia abound. Vagina carvings are so numerous on aspens that one assistant archaeologist thought they were Basque "peace signs." She was promptly corrected: "You mean 'piece signs.'"[112] In the same way, some petroglyphs may have been carved by male Indians longing for females. Depictions of genitalia do not necessarily mean that the artist wanted offspring. They may simply represent ancient sexual fantasies, just as some arborglyphs do.

Campbell Grant, author of the first general study of rock art in North America, believes that a small number

8 and 9. A message of peace: bird flying in the direction of a star. Both by J. M., 1926–29.

10. The primary reason for carving was to record one's name: "Felix Arozpide 29 Junio 1909."

of the petroglyphs are doodlings, copies of symbols and designs done for amusement or for no reason at all.[113] Tree art contains doodlings also. The "Kilroy was here" syndrome seems to be a primordial inclination in most humans. Like the sheepherders, most people carve their names more frequently than any other message. Were the Indians any different? I suggest that some of the petroglyphs are names and clan or band IDs. That appears to be the case for at least one Hopi site.[114]

Did Colonial Herders Carve Trees?

If the petroglyphs were not the inspiration for sheepherders' carvings, did the Hispanic herders in the Southwest provide it? They had been herding in the area since colonial times. It seems likely that Hispanic sheepherders in the summer ranges of Arizona and New Mexico carved aspens then, just as they do today.[115] Even before Oñate colonized New Mexico in 1598, the sheepherders in northern Mexico may have carved trees. But the petroglyphs are older still. Perhaps the question should be, did the petroglyphs inspire the Hispanic her-

11. Uninterrupted script carving, which is more difficult than the usual type carved one letter at a time: "Baptiste Saldubehere 1938," Humboldt County, Nevada.

ders, who in turn inspired Basque sheepherders? Evidence for such a link is lacking, but it is very probable that sheepherders in the American West started carving trees as early as there were sheep to be herded.

It has been firmly established that the blazing of trees began in California as early as the 1850s and 1860s and expanded in subsequent decades as the number of Basques associated with the raising of sheep increased. As noted, Hispanics were among the earliest herders, followed by Scots and Irish and at least some Indians and Chinese. Of all these groups, however, the Basques, in Byrd Sawyer's words, "were the artists of the sheepherders."[116] This may or may not be true, but it is certain that few inscriptions have been found carved by the other groups mentioned, with the exception of the Hispanics.[117]

Tree Art and Graffiti

Some call aspen art graffiti; one author had a better idea and called it "treeffiti."[118] But are the carvings graffiti or merely the musings of men who spent a lot of time alone with their thoughts? A graffito in New York City's Greenwich Village reads: "The only difference between graffiti and philosophy is the word *fuck*."[119] According to this definition, the sheepherders were not philosophers. Besides philosophy and four-letter words, there are several clear parallels between graffiti and the messages on the aspens.

The ancient Romans had a saying: "Nomina stultorum ubique locorum" (The names of the fools are found everywhere).[120] Although creating graffiti is a universal activity, it is often viewed as antisocial behavior and attributed to iconoclasts, people who gain satisfaction from the imagined shock they are causing.[121] Graffiti are written in public places for all to see, whereas most of the arborglyphs are found in very remote areas and were seen only by other herders. If graffiti are, as Robert Reisner put it, "dirty words on clean walls," then at least some of the carvings may be considered to be "dirty" inscriptions on white tree trunks.[122]

Graffiti is often created in the privacy of the bathroom, which becomes a confessional box for the graffiti writer. The exercise is nothing more than a release from social constraints, as explained by the Freudian principles of repression.[123] Decades ago the noted sex researcher Alfred

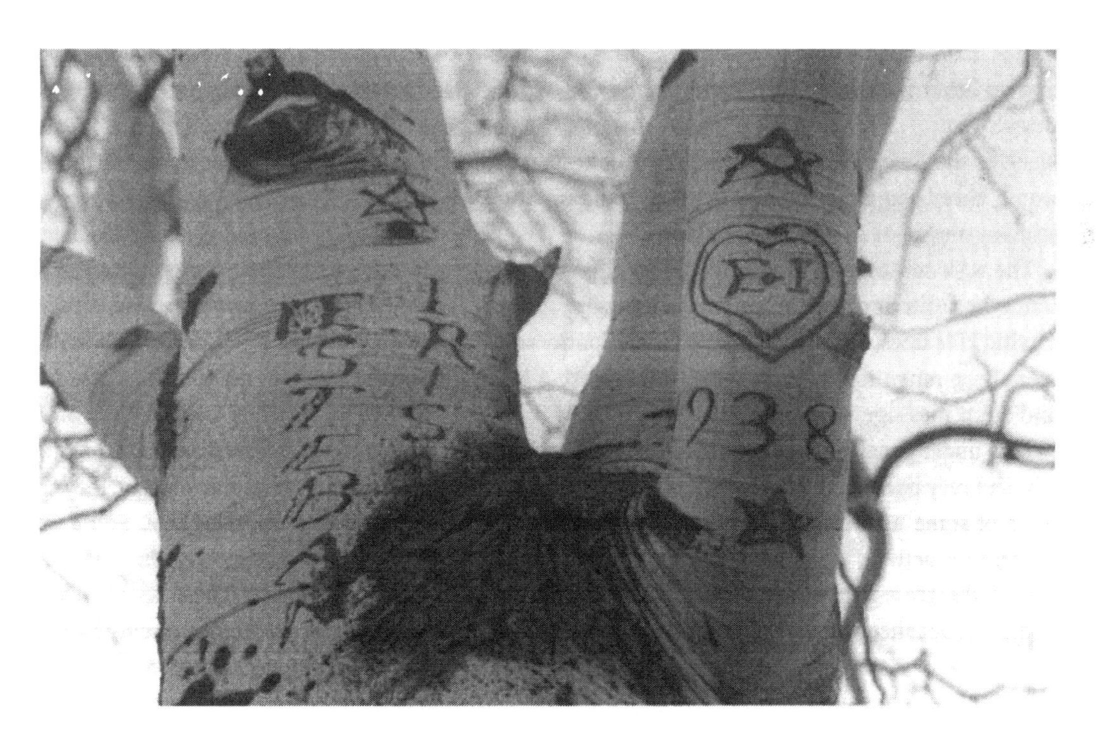

12. "Esteban Irisarri 1938." An attractive arrangement, with the name carved high up on the tree and repeated and embellished with symbols. The star is the most recurrent sheepherder symbol, but of uncertain significance.

C. Kinsey said that "the males who make the inscriptions . . . are exposing their unsatisfied desires."[124] These ideas are applicable to the arborglyph phenomenon as well, for they help us to understand the motives of the herders who carved the more unorthodox inscriptions. To begin with, the parallel between the privacy of the bathroom and the sheepherder's dreadful isolation is uncanny. Further, many of the messages are iconoclastic, deliberate breaks with old taboos, particularly those regarding the sexual mores of the Old Country.

That the herders' sexual desires were unsatisfied is undeniable, given the absence of women. This did not make Basques misfits, however, as psychiatrists often hold graffiti writers to be. Another difference between urban graffiti and "treeffiti" is the relative absence of the word *shit* in the latter. In Germany, for example, defecation is at the heart of graffiti, folklore, and a host of jokes.[125] But not among the Basques. By far the most common "four-letter concept" refers to sexual intercourse, expressed in several languages. The second most frequent is *puta* (whore).

Psychologists believe that graffiti can reveal the truth about the character of the person who creates it.[126] I think the frankness of the carvings authenticates the lifestyle of the sheepherders like nothing else. Arborglyphs bare Basque characteristics and idiosyncrasies like no other cultural manifestation in the Old World. British author Rodney Gallop wrote that the salient mark of the Basque character is its reservedness.[127] He would have been mystified by the utterly candid nature of many messages in the arborglyphs.

The NFS and BLM long regarded arborglyphs as unwanted graffiti and put out posters advising that bark carving kills trees. In a northern Humboldt County summer sheep range someone carved an aspen with a long and stern message warning that carving trees was forbidden under penalty of a thirty-dollar fine. It is in Spanish, and may have been done by a herder on the instructions of some BLM official as an effort to deter further arborglyph activity. Graffiti or not, however, time has turned the arborglyphs into a unique source of information about sheepherders.

Understanding the Arborglyphs

The old theoretical questions of what is art and what does it mean may be asked in reference to aspen carving as well. Art is regarded as a higher form of human activity, as intensified reality. What about arborglyphs? Are they art? history? literature? Reckoning their frequency, the phenomenon can first be classified as strictly ethnohistorical; artistic properties come second. The literary value of the carvings comes in a distant third.

To some extent the Basques may have created and developed the tree-carving art almost absentmindedly, by sheer imitation of previous herders whose glyphs they found in the forest. Someone started carving, and it quickly became a tradition; and tradition is so fundamental to Basque culture that few ever dare tamper with it. That tendency is well explained by an expression Basques often repeat to explain certain behavior: "Gutxiago ez izategatik" (For not wanting to be any less than others). In fact, several herders used those very words when I asked their reason for carving trees.[128] It is the idea of equality, or perhaps conformity, but stated via the negative, another Basque characteristic.[129]

Imagine a newly arrived herder walking the sheep summer range for the first time and coming face-to-face with the names of earlier coworkers. What a shock! His first reaction may have been negative, especially toward the herders who carved boastful or irreverent messages. In the rural Basque Country modesty is a virtue and a sign of wisdom. Bragging is often associated with a lack of maturity. After the initial reaction, and after seeing dozens of inscribed names—some new, some old—the herder would have realized that putting one's name on a tree was in fact a custom; it could not be easily dismissed. Thus, walking up to a nice, smooth aspen, the new herder proceeded to add his name to the list already in the forest. There is no indication of older herders instructing those newly arrived on the ritual of carving trees. I asked several herders about it, and from their responses I gather that most herders discovered the arborglyphs and learned to carve them on their own.

For many, by their own admission, tree carving was a pastime that began as soon as they reached the summer pastures or a few days later. Some carved tree after tree, sometimes day after day, other times several in one day, almost as a ritual. These people usually carved just their name and the date. The carving pattern was not the same for all herders, of course. Most were induced to carve in moments of strong emotions. Although the range of feelings expressed is extensive, the arborglyphs recorded so far tend to feature several dozen major themes, as mentioned above. More sophisticated people might deem the output meager and lacking in variety, but such reasoning is facile. Obviously, the lives and tastes of city dwellers and sheepherders are very different.

What does the arborglyph phenomenon mean? Either the Basques were very aware of their work and its historical value or they were the most self-centered people in the world. The former view is difficult to sustain, because in the last several thousand years Basques, especially rural Basques, have produced very little written documentation. So the carvings must be associated with the herder's self-awareness, his sense of value and worth, a belief in his place in the world and in the community. I believe the carvings prove Salvador de Madariaga—himself of Basque ancestry—right in saying that the psychological center of the people of Spain is the heart; life is reduced to reactions of passion and emotion.[130] The majority of the carvings fall into that category.

Inscribing his name on trees was the herder's response to his environment. Alone in the immensity of the West, many herders felt like kings, lords of vast tracts of lands. Why wouldn't they? They seldom saw another human being, and only rarely did someone else show up to dispute their claim to the land. Loneliness brings out feelings of narcissism, omnipotence, and egocentrism. The lonely person wants to show off before an audience.[131]

When, in the 1950s, an old Bizkaian herder returned home, he told his friends that he had been more than just a sheepherder in America. Among other things, he bragged, he had become the world's boxing champion. People were skeptical, to say the least, but always ready to hear a good yarn, they wanted to know how it happened. "Well," the ex-herder told them, "one day, while I was herding sheep, I climbed the tallest mountain around, and when I reached the summit, at the top of my lungs, I shouted that if anyone wanted to challenge me to a fight, he had better come now. I waited a few minutes but no one answered, so I repeated the challenge, and again I waited a few more minutes. I wanted to give everyone a chance to respond, but no one showed up. Thus I became champion of America."[132]

The strong sense of territoriality expressed here may have its cultural roots in the Basque rural world, where each *baserri* (farmstead) is surrounded by its own fields and meadows. The couple who run the farm, whether they own or rent it, are known as *etxekojaun* (lord of the house) and *etxekoandre* (lady of the house). When, in the morning, these peasant lords walk out the door to do their chores, they are stepping into their fiefdom, however humble it might be. Herders surrounded by the immensity of the western range felt something like that, and it might have motivated them to record their names.

In the Basque Country sheep almost always are kept close to home and the herders have no time or reason to feel lonely and isolated. People would take them for fools if they started carving their names on dozens of trees in the Basque mountains, let alone bragging about their sexual prowess or their forays in the local brothel—if there was one. Carving in the American West reflected the sheepherders' displacement, and leisure time, and the availability of thousands of trees.

Did all the herders carve for the same reason or reasons? Why did a herder carve a deer? And how should we judge the figure of a sensuous woman? Was it artistic drive or sexual fantasy that motivated the carver? (Couldn't we ask the same question of Picasso or Raphael?) When the question is put to ex-herders, they prudently respond that it was just a pastime. Sometimes it appears that they carved simply to express their hatred of the isolation. Others, while also hating their circumstances, felt a certain kinship, a closeness to the range, the mountains, and, perhaps, to solitude itself. The first sentence in Robert Laxalt's *Sweet Promised Land* comes to mind here: "My father was a sheepherder, and his home was the hills."[133]

The location of the carved tree was crucial. The herders did not carve just any aspen; on the contrary, they chose them for a variety of reasons. If a carver had something interesting to say, or if he were a good carver, he would find a tree by the roadside or in some other conspicuous spot that would guarantee "publicity" for his work. The more active carvers vied for strategically located aspens—those by the favorite fishing holes and near sheep camps. Large, smooth trees were the most likely to become "billboards" in the wilderness. The trees beside a popular hiking trail near Reno are heavily carved. One in particular is so totally etched that one herder inscribed: "There is no room" (to carve) on the crowded trunk. He knew only too well that hundreds of nice aspens were available nearby; but no, he wanted to put his name on *that* "billboard."[134]

Herders who retouched carvings or carved over them generally obeyed certain rules. Few were attracted by an ordinary inscription, such as a name—unless they had a bone to pick with that person. They were drawn instead to aspens exhibiting outstanding figures or witty or funny statements, which in turn incited others to carve.[135] Rarely, a herder added his own touch to an outstanding carving as a way of sharing in the experience of the master who had created it. In Tahoe NF there is a very realistic bust of a nude woman (dated 1925) that looks like Jayne Mansfield or Marilyn Monroe with her head turned to the side. A later artist added his own head next to it, facing the woman's at eye level and almost touching it. I found a number of male figures appended to erotic representations of females carved by a previous artist. One of the most interesting female figures found in the Sierra is the work of two carvers some twenty years apart. The original figure is some four feet tall and shows a detailed female head, tiny arms, no breasts—highly unusual—and a geometric vaginal area—like those of the goddesses of Old Europe;[136] she is also wearing pants and high-heeled shoes. A second herder added earrings, eyelashes, and a baby that looks like a lamb in the abdominal area. The second herder no doubt thought that "she" was more attractive, thus more magical, with the new accessories. Herders occasionally criticized someone else's carving skills, most often by adding phalli or written comments to an existing arborglyph.

The trees were the herders' friends, counselors, and confidants. The carving exercise had definite therapeutic effects; for example, when, in a fit of anger, an individual inscribed a curse against a fellow herder or described a quarrel with his boss. By the time he had finished carving, his anger had subsided. The tree was the cheapest and most accessible "psychologist" on the mountain.

One defining characteristic of tree art is its simplicity, but in numerous instances it is plain that the sheepherder sought to give words or letters special flair by carving them deeper, encircling them, or enlarging them. This attention to detail, this purpose, is particularly discernible in the initials, which often are calligraphically styled. I would say they qualify as art—as, certainly, do the symbols and human figures, though they may not be in the same league as Francisco Goya's nudes. There are as many styles of carving as there were herders themselves. Some carved capital letters only, others used script, but the majority mixed them. The first letter of the name and the surname tend to be elaborate; northern Basques were particularly fond of that practice. Some letters may be one foot tall while others are less than one inch. In similar fashion, full human figures range from eight or more feet tall to two or three inches.

13. A fine example of calligraphic signature, complete with rubric, by P. Gortari.

Another characteristic of tree carvings is their perspective. Often the whole message cannot be read or seen from one viewpoint, and it is necessary to walk around the tree. I think most herders had to deal with this problem. Those carvers who sought to affect the reader with a single view remedied the situation by inscribing the words in one or two vertical columns with the letters positioned either vertically or horizontally. Most messages read from top to bottom, but a number are in reverse order, created when the herder was taking a siesta under the tree or was sitting on the ground next to a tree and began toying with the knife at the bottom of the trunk.

The visual limitations were compounded when figures were being carved, especially those involving more than one person. In sexual scenes, for example, often both the man and the woman are standing. Sometimes the woman is above the man, so that both figures can be seen from one viewpoint. One solution was to carve small figures, but most herders preferred large figures to small ones. They knew that time blurred small carvings more quickly than large ones.

The documentary value of the carvings must be understood properly. In careful observation of a grove one quickly learns that the messages are short and repetitive. They do not even come close to covering the spectrum of possible topics regarding the sheep industry or immigration. If it seems that they left out a great deal, it is good to remember that the sheepherders never intended to carve a biography that outsiders would appreciate. We need to appreciate what *they* carved, because that is precisely what is important.

The herders certainly knew what they were doing. For example, I learned, without asking, that many of them knew the proper carving techniques for ensuring longevity and legibility, which says a lot about their concern for readability. More than one herder hoped that a particular tree he carved would grow tall and live a hundred years.[137] But whatever the case, one thing is clear: tree carvings constitute an odd historic chronicle in that they exist not on paper, microfiche, or disk, but on wood. That is odd indeed in the twenty-first century.

Until the 1980s and 1990s Forest Service personnel invariably responded to the term *Basque arborglyph* with a condescending attitude, if not with a knee-jerk reaction against "pornographic" smut and a prompt dismissal as well. The Forest Service employees who recorded one arborglyph grove in California named the site "Sheep Thrills"; another site was called "Little Big Man," after a figure of a man with a large phallus; there may be other colorful names as well.[138] Many of these people now realize that they were misguided and that the carvings are a valuable resource of western U.S. history.

Arborglyphs constitute rare testimony because they are totally uncensored. Mainstream historians of the American West, however, have yet to utilize this massive data bank, which, chronologically and spatially, documents the sheepherders' presence on the land with unrelenting detail. What conclusions can we draw from the thousands of tree carvings so far examined? Scientists love to gather data—the more the better to make their conclusions seem more solid. Much of what we know of prehistoric peoples and cultures is based on the traces and remains they left in the wake of their movements. A good example is the Kurgan culture, which has been used to determine the origin point of the Indo-Europeans and the direction of their movements.[139] If we applied the same methodology to the carvings, we would conclude that the Basques invaded the western United States and for almost a century held a grip over much of the territory. And such a conclusion would be totally wrong. Almost the opposite was true. The herders went through their motions almost invisibly. They affected the land and ecology, sometimes severely, but society was largely unaware of their existence.

The aspen carvings show that it is possible for a relatively small group of people to leave behind in a foreign country numerous cultural relics without having much intercourse with its people and without affecting the majority culture in any significant way. Although it is true that the Basques partially "occupied" the land, they were nevertheless a minority roaming it like ghosts, and when they were in town, they lived in ghettoes centered on boardinghouses. Until they began settling in towns, raising families, and starting businesses, their influence on the majoritarian society was almost imperceptible and difficult to measure or evaluate.

I asked a retired herder in northern Nevada if he thought the Basques had any influence in town. "You betcha your life," he said. "We drank most of the wine in town, and without us, the whorehouses [would have] had to close. Think of all the money."[140] He laughed heartily and, scratching his thick neck, added: "Well, maybe my theory needs some work." His was the standard Basque reaction. This man took it for granted that I would know that the sheepherders' greatest contribution had been in raising lambs and wool. Basques tend to bypass the obvious; some will state the opposite of the obvious in order to get your attention.[141]

A Claim to the Land?

Dates tell a story, too. The earlier arborglyphs reflect the grazing practices before the Forest Service came into being and the Taylor Grazing Act of 1935 regulated public lands. Carving one's name on a grove was a way of claiming a certain mountain range. James Snyder, a Yosemite National Park historian, agrees with this opinion. Although one's name did not carry the legal weight of a mining claim, in no-man's-land, a sheepherder's name carved on a tree was the next best thing; it was tangible proof of being there first, so the carver could claim it again the following year. In fact, many established ranchers used this very argument to claim large tracts surrounding their ranches, thereby excluding the itinerant sheepherders from them.[142]

A number of "John Doe was here"–type messages may be indicative of such assertions, although I found only one arborglyph that may be clearly regarded as a claim. Located above eight thousand feet in the spectacular Schell Creek Range in White Pine County is the following carving in Basque, dated in Spanish and Arabic numerals: "Etcheberrin campua agosto 10 1904" (Etcheberri's camp 10 August, 1904).

It is quite a spot. A trout creek runs through the grove, and enormous mountains rich in lamb's-quarter and other grasses rise skyward nearby. Etcheberri's inscription is significant both for its antiquity and for its message. It pertains to the early period of Basque presence in White Pine County—in fact, during the heyday of the sheep business, when herders were free to roam public lands on a first-come basis. According to one source, the outfit of Adams and McGill bought ten thousand sheep in Bakersfield, California, in 1897, and with them came the first five Basque herders to herd in White Pine County. Their names are not known.[143]

14. A claim to a campsite, when the range was free for the taking: "Etcheberrin campua agosto 10 1904" (Etcheberri's camp 10 August, 1904), located in White Pine County, Nevada.

Today the Schell Creek Range is managed by Humboldt-Toiyabe National Forest, but in 1904 the agency had not yet been organized. The fact that J. B. Etcheberri declared the top of Timber Creek canyon *his* camp and carved his claim on a tree trunk (the sheepherders' billboard) in Euskara (the official spoken language of the sheep range) suggests the presence of other Basques in the vicinity. Indeed, the previous year, 1903, another herder named Salaburu had apparently camped in the same grove, for he left his name on several aspens. Etcheberri may have been a latecomer who was challenging Salaburu, or perhaps he had been there before but had forgotten to carve his name, in which case Salaburu was the challenger. Whatever the case, Etcheberri was putting everyone on notice.

There is no doubt that the carvings supported a sense of "ownership" regarding particular sheep ranges. People who year after year drove their charges into the same canyons and mountain range considered those campsites their own, and some arborglyphs convey this impression. These "sagebrushed" old-timers were set in their ways and unwilling to relinquish their claims to younger herders.[144]

Furthermore, according to Pete Mendiboure of Madeline, California, sheepherders improved their sites. They developed springs, installed pipes to collect water, and did a lot of work that benefited not only them and their sheep but wildlife as well. These improvements gave the herder certain rights. Even the so-called tramp sheepmen felt that they had rights to particular grazing areas. But occupancy was normally the deciding factor. Whoever arrived first was entitled to the pasture. In the spring, competition was so keen that some herders preferred to eat cold food rather than light their stoves and give away their position to rival operators.[145]

Take a Lonely Canyon and Meet Bixenti Astorkia

Suppose you are laboring up a remote Nevada mountainside, overwhelmed by the isolation. You start wondering who else has been there before you—Indians? conquistadors? trailblazers? fur trappers?

As you continue up the canyon, your question is to some extent answered when you encounter tree carvings. You feel history come alive; you feel part of it, as if you were actually talking with the old sheepherders, because suddenly, as you walk around a large tree, you come face-to-face with the herder himself, looking you straight in the eye. He is burly, and under his large nose he has a cigarette in his mouth; he wears a hat and a jacket, and his feet are firmly planted on the ground. In his hand he is holding a sturdy stick. He is not smiling, but neither is he intimidating. You walk some more and you are startled by the figure of a snake. Perhaps the carver had dreams about it or had a scary encounter with a rattlesnake. After a few more dozen yards you see a totally naked woman. Almost certainly the herder had dreams about her. There is no telling what you might find. And every range where sheepherders went contains this type of "living" information about specific herders who stood on exactly the same spot you are standing on now.

If not for the aspens, how else would we know about Bixenti Astorkia,[146] a Nevada laborer of a century ago? If he passed through the port of New York, his name may be found among the records of Ellis Island. But if he jumped ship or crossed the border from Mexico, the chances of his name being recorded anywhere diminish dramatically. And even if his name was registered in New York, what happened to him next? Did he take a train west?[147] What was his destination, and where did he work?

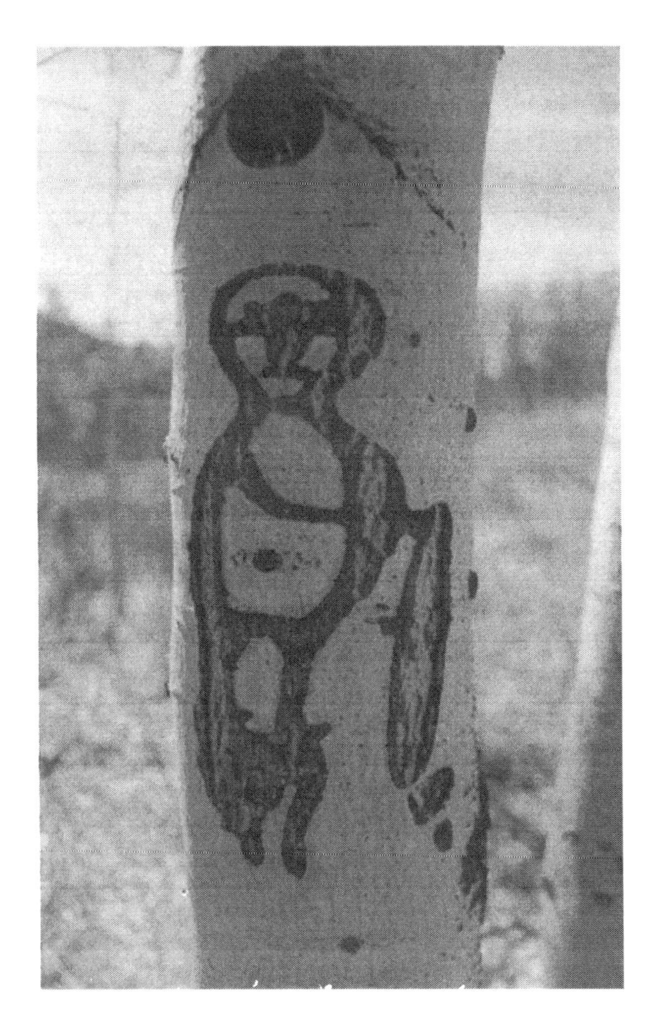

15. Herder with a big stick, Elko County, Nevada.

Possibly the last place anyone might think to find him is on lonely Telegraph Mountain in White Pine County, Nevada. Yet several carvings dated 1901 attest to Astorkia's presence there. We do not know how he found the place. He probably knew someone who worked there. We do know that Astorkia herded sheep on Telegraph for at least two years. We also know what Basque region and hometown he was from, the name of his boss, and how much money he earned. In addition we know some of his thoughts. Somehow these bits of news seem more momentous than similar items about a 1901 immigrant in New York City or San Francisco.

If that range of White Pine County is terribly isolated even today, we can imagine what it was like in 1901. No wonder Astorkia was homesick. To alleviate his misery he thought of the places in Bizkaia that meant something to him. Astorkia found a beautiful aspen growing straight on a shady bank and, stretching his arm as high as possible, he first carved a large cross. Below it, he inscribed: "Bilbao Sorronza Durango Guernica Cortezubi Arteaga Ibarrangelua Elanchove Nachitua Bedarua Ipestar Lequeitio." And then he signed it: "Vicente Astorquia, 1901." The names are of twelve towns in Bizkaia. Most of them are spelled according to the then-official Spanish form; but "Ipestar" and "Bedarua" indicate that Astorkia spoke Basque as his primary language and was born nearby—in Natxitu, in fact—a town located near the Bizkaian coast. On a nearby tree he even left a street address in Bilbao, the largest and most industrial Basque city. He must have been fond of the town because on yet another tree he carved "Bilbao" three times.

All this information is borne on old aspen trees that still stand. Astorkia's example is repeated in thousands of similar carvings, albeit more recent ones, by other herders. The historical significance of the carvings is this: an ordinary sheepherder left his mark on trees, and thereby his memory survived.

16. A carving by Bixenti Astorkia that begins with a cross, followed by twelve names of towns in Bizkaia: "Bilbao Sorronza Durango Guernica Cortezubi Arteaga Ibarrangelua Elanchove Nachitua Bedarua Ipestar Lequeitio, 1901." No doubt they represented the world he had known before and missed now.

A Doomed Resource and the National Register

By the 1930s aspen groves in the West must have been thoroughly adorned with carvings. By their own admission, about 90–95 percent of the herders left their names on trees.[148] This means that tree trunks contain the most comprehensive list of sheepherders ever compiled anywhere in the American West, including by the U.S. Census, the Office of Immigration, or state, county, and municipal records. I suspect that the herders themselves were aware of this. Felix Mendiola of Table Mountain in central Nevada certainly was when he carved: "Felix Mendiola was here, 1912."[149]

According to the guidelines issued by the federal government, arborglyphs should be both protected and recorded. The National Historic Preservation Act of 1966, 80 Stat. 915, 16 U.S.C. 470 and E.O. 11593, as amended, "authorizes the Secretary of the Interior to expand and maintain a National Register of districts, sites, buildings, structures and objects significant in U.S. history, architecture, archaeology, engineering and culture."[150]

Until recently, federal archaeologists and historians made almost no effort to record the arborglyphs, despite the ephemeral nature of aspens.[151] Their failure was the result of a number of factors, not excluding prejudice against minorities and their cultures and wholesale dismissal of the carving phenomenon as pornography or simple doodling. The inevitable result was that a great portion of this massive data bank was lost. There is no reliable method to calculate the total number of inscriptions lost, but very roughly 70–80 percent have probably disappeared. It must be noted that the demise of the carvings is not uniform in every state. For various reasons such as droughts and fires, the decline in some groves is very high; others appear somewhat more able to resist decay. There is an urgent need to record the remaining arborglyphs before we lose them as well.

The news is not entirely negative. Arborglyphs have been recorded in Nevada and California, the Wasatch Mountains of Utah, and northern Arizona. In southern Colorado archaeologist Polly Hammer of Delta and her assistants have gathered and photographed thousands of them.[152] In Cocomino National Forest of northern Arizona, archaeologist Linda Farnsworth and her colleagues, assisted by senior citizen volunteers, have accumulated more than two thousand photographs of carvings.[153]

This much is certain: arborglyphs are subject to nature's law of self-destruction. They are doomed to vanish unless we clone them, that is, unless we copy old and decaying carvings onto the trunks of nearby young aspens. Not all arborglyphs need to be cloned—only the better specimens in each grove, thus starting the carving process anew. Only thus can we ensure the continuity of a vital part of the history of the American West.

chapter one

A Forest of Names

Peavine Mountain and Copper Basin

The sheepherder's primary purpose and ambition when approaching an aspen with his knife was to carve his name and the date. At its heart, the arborglyph phenomenon is the history of individual herders who through their carvings are enjoying their moment in the limelight.

During my presentations of arborglyphs I often ask my audience this question: "If you had only one tree to carve, what message would you inscribe?" The answer is always the same: "my name." Names are essential in our culture. Without them, history would be devoid of interest, for we could not identify ourselves with the players, whether heroes or villains. So the herders were no different from the rest of us in that regard. But, of course, they had other ways to inscribe their identities as well.

Self-identity can be stamped in written form, as a self-portrait, or in the figure of a hand, but each form means the same thing. In the summer of 1925 Manuel Iturregi of Humboldt County, Nevada, put his left hand on a clean aspen tree and with his knife held in his right hand traced around his fingers. In the same drainage, Batita Barrenetche did likewise on 8 September 1956. The hands seem to be saying: "I was here; this is me," and perhaps: "I am a good herder."[1]

Was history per se a concern for the herders? Yes, for a number of them it was. Facundo Amostegui working the Sierra in 1964 wanted everyone to know that he had built the corrals[2] at "Kloba Meros" (Clover Meadows). In Columbia Basin, Elko County, a fellow named Eusebio claimed to be a "famous Bizkaian sheepherder." The herders who bragged about their professionalism or about growing the fattest lambs are also expressing a desire to be remembered by posterity. But even when we are not sure of the carver's intentions, the arborglyphs do answer some of the crucial questions of history: who, when, and where.

J. Saroiberry
Basco Fransis, 1912

J. Saroiberry
French Basque, 1912

Ni nas Jesus
viscaitarra
ni nas arsain pobre bat
1963

I am Jesus
a Bizkaian
I am a poor sheepherder
1963

Pierre Ygoa
Basco frances
Urepel Basses Pyrenees
14 de agosto 1931

Pierre Ygoa
French Basque
Urepel, Lower Pyrenees
14 August 1931

Miguel Recarte
Errazuarra
vasco, 1958

Miguel Recarte
Native of Erratzu
Basque, 1958

Localizing Sheepherders by Range

The carvings are geographical markers, like the soldiers the general places strategically on the map of the battlefield spread on the table. The herders were the soldiers of the sheep business, and if we marked all of their carved names and the dates on the Nevada map, we would have a bird's-eye view of their whereabouts and movements. Let us briefly journey through some of the major sheep ranges that contain tree carvings. (See Appendix 1 for a list of the names recorded in most of the mountain ranges discussed below.)

Yosemite National Park, California

The Sierra Nevada was an early target of sheepherders escaping the crowded conditions in the valleys of California or seeking relief from the drought. Arborglyphs can be found in the Sierra National Forest—south of Yosemite—and farther south in Sequoia National Forest. Presumably some of the older arborglyphs are still there, but no one has recorded them.

According to park historian James Snyder, there were never more than fifty herders in Yosemite in the summer even during the peak of the sheepherding business. Emigrants Pass north of the park, Mammoth Pass, and the headwaters of the San Joaquin River were much more important ranges. Competition was fierce there between cattlemen and sheepmen, and there are several "Graveyard Meadows" where shootings and deaths occurred.[3]

Within the park itself, the oldest blaze outside Yosemite Valley was carved by Antonio GA or EG and is dated 1875. The initial A is more elaborate than the rest, a characteristic observed in many Basque aspen carvings of the present century. Blazes with dates carved in succession are probably the work of sheepherders as well.[4]

Snyder has identified a number of sheep camps, often with crosses carved on nearby trees, a detail that he associates with Basques. In 1881 Col. William Shepherd noted an interesting fact about sheepherding in the Yosemite when he observed that herders felled large trees over rushing streams so that their sheep could walk over them to the opposite side.[5]

Toward the turn of the century, Basque herders accustomed to pasturing their charges in the high Sierra faced the opposition of the cavalry and the new managers of federal reserves, the Forest Service rangers. Snyder told me that the herders knew the terrain better than the federal guardians and often eluded them. According to Jeronima Echeverria, herders sometimes nailed messages on trees to warn fellow herders of the whereabouts of the federal agents.[6]

Most of the Basque herders in Yosemite were from Iparralde, and they inscribed more messages in Euskara than did their countrymen from other regions. Under the circumstances, the Basque language may have become a strategic tool for the herders, who left messages in a language the rangers could not understand. All in all, however, the number of identifiable Basque carvings in Yosemite National Park is negligible.

Eldorado National Forest, California

The archaeologists and historians working in this forest were among the first to record tree carvings. The groves on Packsaddle Road and in the Long Canyon area were recorded by Linda Goddard and Dana Supernowicz from 1978 on. They documented about a half-dozen herders from the 1910s to 1940, most from Nafarroa.

Toiyabe National Forest, Bridgeport, California

Quite a few people have contacted me to report many carvings in the Bridgeport area. I have recorded only a few hundred of them. Today Bridgeport still has some sheep. In the 1930s through the 1950s, when Alfonso Sario and Santiago Presto (they were related) ran sheep, Bridgeport was their summer headquarters. Today the Iturriria Sheep Company leases their land and has taken over their allotments.

Monitor Pass, California-Nevada Border

This forest is rich in information. I have recorded about one thousand carvings, but much remains to be recorded. Especially abundant are artwork and figures, which surpass most other areas. It appears that some of the outfits that ran sheep in this area were based in Gardnerville and Minden, Nevada.

The carvings begin in 1901 with Alfonso Zuricaray, and there are seven more names dated in the first decade of the century. About 70 percent are Iparralde Basques, 25 percent are from Nafarroa, and 5 percent are Bizkaians and others. The Bordas and the Laxalts are represented prominently. The three Borda brothers—Guilen, Batita, and Erramun (Ramon)—were natives of Bidarray (Benafarroa). This large family lived in the Oilaskoa farmstead, perched on a steep hillside. The Bordas carved their names differently, perhaps according to the mood of the moment, in Basque, Spanish, or French. Apparently the first, Guilen, had arrived in Monitor by 1909, but Batita also carved at Bliss Canyon in the same year. Almost all of their carvings are dated between 1911 and 1919, but their carving activity peaked in 1915. In the following decade the Bordas ran their own sheep, and they stayed in business until 1997.[7]

Pine Nut Range, Carson Valley, Nevada

The sheepherders spelled the name phonetically and carved it "Painot" or "Pinot." This is BLM territory, and the Carson City District partially funded the research I conducted on Pine Nut. I have not heard any reports of conflicts between Basques and Native Americans, who visited this range regularly to harvest pine nuts.

The many arborglyphs here (about 750 cataloged so far) indicate that the area was heavily grazed. During dry years dispirited herders would vent their frustration by carving "Pinot is not worth a dick" and similar messages.[8]

This range was used by the same sheepherders who worked on Monitor Pass, and names such as Laxalt, Borda, and Dangberg are repeated in both areas. According to BLM records, Laxalt and Borda quarreled over grazing boundaries on Pine Nut, but such arguments were not unheard of among herders.[9] Even in 1987, an old herder shook his *makila* while complaining to me about the younger herders in his outfit who tried to push their sheep into his territory and feed on the best grass.[10]

Several *harri mutilak* on the mountain ridges on either side of Carson Valley may have been placed there as reference points and range boundaries to minimize disputes over allotments. One on top of Pine Nut looks impressive from a distance, but I never hiked close enough to photograph it.

Pine Nut has relatively few old carvings, possibly because of its proximity to population centers and their mining and woodcutting activities. Incidentally, one abandoned mine and its mill are located right by the aspens, and a nearby tree reads: "Morgan Mining Co." I recorded eight initials and names dated between 1900 and 1910, among them P. Laxalt in 1906; twenty-four for the following decade; twenty for the 1920s; and only thirteen for the 1940s.

Intense carving activity occurred here during the 1940s, much of it the work of a handful of herders who produced considerable sexual material and erotic representation; but the 1960s were the most prolific. By this time four herders from Cantabria (also called La Montaña or Santander, Spain) had joined the Basques, and they carved dozens of trees each.

Hope Valley, California

I found an 1896 carving here, one of the oldest in the Sierra. It depicts a pair of open legs, and below them the word *caca* (shit).

One notable carver is Felipe Errea from Mezkiriz, Nafarroa. He arrived in Reno in 1944 as part of the Nevada Range Sheep Owners Association's effort to recruit Basque sheepherders. All the recruiters were from the highlands of Nafarroa. Errea and his friends Jose Babace, Fermin Jorrajuria, and Raimundo Villanueva came together and worked for J. B. Dangberg of Minden. Babace worked at the ranch, but the others herded sheep and left their names carved in the mountains. Stars were Errea's favorite motif, along with regional references.

Genoa Peak Area, Nevada

As a rule, inscribed names dated during the 1910s far outnumber those from the previous decade. But that is not the case on Genoa. This anomaly appears to indicate that Toiyabe National Forest was already dispensing permits and controlling grazing.

This area contains some prime mountain meadows where a few aspens carved in 1900 still survive. Two of the carvers, Louie Johnson and Emilio Alvarez, were there for several years. Anton Inda's summer grazing grounds extended south to Stanislaus NF and Yosemite, where he was carving several years later. In 1907 he was back on Genoa.[11] Among his contemporaries were John Egoscue and A. Salet.

Several Basque outfits such as the Bordas, Laxalts, Grenades, Hachquets, and Uharts had their *kanpo handia* (main summer camp) on this mountain. Senator Paul Laxalt carved his name near the camp when he was just a young boy.

The legacies of some of the most prolific and outstanding carvers are found here; for example, Batista Amestoy, who imprinted his name calligraphically; Jose Mitjana, the star worshiper; Martin Lanathoua, the snake lover; and Etienne Maizcorena, master of them all, who carved beautiful couples, horses and riders, bucks, and crosses.

Spooner/Marlette Lake, Nevada

This range, located north of U.S. 50, is practically a continuation of Genoa Peak, and thus many of the herders' names occur in both locations. Only the extreme eastern slopes of the Sierra in the Hobart and Little Valley areas contain different names. The groves in Marlette contain older carvings, but fewer than those on Genoa. An article published in 1971 mentions a carving on the Marlette Lake trail dated 1895, but I am afraid most of the carvings from the 1800s have not survived.[12]

One notable herder in these parts was Martin Banca. He must have worked for more than one boss because he traversed much of the northern Sierra. North of Lake Tahoe the documentation on him begins in 1901. In Marlette/Spooner his last date is 1923, but he continued herding sheep elsewhere for another decade. Banca—from the village of the same name in Benafarroa—appears to have been a rather typical figure of the Nevada/California desert, a wanderer and loner, the sheepherder version of Pete Aguer(r)eberry, the Basque miner of Death Valley. This area is significant for Nevada Basques because the old Laxalt sheep camp is located high above Marlette. Although it is snowbound for much of the year, the Laxalts preserve it as a family heirloom and a piece of Nevada sheep history. Robert Laxalt immortalized these ranges in *Sweet Promised Land*, a book about his sheepherder father, Dominique, one of three brothers who came to America in the early 1900s from the mountainous region of Zuberoa. The Laxalts are the best-known Basque family in the state. They include Robert Laxalt, the well-known western author and founder of the University of Nevada Press (1961), and Paul Laxalt, governor (1966–70), senator (1974–86) for Nevada, and prominent figure in the Republican Party.

Unfortunately, few carvings have been found in the vicinity of the camp itself. According to Robert Laxalt,[13] his father did not carve, but his uncle Jean-Pierre left interesting statements, especially in Basque. His preferred subject was birds, of which he left numerous delicate figures. President Ronald Reagan once came to visit the sheep camp, and he, too, carved his name.

In the Laxalt family archive are several outstanding cultural documents in the form of letters in the Basque language that the brothers wrote to each other or to other friends in the American West.[14] These letters are good examples of a type of Euskara Batua (Unified Basque Language) that took form among the Basques in the western United States long before it became a reality in Euskal Herria. Laxalt sprinkled his native Zuberoan dialect with other dialects, especially that from Benafarroa and Nafarroa, the dominant dialects in the Nevada and California sheep ranges.

Mount Rose, Nevada

Between Marlette Peak and Mount Rose are Hobart and Little Valleys. Numerous traces of sheep activity are found here, but few of the carvings are outstanding.[15] Mount Rose is the tallest range in the Reno-Tahoe area, and additional mountain ranges and drainages are covered under this umbrella designation. The area contains large aspen groves that to date remain unrecorded; for example, Hunter Lake, the upper meadows on Mount Rose, and Bronco Creek.

On the western side of Mount Rose I have researched three large groves. The oldest date appears to be 1881, signed by Frank Narajero, a Spanish surname. Other nineteenth-century carvings are: "A. P. 1890" and "1898 Do . . ." (Domingo?). Four names have been recorded for the first decade of the present century.

On the eastern slopes most of the data come from Thomas Canyon. I have found few old dates, with only four names for the 1910s and eight for the 1920s. I have no clue why this is so, but as research continues the picture might change.

Arborglyphs created by Sario, Fermin Gamio, and Santiago Presto in the 1950s abound, and from them we learn that Nafarroan herders were in the majority here. Sario and Presto, both from Nafarroa, owned sheep.[16] In the 1960s Zabalegui, the Sarratea brothers, and Jess Arriaga, who live in the Reno–Carson City area, carved many trees. They worked for Jean Zubiburu, from Esterenzubi, Benafarroa. After Alfonso Sario retired, Zubiburu leased the land from Redfield. The last date carved by sheepherders in Thomas Canyon is 1968.

Mount Rose, rising to 10,778 feet, offers spectacular views of Reno and Lake Tahoe, lush meadows, cold trout creeks, and thick groves. It never gets too hot there, even at the height of summer. The sheepherders left several statements appreciative of the beauty of the mountain, which became very popular with hikers after it was designated a wilderness area.

But beauty alone was not enough to alleviate the boredom of the lonely Basques. One tree says: "This place is very pretty keep it that way," but I doubt that it was carved by a herder. If it was, he was the only ecology-minded sheepherder I know of.[17] There is no doubt about the authorship of the next message: "These places are very pretty, but to stay with a good female." The beauty meant nothing to J. Z., who complained: "Here I am without God," which is the literal translation of "yo aqui sin Dios," but conveys the idea of being utterly alone.

Tahoe National Forest, California

The archaeology department in this forest district began recording carvings in 1979, but it appears that the effort waned or fell short. By the researchers' own admission, many of the sites were "poorly recorded"; even the largest aspen groves were hardly researched until the 1990s, when Tahoe NF archaeologists approached me.[18] But the early efforts paid off; many of the trees recorded in 1979 are gone or their carvings have deteriorated beyond recognition. The agency has IMACS forms on file for thirty-nine Basque aspen sites.[19]

I recorded well over two thousand arborglyphs in the Truckee-Sierraville-Downieville–Nevada City Districts, and although that may seem like a lot, my survey was far from exhaustive.

The carvings would be historically more significant if

it were possible to match the sheepherders with their employers. That information can be gleaned from the grazing maps and the list of grazing permittees in the Tahoe National Forest archives. Unfortunately I have not been able to get copies of them. Records do exist on some of the sheep companies that operated in Tahoe NF. In the area between Truckee and Sierraville, Wheeler was prominent in the first decades of the twentieth century until the Depression. In 1867 Dan Wheeler (Uncle Dan) bought sheep in Lakeview, Oregon, and drove them to Virginia City.[20] Reginald Meaker came after Wheeler, followed by Wilbur Mills, who sold to Abel Mendeguia in 1966, who sold to Victor Arretche and Bernard Etcheberry of Bakersfield. Currently Ray Talbott of Los Banos, California, has the lease. In the Pole Creek area, there were two permittees in the early 1990s: Bernard Etcheberry and Albert Evratchu. According to Tahoe NF sources, Pole Creek was a traditional transhumance corridor even before the federal agency was created.[21]

The forest has preserved a few old names and dates. For example, Beesley and Claytor found arborglyphs dated in the 1890s in Placer, Nevada, and Sierra Counties of eastern California, but that was in the 1970s. Between 1910 and 1920 no fewer than twenty-five herders worked in or passed through this forest. The carved evidence increases significantly for the 1920–30 period, then decreases to just twenty-two herders between 1930 and 1940, when it would have been expected to hold steady or even increase. The 1940s yielded considerably fewer names, chiefly because of World War II.

During and after World War II sheep operators made strenuous efforts to import Basque herders, and in the following years the industry strengthened somewhat. In most ranges there are many more names for the 1950s than for the previous decade, but research conducted thus far in Tahoe NF does not bear this out. And the tally for the 1960s appears to be far below what it should be. In the 1980s and 1990s most of the herders in this area were Mexican, with some from Peru.

Among the "professionals"—that is, those who dedicated a lifetime to the sheep—were Martin Banca, 1901–30s; Jean Barbieri, 1919–50; Cruz Delgado, 1937–54; and Charlie Paul, 1911–50. One of the most celebrated carvers in this part of the Sierra was D. Borel. In August 1925, nine years after his first stint as a sheepherder on Mount Rose, he was south of Sierraville, California, where he spent the summer and part of the fall. According to Albert Gallues of Reno, Borel was a little Frenchman with a large moustache who did not speak much or any Basque. He was employed by Martin Gallues, manager of the Wheeler Sheep Company.[22]

Plumas National Forest, California

This forest contains large groves, some old trees, and a few of the oldest carvings found in the northern Sierra. They are dated from 1887 to 1892, but the carvers' names are not clear.

The earliest arborglyph sites recorded in Plumas National Forest date from a 1980 survey conducted by archaeologist Dan Foster in Grizzly Valley, Beckworth Ranger District, California. Foster today is the supervisor of the archaeology program at the California Department of Forestry and Fire Prevention in Sacramento.[23]

I have not researched Plumas NF as much as I have wanted to. In the early 1990s I visited the Jenkins Sheep Camp by Davis Lake near Portola, California, and Poco Valley Sheep Camp at the invitation of archaeologist Cathy Sprowl. The first site, which needs repair, is impressive; it contains a large cabin and a large oven surrounded by a protective wooden fence. The oven, built by a herder named Urrutia, is partially crumbled.[24] The name of the second site, Poco, is intriguing; the Plumas NF personnel did not have an explanation for it. There was a Basque sheepman by that name in Surprise Valley, California, but I do not know if the two names are connected. The cabin in Poco Valley is in good shape, but the oven is in disrepair.

Plumas NF officials from Milford, California, have told me of several sheep camps and ovens in their district, but I have not yet visited them.

For the data on Plumas I am indebted to E. Loren Kingdom of Greenville, California, who graciously sent me two videotapes of arborglyphs he recorded in 1992–94 in the Quincy area. In fact, he recorded the oldest date on an aspen ever videotaped, 1870, which is still clearly readable and appears to be authentic.[25] Even though there is no name associated with it, the carving proves that some aspens live considerably longer than sixty to eighty years. The carvings in Plumas NF feature a lot of art, human figures, and animals as well as messages and statements. It is a rich forest indeed.

South Warner Mountains, Modoc National Forest, California

Geographically speaking, the research conducted here in 1995 covers but a small part of the Warner Range, but it proved that rich, history-laden groves endure in isolated ranges. The State of California owns some timber in the South Warner Mountains, and in 1995–96 Daniel R. Sendek, forestry's licensing executive officer, photographed some arborglyphs in the Blue Lake area. One photograph showing three aspens carved by Marcos Yrigoyen in 1933 appeared on the cover of the California Board of Forestry and Fire Protection's newsletter.[26]

In the 1990s two sheep outfits still operated on the South Warner Mountains.[27] This border area of California and Nevada has seen a lot of Basque history. Below the mountain I researched, near the tiny, picturesque town of Eagleville, California, are buried three Basque sheepherders—Bertrand Indiano, Jean Battista Laxague, and J. Pierre Erramouspe—and an Englishman named Harry Cambron. All four were killed in the winter of 1911 in northern Washoe County, Nevada, touching off the episode known in Nevada history as "the Last Indian Massacre." The Indians who were blamed for the crime were chased by a posse and slaughtered almost to a man.[28]

Not all sources are in agreement regarding what happened. One writer cited abuse of Indian women by Basques as the motive for the killings.[29] Nevada historian Phillip Earl believes that the killers were cattlemen from Surprise Valley, California. He told me that the Indians' guns did not fire the bullets found in the bodies of the sheepherders.[30]

I have found no carvings by Indiano or Laxague, although his nephew Gratien Laxague left many inscriptions, and Erramouspe may have, of which more later. The Warner Mountains in northeastern California and aspen groves located in the Summit Lake area of Humboldt Country, Nevada, must be thoroughly researched for further carvings by these men or other herders who might have alluded to the killings. In the general vicinity where the sheepherders were massacred I found several carved epitaphs, one of them in memory of a non-Basque: "Here died Bob Makale, 1956." The phonetically spelled name of Bob McCully and the message are carved under an elaborate cross and repeated on a couple of nearby aspens. If specifying the spot where McCully died was so important to this carver, it seems likely that someone carved a comment in memory of the slain Basque herders.[31]

In the winter of 1990 I spoke with Mary Jane (Erramouspe) Cook,[32] who told me that Jean Pierre Erramouspe came to Bakersfield in 1888 and homesteaded in Badger Mountain, Humboldt County, Nevada, and later moved to Cottonwood Ranch in Eagleville, California. He married Dominika Itzaina (usually misspelled "Itciana"). After he was killed, his brother Jean Erramouspe (known as "Manes Ttipia") married Dominika in 1913. Mary Jane said that in the early days, most of the Basques lived in Fort Bidwell and Cedarville, California.

The Basques came early to California's Surprise Valley. According to the 1900 Census, James Dunn of New York, one of the area's largest sheepmen, hired Basque herders. By 1885 Baptist Alsaga (could be Latsaga, the Basque spelling for Laxague), Antone Ossafrain, Pedro Ossafrain, Eugene Salet, Jose Refugio, and Weper Madira[33] worked for Dunn, and a few years later (1894) Pierre or Pete Erramouspe arrived. I found several carvings by Eugene Salet dated from 1897 to 1902.

One of the early Basques in the area was Jean Pierre Duque, who herded sheep between the Warner Mountains and Oregon. According to his niece, he was one of the first herders in Nevada to use burros.[34] Another early herder not listed in the Census was Batita Poko. Poko/Poco (he spelled it both ways) was herding sheep in the South Warner Mountains as early as 1892, according to a large aspen. He carved several trees in 1895 as well, and many more in 1911. Poco and Duque herded sheep for the Irishman Pat Flanigan, and in about 1900 they bought his Smoke Creek Ranch.[35]

Granite Mountain, Gerlach, Nevada

I recorded only a few dozen arborglyphs here. According to Leandro Arbeloa of Reno, most of the carvings are high up on the mountain.[36]

The Espil Company is one of the most recent outfits to graze sheep on this range, but Granite has seen feverish sheep activity in the past. According to one report, a Basque named Anchordoqui homesteaded in this area, and in the 1890s Louie Gerlach bought his place (Gerlach is a railroad town, dated to 1910).[37]

Basque sheepmen from California's Surprise Valley, such as Laxague and Urrelz, used Granite Mountain for summer pasture. In 1925 Joe Ugalde arrived from Idaho and formed a partnership with Pedro Camino, Pierre Camino, and Lawrence Holland. Ugalde was in business until 1948. Another Basque sheepman in the area was Martin Lartirigoyen.

The Homestead Sheep Camp is on the northeastern side of Granite Mountain. There is a question as to whether it is the original Anchordoqui place.[38] The camp comprises a cabin, an oven, and a unique stone cellar, still in working order if only it were cleaned of the debris the packrats have accumulated inside.

In nearby Smoke Creek, soon after the turn of the century, Dominique Laxalt, a youth of eighteen, was breaking mustangs for John Jaureguy and driving teams through Gerlach and the Black Rock Desert. His son Robert Laxalt was born in this area.[39]

Summit Lake Area, Humboldt County, Nevada

This area is managed by the BLM,[40] and the archaeology department of the Winnemucca District made my research possible by providing transportation, maps, and film. In addition, department personnel joined me and others in the field, helping to record the carvings in the vicinity of Summer Camp, which used to be called Sheep Camp.

The Black Rock area of northern Nevada and its environs constitutes a unique microecology whose significance transcends Native American history and the earliest pioneer roads that crisscrossed its inhospitable immensity. The region also saw the greatest concentration of Basque immigrants in Nevada. It stretches from Surprise Valley, California, and Smoke Creek in Washoe County in the west to the Mary's River and beyond in Elko County in the east. The northern boundaries of this sheep haven extended from Steens Mountain and Jordan Valley in Oregon in the north to Paradise Valley in Humboldt County. Jess Goicoechea of Elko, himself a lifetime sheepman, vouched for the high ovine population density in this area.[41]

Although the region is too dry to sustain many aspen groves, a surprising number of older aspens still stand in canyons east of the Black Rock Desert in the vicinity of Bartlett Butte and Summit Lake. The ecological contrast between the flat, barren, alkali bed of the desert playa and the moist, grassy earth in the shady aspen forest is incredible. But that is Nevada.

Hundreds of Basque sheepherders worked here for more than a century, beginning in the 1880s. There are two aspens in the foothills of Bartlett Butte carved by PE and dated 1895 and 1896. A cross can be seen on a nearby tree, and an E (or P) followed by the oldest date ever recorded: 1830. Since the aspen does not appear to be *that* old, 1830 may refer to someone's birth date. In any case, we have no historical information about white people being here in 1830.

The question is, does PE stand for Pierre Erramouspe? His name was Jean Pierre, but he went by just Pierre or Pete. Recall that he started herding for James Dunn of Alturas, California, in 1894, and although Alturas seems a long way off, we know that sheepmen traveled hundreds of miles in search of pasture. Another coincidence is the matching of the dates. PE continued working and carving in the area until 2 August 1907, and another distorted carving of his appears to be dated 1909, two years before Erramouspe was killed. It is the last known carving by PE. According to the available information (U.S. Census and oral interviews), no other herder with the initials PE was in that area.

This remote part of northwestern Humboldt County has yielded the longest list of early Basque herders compiled to date. Additional names will no doubt be collected as research continues in the still unrecorded large groves nearby.

Certain aspects of the findings are puzzling. For example, why are there so many carvings dated 1900–1901 but so few—or none—dated 1899? Was it because the herders wanted to record the beginning of a new century? Perhaps. It certainly does not mean that between 1899 and 1902 the Basque presence in Nevada and California experienced a peak followed by a sudden drop. Quite the opposite; Basque involvement in Nevada's sheep business experienced a high during the first decade of the century.

In the Mahogany and Wood Creek drainages there are about thirty names dated between 1900 and 1910. Some are repeated and others are difficult to read, but in any case thirty names is more than I have found anywhere else.

Worth mentioning is the predominance of the Uriartes, four different herders with the same last name who may have been related.[42] They continued herding well into the next decade. George Itçaina was recorded here as well (he is surely the George Itciana seen in the media photographs of the posse that pursued Shoshone Mike and his band of Indians in 1911, although the caption says he was from Elko).[43] In the 1910s there were twenty-eight herders, an average of three per summer in two drainages alone. However, during the 1920–40 period the number of herders fell by one-third.

17. The sign over the cabin's door says "Summer Camp," but this used to be a major sheep camp. The cellar—not shown—is at the left. The photograph shows BLM archaeologists from Winnemucca, Nevada; members of the AM-ARCS organization in Reno, Nevada; and other volunteers. From left to right: the author, Annaliese Odencrantz, June Rivers, Erik Mallea, Barbara White, Regina Smith, Kirk Odencrantz, and Peggy McGauckian.

The three major emigration zones in the Basque Country—Iparralde (Benafarroa and Zuberoa), Bizkaia, and Nafarroa—are evenly represented among the herders in this part of northern Nevada. In many other groves of the state and the Sierra, Bizkaian herders were in the minority.

I was amply rewarded by the carvings in this remote part of Humboldt County; they were the clearest vindication yet of the validity of the endeavor. The carvings are remarkable for their number and their cultural richness; they shed light on previously unknown historical aspects of Nevada's past, Basque immigration, and the sheep industry.

Pine Forest, Humboldt County, Nevada

The data in Summit Lake complement those of Pine Forest, located south of Denio, because a number of the same herders worked in both mountain ranges.

In this area are two ranches owned by Basque cattle- and sheepmen: Leonard Creek Ranch and the Dufurrena Ranch. Only Dufurrena is still in the sheep business. Linda Dufurrena—a non-Basque—who lives on the ranch, is a well-known Nevada photographer specializing in sheep, Basque, and Nevada wilderness scenes. The Leonard Creek Ranch—owned by the Monteros and the Bidart brothers since 1926—is notable for having had a public school functioning there since that same year.[44] Mitch Bidart, Louie Bidart, Mike Montero, Ray Montero, and Frances Larragueta made up the first class. A minimum of five children were needed to open the school, but at first there were only four, so Ramon Montero, a native of Nafarroa, brought Frances Larragueta from Winnemucca.[45] This was very probably the only public school in the United States with a 100 percent Basque student body.

Pine Forest is the highest mountain range in this part

of the country (except for Steens), so it is no wonder that it experienced intensive sheep grazing. Winnemucca BLM archaeologists have collected dozens of arborglyphs on Pine Forest, and I am using their material here.[46]

One thing is immediately clear: Pine Forest contains fewer older carvings than the Summit Lake area. The 1900–10 decade yielded just four clear names, and the 1910s yielded fourteen. After that, however, Pine Forest catches up and even surpasses just about every range I visited. In 1923 alone eighteen herders were competing for water and pasture. Forty-two herders have been counted for the 1920–30 decade, a number surpassed only by Cherry Creek, Nevada. Oddly, no carvings were recorded for the years 1920 and 1929.

The yield for the 1930s is only fourteen names, an indication, perhaps, of the severity of the Depression and the sharp decline in the sheep industry. According to rancher Buster Dufurrena, however, in 1932 there were twenty to thirty sheep camps in Pine Forest.[47]

Steens Mountain, Southeastern Oregon

I spent a few days on Steens Mountain. On 15 September 1990 I ran into two old-timers who were fishing at Fish Lake. It was about 9:30 A.M., and these fellows were already eating sardines and cold salami and washing it down with beer. I approached them with my camcorder and asked if they minded answering some questions. The following is based on what they told me.

The earliest sheepherders in the Lakeview area of Oregon were Irishmen who ranged their charges on Steens and the Hart Mountains and in Fremont National Forest. About half of the Irish may have returned to Ireland, but others stayed, and today about 50 percent of the population in the area is of Irish ancestry. The Irish turned to cattle in the 1950s or earlier when they could not get herders.

At least in the 1930s, and probably earlier, Steens Mountain and the Pueblo and Trout Creek Mountains were taken over by Basques. One of the longest-lasting old-time herders was the Bizkaian Tomas Zabala, who carved a lot and built several *harri mutilak* in the area. He retired, and later died, in Winnemucca, Nevada. Many Basques who worked with sheep on Steens settled in Burns and Hines, Oregon, and were employed in the lumber mill.

Nevada has Porno Grove, but Steens has Whorehouse Meadows, so named because prostitutes drove up and parked their wagons under the aspens in the nice meadow near Fish Lake. Lakeview had a red-light district until 1962, and I found a carved reminder of that little-known fact in the South Warner Mountains. In Whorehouse Meadows and its environs I recorded about seventy-two arborglyphs. Two of them, dated 1933 and 1937, alluded to prostitution activities, including female names, one of which appeared to be Basque. Most of the sheepherder names I recorded were Bizkaian; the oldest was Jose Egurrola, 1923, and the most recent was Resu Uriarte of Arratzu, Bizkaia, dated 1981. Interestingly, I saw no Hispanic names. Another anomaly on Steens was the abundance of carved figures of tea- and coffeepots, glasses, bottles, and even a fancy detailed figure of a steamship.

Steens Mountain features long, deep canyons populated by aspens, which I presume contain thousands of additional carvings.

Columbia Basin, Elko County, Nevada

In the old days, when the green aspens bearing markings dated in the early 1900s were young, the sheepherders used to catch Pacific salmon there, but nowadays they are happy to get trout for dinner. The old carvings went the way of the salmon, in part because of the beaver activity. A herder who tended camp in this paradisiacal place remembers seeing a carving located, ironically, on Beaver Creek, carved by Vicondoa in 1889.[48] No further information is available about this ancient carving, the oldest dated arborglyph in Elko County (but there was a Vicondoa Sheep Company operating in the 1950s and 1960s in the South Warner Mountains).[49]

The results of the research in the Columbia Basin have been meager so far, although the names of sheepherders increase from the 1950s onward. The Allied Sheep Company leased the Columbia Basin forest reserve, and in the 1950s and 1960s a dozen or more Basque herders worked there. Manuel Arbillaga, born in Argentina, was the manager for many years until Fermin Alzugaray from Nafarroa took over. The main summer camp contained several buildings and a large oven. The I & L Ranch in Tuscarora, Nevada, was their winter headquarters, and the sheep ranged in the Midas, Valmy, and Golconda areas. The *trela* (stock trail) from Pumpernickel Valley (the herders called it "Panpanika") to the high country took thirty-two days to traverse.

While the sheep grazed in the Valmy area, the camptender had an additional duty: fetching water. Although water was readily available in the hot springs that abound in the valley, it did not cook beans—not satisfactorily, anyway—and the sheepherders were very fussy about their beans.[50]

Independence Mountains, Humboldt National Forest, Elko County, Nevada

There is still a chance that the Independence Mountains will yield the names of the Basques who worked at the Spanish Ranch after Pedro Altube settled there in the 1870s,[51] but the longer we wait to look, the less chance we have of finding them. The elements and time are not the only threats to the survival of the trees. In the past, entire sections of groves have been bulldozed by mining companies, and others may still be threatened.[52]

Initial recordings in the Toe Jam Mountains yielded several carvings by a "Mr. Pete Elgart," the oldest of which is dated 1913.[53] Other obscure names dated in the 1910s are located in Jack Creek Canyon; perhaps their authors worked for Altube or the YP.

The North Fork Canyon, traversed yearly by the herders of Ellison Sheep Company, is one of the few areas in this mountain range that I recorded thoroughly. It was here that I found the earliest carvings by Frank Rodriguez, dated 1913 and 1918. Otherwise, the canyon contains few older arborglyphs and is not unusual in any way—except, perhaps, for several anguished messages whose wording suggests that they were authored by an early Peruvian herder. One Basque herder carved "Coward" next to one of them.

The ecology of North Fork Canyon has been profoundly affected by the mining operations in the area. The new mine road was bulldozed right through the middle of an aspen grove, and I do not know whether the carvings were recorded before they were destroyed. The aspens that flank the creek bed have been spared, but when I was in the area the water ran muddy with silt. The most impressive sight is the bridge that the mining company put into place so that the monster trucks could travel from one side of the canyon to the other without having to descend into it. They accomplished it by filling the canyon with dirt.

Ruby Mountains, Elko County, Nevada

The rugged and spectacular Ruby Range in Elko County is under the management of the Humboldt and Toiyabe National Forests, and their personnel provided transportation, maps, and film for the research conducted there. Initially I was greatly disappointed because some of the canyons I visited in August 1990 contained fewer carvings than I expected, and not very old ones at that.

The sheepherders I interviewed indicated that there are numerous carvings in the groves in the upper elevations, which are difficult to reach even on horseback. According to one herder, "Up there, with all the loneliness, is where the sheepherders carved their hearts out."[54] Those carvings are in extreme danger of disappearing. I want to thank Cody Krenka of Ruby Valley for hiking up to Cold Lake, high in the Rubies, and photographing the carvings that are still there.[55] He found some carvings from the 1920s, which seems to indicate that the groves at high elevations contain information on earlier herders.

Lamoille Canyon is said to be one of the most spectacular canyons in all of Nevada. One ex-herder who worked there in the 1960s saw carvings dated in the 1890s. Another area of the Rubies that some Elko sheepherders suggest might be rich in carvings is the range used by Pete Itçaina's herders. It seems that a great deal of recording remains to be done in the Ruby Range, which is more than eighty miles long.

Toiyabe National Forest, Central Nevada

The central part of the state encompasses several large sheep ranges, among them Toquima, that remain totally unrecorded. The area treated under this heading is located south of Austin, mostly on the western side of the range, and it was researched in 1991 thanks in part to the cooperation of Toiyabe National Forest, Austin District. Paul Ynchauspe, a Basque rancher north of Austin, used to run sheep on the Toiyabe Arc Dome area.[56]

One anomaly I noticed was the absence of non-Basque names; another was the low number of carvings for the 1930–40 decade—only two. While recording the arborglyphs I met a doctoral researcher from the University of Washington who told me about several *harri mutilak* he saw on the trail to Arc Dome. The skies were threatening rain, but I had to walk up the big mountain and videotape the stone monuments. I found two. One was typical: a large, squat pile of stones about four feet high. The other was smaller but altogether unusual with a long, large stone leaning upright against the main body of the *harri mutil*.

Table Mountain, Toiyabe National Forest, Nevada

This range, predominantly covered by aspens, is unique in Nevada, where normally the groves cling to creek banks and springy meadows. No motorized vehicles are allowed here, so I am grateful to Toiyabe National Forest, Tonopah District, for organizing a four-day field trip complete with horses, pack animals, and first-class accommodations under the aspens near the mountaintop in the summer of 1991. I was invited by Toiyabe National Forest archaeologist Arnie Turner and the Tonopah District to ride with a dozen other people—officials, rangers, conservationists, researchers, and supervisors.

I became intensely aware of the remoteness of Table Mountain when the horses began crossing the creek and suddenly the water around us churned wildly. It was full of fish! The long train of pack mules and riders moving slowly up the steep trail was quite a sight. From the back of my horse I had the opportunity to reflect on the sheepherder's view of the environment—his dependency on animals (horse, burro, dog) and his life conditioned by the sheep, coyotes, and other predators. The awareness of the immensity of the land was a particularly unsettling feeling.

Table Mountain seemed friendlier—more greenery, fewer rocks—than most other Nevada ranges, but it was nevertheless an alien world that belonged to the deer and elk, sage grouse, eagles, and fish. At one point we roused a herd of fourteen bucks and watched, motionless, from a distance as they jumped one by one over a fence line. They seemed to be running on tracks, up and down, single file, like on a roller coaster. Then, safe on the other side, they stopped and turned, and displaying their magnificent racks against the afternoon skyline, watched us.

Any "connoisseur" of the outdoors knows that Table Mountain is home to Porno Grove. It was my first visit to this notorious slice of Nevada history. I deal with Porno Grove in chapter 4; here I will simply say that we recorded it in its entirety. Table Mountain makes a good argument that abundance of trees does not guarantee an abundance of arborglyphs. The mountain did not yield a lot of data, one reason being that supposedly there have been no sheep on Table Mountain since 1948.[57] Some sheepmen must have sneaked in, however, because I recorded carvings dated in the 1950s.

A noteworthy herder here was Pete Elgart, whose first carving is dated 1907. After trekking through northern Elko County and north and south in the Sierra, he returned here—where he had started—and in 1948 carved "adioz." He must have been in his seventies by then, and this time he meant it.

Desatoya and Clan Alpine Mountains, Nevada

These two mountain ranges, located in central Nevada between Fallon and Austin, on either side of U.S. 50, are under BLM management. My research was conducted thanks in part to the archaeology department of the Carson City District.

The headwaters of Edwards Creek and Topia Canyon in the Desatoya Range are known as "Basque Summit," which suggests intense sheepherding activities in the past. Lack of aspens is the chief reason little carved evidence of sheepherders remains. The groves now standing are in poor shape because of canker, drought, or other causes, taking a severe toll on the arborglyphs.

In both mountain ranges most dates begin in the 1920s and multiply in the following decade. In 1993, when I researched this area, I did not record upper Topia Canyon because old aspens knocked downed by the wind were blocking the road and I did not have time to walk.

Traversing Basque Summit from north to south, a side jeep road takes you to Billy Canyon. Here the trees are hemmed in by the mountains and have survived the wind, which explains the older carved dates. One of the oldest carvings anywhere is here, dated 1872, but it is not totally clear. Another nearby tree is inscribed "1899 E[r]fil Uxine," but it is not clear either.

On the north side of U.S. 50 in Clan Alpine I recorded only a few carvings, but additional information may exist in two groves with older aspens that I saw way down the canyon on the northwestern side. It was a wonderful sight, beckoning me with its dazzling fall colors, but in November the sun sets in a hurry in Nevada, and I resisted the temptation to walk for several miles down the goat-steep and rocky terrain.

During the Depression, when the banks took over the bankrupt sheep outfits, it was common for banks to hire an experienced sheepman to run the business. Bert Arambel of Fallon was one of them. He was based in Whitaker's Alpine Ranch, and he hired several countrymen from Iparralde who were related to each other, namely Pete Etcheverry of Arnegi and his brothers-in-law, Coscarat and Goyhenetche of Urepel. I found one tree mark by P. Etcheverry. Pete was known as "Pete Zaharra" (Old Pete), according to his son Jerry Etcheverry of Winnemucca. He came to Eureka, Nevada, in 1913 and did not go into town for seven years. The first winter he slept in a bedroll. The following winter he bought a tent for $4.75. In 1920 he and two other partners bought sheep; Etcheverry paid cash for his share. During the Depression, the four partners went broke so Old Pete went to work for Arambel.[58]

Another Basque, Tony Ormaechea, ran sheep on the nearby Clan Alpine Ranch; in 1942 he sued Whitaker over water rights. It was decided that there was not enough water for both operations, and Ormaechea took over both ranches. Even today, several beautifully preserved stone buildings can be seen among the outbuildings of Alpine Ranch, and there is another stone cabin high on the mountain in the Cherry Creek meadows owned by Ormaechea. When I visited it in the fall of 1994 part of the stone wall and roof had collapsed. Inside and on the doorway I could read Anglo-American and Basque names, including Joe Onaindia's, dated in the 1930s. Today it is cattle country up there.[59] Angelo Mendiguren was another sheepman who ranged on the Smith Creek side of Desatoya. Most of his herders, who were all related to each other, were from Aldude.

John's Wash, White Pine County, Nevada

The BLM, Ely District, provided the transportation for my party's fieldwork in White Pine County. Four of us researched this section of the Wheeler area for a good part of a day.[60] Great Basin National Park is located in this area also, and I understand many carvings can be found within its boundaries, but I do not know of anyone having recorded them.

According to published sources, the Basques reached White Pine County somewhat later than Reno, Winnemucca, or Elko. The Mormons were the earliest white settlers here, and some of them ran sheep. One sheep owner of the area was Jean "Manes" Auzqui, who carved his name but did not date it. Jean Ithurralde of Eureka came to Nevada in 1929 and worked for Auzqui for more than ten years. Auzqui was a big man, so the Basques called him "Manes"; Ithurralde, being smaller, was called by the diminutive, "Mañiz." Auzqui's sheep outfit was the largest in the Ely area with fourteen thousand sheep and twenty employees. He had five other partners: Paris, Uhalde, Etchegaray, Labarri, and Urrizaga. Ithurralde said that during the 1930s and 1940s there were more than one hundred Basques in Ely.[61] Many of the names I recorded in the Ely area were of herders employed by this sheep company.

Another well-known personality of the Ely area was Jean Goyhenetche, recently deceased, a sheep owner and lamb and wool buyer. He carved extensively in the Wheeler and Cave Lake areas. His father trailed sheep from Bakersfield, California, to Jarbidge, Nevada, in 1894 or 1895.[62]

John's Wash offers the curiosity of trees carved by three groups of herders: Indians, Mormons, and Basques. More arborglyphs are dated before 1910 than between 1910 and 1920. The dates from 1899 are probably the work of Mormon herders or cowboys. I did not find any human figures carved by them, although I did discover one biblical quote by Swallow.[63]

The low number of carvings for the 1920–40 period—only twelve—indicates that this was not good sheep country. It is very rugged, and the dense thickets would be a nightmare for the herder.

Cave Lake, White Pine County, Nevada

This area and Copper Canyon should be studied along with the data on nearby Loop Road, which comprises one sheep range that remains largely unrecorded. Like many other ranges, the earliest date here is 1900. I think sheepherders wanted to commemorate the turn of the century and carved more often in that year. Once again we find several Anglo-American names, presumably of herders, one of them claiming to be an Indian.

Berry Creek and Timber Creek, Humboldt National Forest, Nevada

Both drainages are north of Ely in White Pine County. The whole Schell Creek Range, Moriah, Wheeler, and environs probably contain thousands of unrecorded arborglyphs.

In July 1991, with the cooperation of Ely BLM District's archaeology department, I spent two days on Berry Creek. I drove up the north fork and could see a profusion of carvings right from the truck. I only partially recorded two branches of the south fork, so this is a meager sample of what might be found in the area.

The names of the herders are sometimes duplicated in Berry Creek and Timber Creek, which are not far apart. I spent only a day here and have just a vague idea of how much remains to be recorded. I could see additional groves further up the mountains, and I would expect them to be carved as well.

Considerably more names and older dates were recorded at Timber Creek than at Berry Creek. Several herders apparently received a great deal of satisfaction from carving figures of stars, of which there are many. But Jean

Baptiste Etcheberri became a star of greater magnitude when in 1904 he carved his claim to the campsite at Timber Creek, and in the Basque language, a good indication that the anticipated readers were Basque speakers.

Pete Elia, who one day would become a sheep owner and lamb and wool buyer in Elko, began his herding career in White Pine County and carved his name at Timber Creek.[64]

Telegraph Mountain, White Pine County, Nevada

About 150 arborglyphs were recorded here, a modest figure as sheep ranges go, but the many old carvings set this site apart, as do the erotic and sexually graphic glyphs, a match for central Nevada's Porno Grove.

Telegraph suffers from the same handicap as Yosemite National Park in that the oldest arborglyphs are just initials or are difficult to read. Telegraph is also anomalous because several herders from Spain had already arrived here in the early 1960s. In many other groves the changing of the guard from Basques to non-Basques occurred in the late 1960s and continued throughout the 1970s.[65]

The Bureau of Land Management should be encouraged to look after Telegraph Mountain and manage its cultural resource according to the mandates of the National Register of Historic Places. I hope that the BLM and the White Pine Historical Society, in cooperation with local Basques and the North American Basque Organization (NABO), will preserve some of the best specimens here and in the nearby Cherry Creek Range (see below) for future generations.

Cherry Creek, Elko and White Pine Counties, Nevada

This area is also managed by the BLM, and the archaeology department of Elko District funded the weeklong research I conducted there. It was one of my most rewarding field trips. When my son, Erik, and I arrived in the first grove, we were greeted by a big herd of deer napping in the shade.[66] After they scampered away, a huge, black old bull showed up and would not go away. He moved about so slowly that we thought he was either retarded or close to dying.

Cherry Creek is one of the most isolated sheep areas in northern Nevada. It runs north-south and straddles Elko and White Pine Counties. The area appears to be rather dry, but that did not keep the sheepmen away from it. On the contrary, in the early 1990s there was still one Basque sheepman there: Bert Paris, son of Beltran

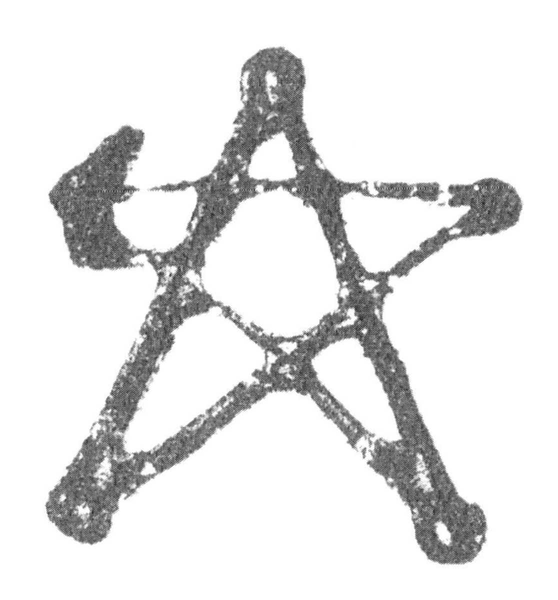

18. A fine example of a star in White Pine County, Nevada, dated in the 1920s.

Paris.[67] His nephew, Pete Paris of Lee, Elko County, is also a sheep owner. Bert Paris's carvings are dated in the 1940s. He and his brother used to own several ranches in the western foothills of Cherry Creek.

This range radiates Basque history. On the eastern side is Ordoqui Creek (also known as McDermitt Creek), named after Pedro Ordoqui from Lesaka, Nafarroa, another sheep owner, who operated from 1939 to 1955. Ordoqui was a large man, which is well attested by a portrait of him on an aspen. He had a brother, Agustin, and another partner nicknamed Xubero.[68] Ordoqui's earliest carving is dated 1912. In the meadow below the canyon Ordoqui built a beautiful home that today is in ruins, overrun by rats and mice and full of dirty mattresses and bed frames left by deer hunters.

Manes Auzqui, mentioned above, was in Cherry Creek in 1922. Pedro Camino also owned a sheep outfit, but I have not seen any carvings by him; and there were other sheepmen there, too, of course.

The Cherry Creek Range comprises some half-dozen heavily carved large aspen groves. I have recorded perhaps 80 percent of the arborglyphs. In order to economize on research time, smaller groves far from jeep roads were not surveyed. I encountered a familiar problem at Cherry Creek: the pre-1910 carvings are few and the readable names even fewer. After 1910, immigrants from Nafarroa dominated sheepherding in this part of

19. The old Ordoqui Ranch, west of Currie, Nevada, was unoccupied when this photograph was taken in 1991.

Nevada, but by the mid-1920s the Iparralde Basques constituted the majority. Bizkaians can be counted with just one hand. Also, I recorded one Portuguese and several Spanish surnames.

I recorded six herders and five Anglo names for the 1900–10 period. That meagerness contrasts dramatically with the data from the next decades. The large older trees lie in aspen "cemeteries," the bark peeled bone clean. Cherry Creek has one such cemetery at a high elevation that is especially strange. Every aspen is blistering white and dry, and all of them lie in a big pile. The place is too high for beavers, and the trees were not large enough for logging. An invisible hand or high winds seem to have suddenly struck them down.

I recorded more than twenty-eight names for the 1910–20 decade, eleven for the 1910–12 period alone. In the next decade, however, sheep activity on Cherry Creek virtually exploded; I recorded fifty-five different names, more than for any other mountain range I surveyed. I spent a week on Cherry Creek, however, and only a day or two in other ranges. If other sheepherder mountains were as thoroughly recorded, the tally of names would certainly be much higher.[69]

In the 1930s the sheep business was hit hard by the Depression, but the Basque presence on Cherry Creek continued unabated. I recorded more than fifty names for that period, and many for the next decades as well. When I visited the place in 1990 I saw a strange site: a sheep wagon. There must have been quite a few of them around once, but you rarely see one being used on the range anymore. Most outfits today house the herders in small trailers that are lighter than the old sheep wagons. The only other place I saw sheep wagons was in Idaho.[70]

. . .

The arborglyph areas described above constitute the principal sites I recorded. There are many others not yet recorded. A few possible sites are the Strawberry Mountains north of Burns, Oregon;[71] Mount Jefferson by Tonopah, Nevada; Soda Springs, Pocatello, Idaho; Cold Canyon, Donner Lake, California; North Star ski area, Truckee, California; Horsethief Canyon, Woodford, California; and the Bitterroot Mountains, Montana.

Arborglyphs and U.S. Censuses

One obvious result of this research is the preservation of the names of many herders who lived and worked, virtually unknown, in Nevada and California. In some instances the data retrieved from the trees tell us not only who they were but exactly where they worked and how long they stayed with the sheep.

According to the 1900 Census there were 180 Basques in Nevada. Almost half (84 individuals) lived in Humboldt County, followed by Elko (57) and Washoe (18) Counties.[72] The same statistics reveal that 73.4 percent of Nevada's Basques were from Spain. If the tree carvings are taken into account, however, as indications of the presence of Basques from other areas, the percentage favors Iparralde Basques. This is an important clue that appears to cast doubt on the census figures.

According to Marie-Pierre Arrizabalaga's analysis, as many as eighty Basques may have immigrated into Nevada between 1906 and 1907.[73] If most of them became sheepherders, as is likely, it follows that we should have an increased number of arborglyphs for the 1907–8 period; but we do not. My own recordings of carved names indicate an upsurge in 1909, but given the number of variables I am reluctant to draw any conclusions from that.

If all the still-readable names carved on trees were read and tabulated, even disregarding the lost data, the numbers for Basque herders in Nevada and eastern California would increase considerably vis-à-vis those who appear in the census.

Non-Basques

The information on the trees indicates that a number of the sheepherders were Spaniards from Spain, Mexicans, and Hispanics from the American Southwest and more recently South America.

The best calligraphic carving found so far was inscribed in the South Warner Mountains of California by a Hispanic herder named Elias Martinez, who gave two addresses: Pagosa Junction, Colorado, and Santa Fe, New Mexico. Today most herders are from Peru, Chile, and Mexico. Interestingly, it is not always their last names that give them away. I recorded a number of Mexicans from Zacatecas and Peruvians with Basque surnames and non-Basque first names. Herders named Anibal, Ivan, Gonzalo, or Rosario are not Basques because those first names do not exist in rural areas of Euskal Herria.

Roughly 10 to 15 percent of the glyphs were carved by non-Basques—excluding Anglo hunters, fishermen, hikers, etc. Not everyone with a Spanish or French surname is automatically non-Basque, however, just as there are a lot of Hispanics with Basque surnames such as Ibarra, Aguirre, Garcia, etc., who are not culturally Basques. A number of the Nafarroan herders had Spanish surnames (e.g., Villanueva and Perez) but spoke Basque. Others from around Pamplona had Basque surnames but spoke only Spanish. Some herders from Iparralde had French-sounding surnames such as Grenade, Paris, Lacroix, and Brust but were culturally Basque.

Case Studies: Peavine Mountain and Copper Basin

How much information is there on a given mountain? In order to answer the question and assess the true dimensions of this cultural phenomenon, let us scrutinize the arborglyphs in two sample mountain ranges: Peavine and Copper Basin. I am inviting you, the reader, to hike and explore with me these two areas of Nevada brimming with sheepherder history.

I have two reasons for choosing these two sites. First, both are in Nevada but they are 380 miles apart; one is a few miles from the California border, and the other is in the northeastern part of the state near the Idaho border. This separation and the proximity to two other states lend credibility to the data. Second, I know Peavine and Copper Basin better than other sheep ranges because I have researched them more thoroughly. While the first qualifies almost as suburbia, the other is as remote as can be. Let us take a hike.

Peavine Mountain (Washoe County, Nevada)

Peavine is situated on the northern fringe of Reno, "The Biggest Little City in the World." As Nevada mountains go, it is average to small; its windswept peak reaches an elevation of 8,240 feet.[74] The once-lonely peak that the sheepherders knew is today studded with antennae and satellite dishes, and the traffic up and down the main dirt road would certainly baffle any sheepherder from the 1920s and 1930s. The mountain has a steady clientele of hikers and bikers.

On a southeastern foothill of Peavine overlooking the city of Reno stands the national monument to the

Basque Sheepherder, a twenty-three-foot-tall abstract bronze statue dedicated on 27 August 1989. It is the work of sculptor Nestor Basterretxea, who appropriately entitled it *Bakardade* (Solitude). Etched in the plaque in front of the sculpture he wrote this poem, which appears in English and Basque:

A FIGURE
AS IF SCULPTED BY THE WIND ITSELF,
A MAN SOLITARY AND STRONG,
HELD STRAIGHT BY HIS OWN WILL,
PATIENT LABORER,
FACING UNTO THE UNCERTAIN HORIZON
OF ADVENTURE.
ENDLESS STRETCHES OF SILENCES OF MOON
AND STARS,
THROUGH MOUNTAIN TRAILS:
THIS MONUMENT IS
ETERNAL HOMAGE AND MEMORIAL
TO THE BASQUE SHEEPHERDER.[75]

It must be stated that the abstract style of the monument did not please the sheepherders themselves, who would have preferred something more realistic[76] (perhaps something resembling tree art?). Apparently the promoters failed to consider the fact that the herders had created their own art style.[77]

John Ascuaga, owner of the Nugget Hotel-Casino in Sparks, Nevada, commissioned another, more conventional, statue to honor the Basque sheepherder. This one, sculpted by Douglas Van Howd, stands seventeen feet tall. The bronze monument captures a vignette from spring lambing in Nevada: a herder holds a newborn lamb in the warmth of his coat. The faithful dog, of course, looks on approvingly. The *txapela*-clad herder has a long sheep crook in his left hand. Ascuaga unveiled the statue on a sunny afternoon on 4 August 1998, amid Basque dancing, aperitifs, and wine.[78]

Just as Gernika is the sacred city of the Basques in Europe and Aralar their sacred mountain, Peavine is now the sacred mountain of the American Basques. Thousands of Renoites each year hike up to Peavine to enjoy the magnificent views of their city. The trek can be very rewarding and surprising. The fact that many people go to Peavine and return without discovering its cultural cache—the Basque arborglyphs—is a statement about the elusive nature of the resource.

Peavine Overview

Peavine was heavily used by sheep operators, as attested by the seventy-one-plus sheepherders who recorded their names on Peavine's aspens from 1901 to the present. There are four main groves. I have cataloged more than five hundred carvings on Peavine so far, and the final count may be even higher, for it seems that on every visit I find something new.[79]

The testimony of the trees begins in the year 1901 with C. Paul, who may have been a lifelong sheepherder in northwestern Nevada and eastern California. He was a prolific carver, too, beginning in 1899, but laconic, even for a Basque, and monotonous. He never supplied any more information than his first initial, last name, and the year. Almost a half a century later he was still on Peavine, still busy with his knife, but the message was still the same: "C Paul 1947."[80] He never deviated from that pattern, so for years he frustrated me. I had no idea of his identity or nationality—until one lucky day.

Peavine has a long history of sheep grazing. The first herd of sheep in the Reno area probably grazed the Peavine foothills. The Peavine Ranch, located at the junction of Red Rock Road and U.S. 395, included all of the Stead airport from Horse Mountain to Lemmon Valley Road. Sometime in the 1910s a Danish fellow by the name of Anderson and his father-in-law, Andrew Nielsen, also from Denmark, bought the place. Anderson was a banker in Reno and set up Nielsen—an 1899 immigrant—in the sheep business.

Andrew Nelson Jr. (Nielsen was Americanized to Nelson) remembers tending camp for his father in the 1920s in the summer range of the Hell Hole area, Diamond Crossing, and Five-Lake Canyon. He does not remember if his father had an allotment on Peavine after the U.S. Forest Service took over the high country. He thinks most of the mountain land was private.

"There was this fellow, C. Paul, who carved a lot on Peavine," I said.

"I knew Charlie Paul," Andrew answered, almost automatically.

"You knew C. Paul?" I asked, incredulous.

"Yes, I always remember him working for my father."

His words were a dream come true.

Nelson continued: "Charlie Paul was German and came [to the United States] before I was born. He was chunkily built, a Squaw man. When I married in 1945 and brought my wife over to the camp he said: 'Andrew, she's too skinny.' Charlie was tough. One day a horse kicked him in the chest and it sounded like a drum, didn't hurt him a bit. . . . Charlie's wife was often [in the range] with him. After my father lost the sheep in 1932, Charlie may have worked for my uncle James Nelson, who ran sheep around Independence [Lake, California]."[81]

Finally, the enigma of C. Paul was solved.

Eighteen years are unaccounted for in the Peavine carvings: 1903–6, 1908, 1910, 1912–14, 1916, 1944, 1948, 1966, 1979, and 1983–86. The voids must be regarded as the consequence of destruction of trees and loss of data rather than absence of carving and sheep activity. Most of the missing years are pre-1920, which is explainable by the fragile nature of aspens. The 1944 hiatus may have been caused by World War II, when many herders went off to fight. During most years there were several herders on Peavine.

After the 1940–45 decline—when seemingly Charlie Paul had Peavine all to himself—the sheep business, under the umbrella of the Western Range Association, experienced a comeback in the 1950s. The aspens have preserved most of the names of the herders for this period. In 1959 seven herders (four from Nafarroa, two from Benafarroa, and one from Bizkaia) spent time on various parts of Peavine, although, strangely, no two sheepherders ever carved on the same day. During the ninety years of herding history on the mountain there is only one exception: Magencio Valencia and Juan Zabalegui (both from the town of Lerga, Nafarroa) both carved on 10 October 1960.

In 1989 I met two herders on the mountain: a Peruvian and his Basque camptender. A third herder, a Mexican, was stationed at the base of the mountain by the corrals of what used to be Peavine Ranch, guarding sheep that were about to be transported. This was the traditional shipping location of lambs that summered on Peavine.[82] The camptender told me that he never carved his name, and neither did the Mexican. They worked for Koffman from Lincoln, Nevada, but later he moved to Oregon. After Koffman, Ray Talbott of Los Banos, California, ran sheep on Peavine for a few more years, but according to Toiyabe archaeologist Terry Birk, the forest allotment on Peavine has been inactive since 1991 or so.[83]

The Landa Sheep Company

Even though only Steve Landa's carvings have been recorded on Peavine, the Landa family has a long association with the area. The Landa brothers—Steve, Marcelino "March," and Anastasio—operated a sheep business on Peavine from the 1930s until 1959 or so. I have been informed that Steve Landa and Frank Erro were the two foremost bread-oven builders in the entire region. Erro worked as a sheepherder and then as a government trapper in the Winnemucca-Gerlach-Cedarville areas.[84]

The history of the Landa Sheep Company is complicated but typical enough when one realizes that in the 1930s a multitude of Basques were in the same business, and everyone seemed interrelated by blood or marriage. Jeannette Landa, for example, married "March" in 1949. She was from the Etcheberry sheep family. When Jeannette's father was killed by a drunk driver in 1933, her mother married a Nafarroan named Azcarate, also a sheepman. In 1935 Azcarate bought the Mustang Ranch, located east of Sparks, Nevada, which later, under different ownership, became a famous bordello.

The Landas, the Etcheberrys, and the Azcarates shared their Sierra summer range and wintered their sheep in the Truckee Meadows, the Virginia City foothills, and Largomasino Canyon. The Landas hired about ten Basque herders, and most of their names appear in the Peavine list. They usually ran three bands on Peavine and the rest in Blackwood Canyon, Tahoe National Forest. Jean Borderre, a prolific carver, was their camptender and a proud *aldudarra* (native of Aldude), as he liked to carve.

A Puzzle

The history books say that during the Great Depression of the 1930s many or most of the sheep operators in the West went out of business. According to Alberto Gallues, son of Martin Gallues, who ran sheep first for Wheeler and then for Reginald Meaker, four-fifths of the sheep owners "lost their shirts."[85] At the time, Nevada had 2 to 2.5 million sheep, according to one sheepman.[86] Textbooks and many writers reiterate the losses, and it is difficult to argue with them. It is said that when the financial empire of George Wingfield crashed in 1932, most of the banks in the Reno–Carson City area closed and many livestock operators went out of business. Such assessments may lead us to think that when sheep operators went bankrupt, sheep also vanished.[87]

The aspen carvings may be better barometers of sheepherding activity than general history books, and they tell a different story. The number of herders who continued carving their names in the 1930s leads to the conclusion that sheepherding continued unabated during those years. I recorded a total of thirty-two names (carved by twenty individuals) on Peavine for the period 1930–40, which hardly indicates a sheep debacle. The recordings in other groves tell a similar story.

How do we explain the anomaly? Ambrose Arla of Reno gave me the clue. I found Arla's name on Peavine dated in the early 1930s, so I asked him about the herding situation during those years. Here is a part of his story: "I came to the United States in 1929 to escort my sister who was getting married.[88] I did not have enough money to return home, so that year I started working near Reno. They gave me a herd of sheep, two dogs, a tent, a rifle, and a donkey. During the Depression the sheep boss went broke and . . . the bank took over the sheep. In 1933 the bank paid me fifty dollars a month. Later I worked for other sheepmen."[89]

During the Depression sheep numbers dwindled somewhat, but the sheep did not disappear. Mostly they changed hands. Since few people had any cash, most animals ended up in the hands of the bank, and the banker became the sheep boss. One could argue that some herders actually benefited from the Depression. The banks were desperate with so many sheep on their hands, and they began to pay higher wages. The standard pay was about thirty dollars a month, but Ambrose Arla was getting fifty dollars a month right in the middle of the Depression.[90]

Perhaps this explains better than anything else the strange message carved on Peavine by Frank Rodriguez, a restless spirit from Galicia, Spain, who inscribed an incredible statement on the northwestern slope of Peavine: "Long live the depression of 1932."[91] What prompted such a message? Either Rodriguez received a raise, or it was his way of saying, "It's all the same to me, folks." Of course, he may have had something nastier on his mind, like "Screw you people in town, without jobs and money." Some of Rodriguez's statements are reminiscent of personality traits exhibited by typical western characters—loners, prospectors, and the like living on the fringe of society. Rodriguez was a talented and prolific carver, and we will have opportunity to pick up his trail again later.

The Men on Peavine

No fewer than 105 people left their names on Peavine.[92] The surnames often allow a fairly precise determination of the carver's region of origin in the Basque Country, and sometimes even his hometown. Some last names are too common to be pinpointed to a particular area, but it is clear that a herder named Echebarria/Echevarria was a Bizkaian, Echeberria/Echeverria was from Gipuzkoa or Nafarroa, and Etcheberry or Etcheverry was from Iparralde. This surname is as common in the Basque Country as Smith is in America, but the different spellings reflecting dialectal differences allow us to deduce the regional origins. Sometimes no guessing is required because the herders themselves are still around to tell us where they were born.

On the strength of the last names, the regional breakdown of the sheepmen on Peavine is as follows:

Nafarroa	25
Iparralde	16
Bizkaia	9
Non-Basque surnames	11
Unidentified (initials, etc.)	44

The "unidentified" group is large, of course. All the carved initials (e.g., T.S., 1972) are included in it, and some of them were surely sheepherders. Other surnames, such as Leiva, Aliaga, and Suqylvide, are Basque, but their regional origin cannot be determined with accuracy. (See appendix 2 for a complete roster of the names on Peavine.)

Noteworthy Herders

Without detracting from the democratic spirit of the arborglyphs, I would like to mention a few individuals who left their mark on Peavine and later went on to settle in northern Nevada communities. John Etchemendy was the original owner of the Overland Hotel in Gardnerville, Nevada, which is still in operation. John Etchart's relatives ran the Star Hotel in Reno. As recently as the 1980s there was a sheepman in Winnemucca, Nevada, named Etchart. Ambrose Arla, too, carved several trees on Peavine, though in an interview he never mentioned working there. Arla lived in Reno for many years. Another name many Basques recognized, especially in the Boise, Idaho, area, is Justo Sarria. He never worked in the Peavine area, but he left several carvings there while visiting sheepherder friends.[93]

Also worthy of mention are the Lanathouas—Charles and Jean Pierre, father and son—who worked for the Landa Sheep Company of Reno for more than ten years. Jean Pierre carved a lot, all up and down the Sierra Nevada. Several Lanathouas reside today in Bakersfield,

California. Another frequent carver on Peavine and its environs was Ynacio Arrupe Acorda, who was related to the Landa family.

These and their fellow herders addressed a multitude of thoughts and feelings in their tree carvings. When I asked Jean Lekumberry what types of carvings he made, he answered: "Things that happened on that day, religious things."[94] You might say that he was keeping a diary. But on Peavine as elsewhere, the primary message is the identity of the herder. A few carvers embellished their identities with individualistic details. One identified himself as *euskotarra*, or Basque by race and nationality. Several others simply carved "Basqo/vasco" or "espanol vasco" (Spanish Basque). One Nafarroan proclaimed himself "the Spanish gentleman," while a herder from Arrosa, Benafarroa, perhaps more humble, dubbed himself "Martin Tipy" (Little Martin), probably a nickname.

Jose Otazu carved his age: "28" (in 1959). J. Urreaga indicated that he was from Irurita, Nafarroa, and, in a rare instance, carved the date in Basque. Several others revealed their hometowns. Juan Zabalegui was, like Otazu, from Lerga, Nafarroa. In the 1950s and 1960s there were sixteen herders in the Reno area from this tiny Nafarroan locality.[95] Zabalegui was a devout man who carved several crosses and statements of a religious nature. He apparently did not think America was a God-fearing country, for in one instance he appealed to the Virgin Mary for protection "in this land without faith."

Peavine contains patriotic expressions, too, although fewer than most groves I recorded. One "Long live Spain" provoked someone to carve a comment underneath: "This guy up there is an s.o.b."[96] A herder from Bizkaia hailed ETA (Euskadi Ta Askatasuna, Independent Basque Country and Freedom), a nationalistic Basque armed guerrilla group. Others, like Zabalegui, being more traditional, hailed "the sheepherders, Spain, Christ, and the King."[97]

Life with Sheep and Other Matters

Someone happy to reach summer pastures and end his daily trekking across valleys and roadways carved "Adios trel" (good-bye trail) and underneath added "1444," which may refer to the number of sheep in his band.

On 9 September 1954, J. U. noted that he had arrived from California. Two years later it must have been a dry year on Peavine, because on 7 July he said he was leaving for the Sierra. Such news usually meant that a herder had run out of pasture. In 1957 J. U. was still herding on Peavine but had grown tired of it and carved that he had been deceived regarding sheepherding in the United States. Like many others, he came to Nevada because he had acquaintances there—in his case, his uncle, who may have painted life on the range a bit too rosily.

On the other side of the same trunk, without interruption or commas, J. U. continued his diatribe on another topic that was bothering him: "It would be better if the sheep bosses paid once a week." This is the only such carving of which I am aware. Most herders preferred not to see their money and asked the boss to deposit their wages directly in the bank.[98] Why did J. U. want to be paid weekly? What would he have done with the money? If he had a horse, it is conceivable that he might make it down to Reno for some drinks and company and back up the mountain again in time for a snooze before Mr. Coyote became hungry or the sheep began to stir at dawn. The carving makes sense only if J. U. was a camp-tender and had mobility, perhaps a truck, and was more interested in spending the money than in saving it. This is just guesswork, of course. However, some herders did not trust the banks and preferred to keep the checks or the cash with them, stashed in their bags.

A large number of messages in groves are too vague to be put into a definite context. One on Peavine by J. S. hints only at some kind of a problem: "There is sweet talk here."[99] J. P. Lanathoua apparently had no problems. If a herder learned to deal with the isolation, his job was the best in the world. A lot of free time, a lot of great country to enjoy, the best fresh air to breathe, and plenty of food. One day Lanathoua was enjoying the coolness of the aspens and recorded the simple feeling right then and there: "It is nice in the shade."[100] One fellow, who perhaps ran out of wine, appreciated the springs on Peavine and carved the prosaic remark: "Juan drinks cold water."

Lanathoua really appreciated the United States. After a few years as a herder in Nevada he returned home, planning to stay, but things did not turn out as he had expected. He took the plane back to America and, when he finished descending the stairway, knelt down and kissed the soil, saying: "This is my home now, and I shall never again return to Europe." He did not.[101]

When it came to griping and criticizing, no one could top Frank Rodriguez. He had a knack for it. He worked on Peavine between 1928 and 1939, and sometime during that period he carved on two adjacent aspens: "Pradera's sheepherders, sons of a *chingada*, they have no shame"[102] and "The men who work for Pradera are fags, they have no shame." Both appear to have been carved on the same day. Pradera very probably was Rodriguez's sheep boss, although he may have been the owner of another sheep company with whom Rodriguez had a bone to pick.[103]

It is unclear if the following unsigned message in Basque indicates a complaint by another dissatisfied herder or is a mere sexual fantasy: "If I could catch someone I know, what would I do?"

At the same time that Ynacio Arrupe was drawing pictures of airplanes with four propellers, he was mulling over his future. He had been a sheepherder long enough. On a tree near the airplane carvings he registered the five years he had worked on Peavine. Above the list he carved a modernistic house, complete with shrubbery and a tree, with the caption "American." He may have been thinking of buying such a house in town, but in the end he took the airplane back to his native Bizkaia.

Life in the United States held a great attraction for Basques in the 1940s and 1950s. The United States was part of "Amerika," a word that for centuries had a definite meaning in the Basque County. Many people had at least one uncle somewhere in America, and if a man wanted to see the world or make his fortune, he went there. After arriving in the United States, the sheepherders saw many new and unexpected things, as the following message on Peavine, carved by a no doubt wide-eyed man, demonstrates: "Yesterday I saw Russians on TV." This carving has no date and no signature, but it must be from the early 1950s. Was the herder surprised to find that Russians looked like other people? It is just a small detail, but one pregnant with historicity. Probably the herder grew up in Franco's church-dominated Basque Country, where Russians and Communists were equated with ugliness and evil. Then he saw them on television . . .

The Eternal Female

The aspen groves on Peavine contain the usual sexual and female figures, although they are neither numerous nor overly artistic. There are several plump women, roughly sketched, as well as couples copulating. One of these was axed by a self-appointed censor, almost certainly not a sheepherder.[104]

The proximity of brothels in Reno—there were many in the 1940s and 1950s[105]—is reflected in this message carved in a very tight Nafarroan Basque dialect: "I cannot catch any of those prostitutes in Reno in order to do it with them. What a misery! I must stay in heat. Let us have patience, better years will come."[106] The statement is not signed. It appears to be from the 1930s and was probably carved by someone who seldom went to town. In those years herders rarely took vacations, and a once-a-year visit to town was fairly common. This explains the carver's "better years will come."

In the solitude of the forest the herders could afford to be very frank about intimate sexual details. They did not care if other herders read them. One said that he was horny and was going to ejaculate right there. Another disclosed his sexual preference for hairy, redheaded women. The lack of female companionship is conveyed succinctly by another herder: "I have no other pain [but] that of [lacking] a girl."

Several Peavine aspens are carved with "hurrahs" to pretty girls. J. B. missed his loved one in Benafarroa and wished she were available for sex.[107] Another herder was bold enough to cheer two lovers: "Hurrah for my wife and my girlfriend." And he signed the carving!

The arborglyphs on this mountain indicate that some herders had girlfriends in town who, the distance being relatively short, occasionally paid a visit. It seems that once one brought along her girlfriend, and the herder carved about her: "The blond from Verdi whom Linney brought over is, as far as I am concerned, a knockout."[108] In all fairness, it did not take a Hollywood bombshell to impress a lonely sheepherder.

The Carved Media

Herders also used trees for "internal memos" within the sheepherder community. Peavine has a number of these "interactive" carvings, of which I was able to read some, though not all. No one rushed to answer them, for the sheepherding world ran according to a different clock. Herders might leave a message knowing full well that a response might take years. Or they carved a message to no one in particular, but to all in general, so any herder could respond with a remark of his own on the same or a nearby aspen. For example, if a herder carved an offensive or controversial statement, years later someone might respond, agreeing with him or defending the accused. Thus with questions and answers a consensus might be built about a number of topics important to the herders; for example, women and sex, wine, money, sheepherding, and life in Euskal Herria and in America.

One thing was obvious: the tree, simply by standing there, delivered the carved message to every herder who passed by. A verbal statement, a curse directed at the camptender, could be forgotten five minutes later, but if carved, it remained for all to see. The audience was small, but the medium carried a punch.

The nature of these "memos" varies greatly, from complaints to jokes. The following three are illustrative:

"Someone might say that the guy who wrote this was dumb, but isn't it true that whoever reads it is also dumb?"

Someone else carved below: "Whoever reads this is a fucker."

A third response: "Whoever wrote it is a fag."

Two compatriots from Nafarroa, Juan Zabalegui and Magencio "Mac" Valencia, carved a number of messages to each other, but it is not certain whether they should be taken seriously. It appears that sometime in 1960 Mac either left the sheep or was contemplating it. But he could not decide whether to live in Nafarroa or in Madrid. In the meantime he went to Mexico, leaving his friend wondering. The move must have been big news, because Zabalegui announced it to the rest of the sheepherders, not only on Peavine but on Mount Rose as well: "Valencia is in Mexico." If he did go, he returned to Peavine to carve the following (transcribed literally): "Hi de Mexico for you salud and suerte los desea buestro amigo a Jose and Juan" (Hello from Mexico, I wish Jose and Juan health and good luck, your friend [Mac]). Underneath it says: "Le das gracias" (Thank him).[109]

Zabalegui carved two more trees with messages that denote curiosity concerning his friend's whereabouts: "Let us see whether Mac opens a business in Pamplona, 20 June 1960." On the back of the aspen carved by Zabalegui, Mac appeared to hint at a move away from Pamplona: "Magencio Valencia, Spain C Madrid, 27 June 1960." I do not understand what he meant by this—it does not appear to be a street address—but apparently Zabalegui did, and, itching to know what his friend was up to, he carved: "Mac, are you going to live in Madrid or what? your friend Johnny, hi."[110]

Such messages are an aspect of the sheepherders' world revealed for the first time by the carvings. They suggest that the isolation endured by the herders during the summer was not as relentless as the previous literature suggests. That the herders were lonely is undeniable, but the messages show that from time to time the physical isolation had moments of reprieve (I am aware that Peavine is not the best example to make this argument).

Nobody Lonely

The strangest thing about Peavine, the thing that sets it apart from all other groves, is that not one of the five hundred recorded arborglyphs refers directly to loneliness, isolation, or boredom. There must be an explanation for this, since the topic is one of the most common on every sheep range.

I think the answer lies in Reno's proximity. Whether the herder actually went to town or not, he did not feel isolated on Peavine. At night he could see its lights fairly close by. During the day he could sit on a hill and watch the traffic in and out of the city and even hear its hustle and bustle. If he became desperate, in the evening, the herder could almost walk to town (depending, of course, on where on the mountain he was camped). After World War II the sheep operations rapidly became mechanized, and Peavine was just a short ride away from Reno in a pickup.[111]

Copper Basin (Elko County, Nevada)

Copper Basin is one of the most spectacular and lush drainages in Nevada. An arborglyph on the southern flank of Copper Mountain peak (elevation 9,912 feet), Humboldt National Forest, reads:

Lenago neskatan	Before I used to chase girls
oin arditan	Now I chase sheep
beti amesetan	[I am] always dreaming
J. Z. 1915	J. Z. 1915

It is one of the most dramatic and historically accurate arborglyphs I have come across. I don't know who J. Z. was, but his laconic statement speaks chapters on sheepherder life. Especially true is his "I am always dreaming." For most herders life *was* a daze. With hundreds of square miles to roam yet, like J. Z., imprisoned by that very immensity, dreaming was the only escape.

The first arborglyphs I ever saw (in 1968) were in the Copper Basin. Almost thirty years later, in 1997, under the auspices of Humboldt National Forest, Elko, Nevada, I returned there to do fieldwork. Along with forest historians and archaeologists, there were twenty or so PIT (Passport in Time) volunteers from as far away as New York and Florida, and as close as Elko and Sparks.

Appendix 3 lists the names and dates found in Copper Basin.

Indeed, our effort was an authentic Passport in Time project because we stayed in a typical 1950s sheep camp, which closely resembled a 1930s or 1920s one. For decades the site had been the *kanpo handia* (the main [summer] camp) of the Goicoechea Sheep Company of Elko. The company went out of business in 1976, but the remote camp has not changed much in the twenty odd years since.

It still has two cabins, two fireplaces (one of which is a typical sheepherder brick oven), running water (collected in a ceramic full-size bathtub), and the remains of old corrals and fences. The old two-seater outhouse is now fallen and in pieces. When the outfit was active, the camptender lived in this camp from July to September. After that, the threat of snow drove the herders down to lower elevations.

Bits of History

I mentioned earlier that Pedro Altube and his friend and sometime partner, Jean Garat, arrived in northern Elko County in the early 1870s. They were successful ranchers, and Altube, a colorful character of the Old West, has been called "the Father of the Basques in the American West." In 1960 he was inducted into the Cowboy Hall of Fame in Oklahoma City, Oklahoma.[112]

Elko County was cow country first, some say, and only later was invaded by sheep. It may have been the hard winter of 1889–90, when 134,000 cattle died, that turned many cattlemen into sheepmen.[113] Northern Elko County is well suited for sheep range, and the Jarbidge area in particular, including the Mary's River drainage, may be some of the best grazing anywhere. Tax records for Elko County—except for the period between 1870 and 1889—provide the most readily available information on the history of sheep in the area, although it may not always be completely reliable.[114]

According to the U.S. Census of 1900 there were 887,110 sheep in Nevada in 1900, and 121,945 of them were in Elko County.[115] By 1910 the number had climbed to 1,154,795 for Nevada and 247,913 for Elko, and it continued to rise for at least twenty years.[116] In 1930 Nevada's sheep population had grown to 1,201,837, of which Elko claimed a sizable share: 358,262. The sheep industry experienced a drastic downward trend after that from which it has not yet recovered. In 1992 sheep in Elko County numbered just 56,424 head.[117]

These figures may not be accurate. According to an Elko sheepman involved in the business since the 1930s, one could legally report just 60 percent of the total number of sheep one owned, and everyone followed that practice. If this is true, then the census figures are too low by some 40 percent.[118] In any event, sheep outnumbered people by a wide margin. There were only fifty-seven Basques in Elko County according to the U.S. Census of 1900; by 1910 the number had increased to 194, and not all of them were sheepherders.[119]

Early Grazing Records

It is very important to know the history of the livestock companies so that the herders' carved names can be matched with those of their employers, which in turn opens up a whole new avenue of investigation.

The Jarbidge range was incorporated into the U.S. Forest Service system on 20 January 1909, but by then the area had been used for sheep husbandry for well over twenty-five years. George B. Williams was the first to hold a grazing permit in Copper Basin, sometime around 1910.[120] His half brother, state senator Warren Williams of Fallon, had been running sheep since the 1870s, and by 1889 had expanded into northern Elko County.[121] According to Beltran Paris, who started working for Williams in 1912, George had thirty thousand sheep and Warren had forty thousand. The herders who worked for George Williams, twenty-seven of them in all, were Basque.[122]

It was somewhere in the Jarbidge summer range in 1905 that Simon Salas, a herder working for George Williams, was shot and wounded by Dan Wallis (or Wallace), a homesteader. The trial in Elko attracted a lot of attention and served to highlight the line between herders and cattlemen. Salas spoke no English, and even though his account was more believable than Wallis's, the hearing ended in a hung jury. As often occurred in trials of this nature, the cattlemen appeared to have the upper hand. Wallis went home a free man.[123]

In 1918 the Ellison Ranching Company ran 10,000–12,000 sheep on its allotment near the southern end of Copper Basin (Badger Creek). The earliest Basque presence in Copper Basin was also in 1918. According to U.S. Forest Service grazing maps, M. Lostra and M. Bastida had a small allotment just east of Jarbidge Road where it rounds Copper Basin (near the headwaters of 76 Creek) on which they ran 1,250 sheep. Their names are absent from the 1929 grazing map. Farther east, in the same year, Fernando Goicoechea ran the same number of ovines. According to Dave Goicoechea, his ancestors homesteaded in Copper Basin.[124] (The Goicoecheas are a well-known clan in Nevada with members in several other states.)[125]

In 1918 Fernando Goicoechea's neighbor to the south was Pete Itçaina, "the sheep baron of Elko," according to author Clel Georgetta, but the range maps do not indicate the number of Itçaina's sheep.[126] The Goicoecheas had secured additional leases by 1929—one in the Miller Creek area, which was enlarged to north Copper (Coon Creek) in the 1950s. By 1929 Goicoechea had Basque "neighbors" to the north of Miller Creek—Joe Saval and John Belaustegui, at Corral Creek and Deep Creek, respectively—while Pete Olabarria ran his sheep in the nearby Red Bluff and Rattlesnake Creek areas.

Copper Basin must have been a busy place during the summer months. It is clear that competition for grazing permits was keen. Two outfits either disputed part of Ellison's allotment or were granted leases in areas the company had grazed earlier. One was John Belaustegui, to whom in 1937 the U.S. Forest Service may have awarded Badger Creek. But a year later the place was leased to a man named Chevalier.

The Goicoechea Sheep Company built its main camp in the northwestern corner of Copper Basin, an area that, according to the range maps, was part of the Ellison allotment.[127] Often it is difficult to pinpoint an allotment's boundaries, which sometimes makes it impossible to determine which herders worked for which outfits. In most cases, however, I think the data allow us to draw fairly accurate conclusions.

Copper Basin Findings

Copper Basin, surrounded by towering mountains and covered with aspen groves, is a pristine, lush corner of Nevada. The array of wildflowers, plants, grasses, and bushes is extraordinary. I certainly found it hard to believe that year after year this basin had been grazed by thousands of sheep.

The Ellison Sheep Company still keeps an allotment at Copper Basin. The outfit's main camp used to be located by the 76 Creek Cabin where in 1990 I recorded two structures (the Forest Service burned them not long after that).[128] Today, a supply camp is not really needed to run a sheep operation; outlying herders can be furnished with provisions via pickup trucks. In front of the 76 Creek Cabin there is a brick oven that looks "classic" and is still in good condition. Another oven also used by the Ellison men is located below Goicoechea's camp, in the heart of Copper Basin, several hundred yards northeast of the old mine. This used to be a sheep camp.

Other structures of the sheepherder legacy are the *harri mutil* found on summits and strategic points. The one standing on top of rocky Copper Mountain peak is almost six feet tall and is visible from miles away. It was partly built by Erremon Zugazaga, present operator of the Biltoki Restaurant in Elko, when he worked for the Goicoecheas in the 1960s. Copper Mountain peak was the boundary between the Goicoechea and Ellison allotments, and that is why the *harri mutil* is there. At the extremity of the basin proper, two *harri mutilak* stand together, seemingly one male and one female, the latter being the larger.

Today all the herders employed by Ellison are from Peru except for one Mexican old-timer. The camptender is Canadian. According to one of the Peruvians who volunteered to answer questions, the salary of the sheepherders, new and old alike, is seven hundred dollars a month in addition to room and board. Unlike Basques, who would not stay with a company that did not provide wine—one gallon a week, please—Ellison's men get no wine ration.[129] More surprising, no rifles or guns are issued to them, something unheard of a few decades ago.[130]

Of the twelve hundred arborglyphs we recorded in Copper Basin, many are repetitious or consist of just initials. About two hundred may have been carved by nonherders such as hunters and hikers. In all, we recorded no fewer than 1,003 names and 203 initials, each with their dates, constituting 154 different names for herders and 103 different initials (many belonging to nonherders). By means of the recorded carvings we thus rescued for history the names of 257 individuals.[131]

Names and Themes

The Copper Basin arborglyphs contain fewer messages in Basque than those in other Nevada sheep ranges. One reason, I presume, is that many of the herders here were Bizkaian, and they tended to carve in Spanish. As far as usage of the English language, the most common examples are the herder's name rendered in English rather than in the original: John for Jean, Frank for Francisco, Pete for Pedro, and so on.

Carved messages and narratives are also relatively uncommon here. The carvers in Copper Basin seemed more artistically oriented. We observed a number of calligraphic initials rendered with considerable flair. That is because Basques learned calligraphic writing in primary school. Carved statements began to increase in the 1960s, when the herders were more educated than before.

As is true for all the groves researched thus far in the American West, the number-one theme of the carvings is egotistical. Every herder wanted to leave his name so that others could find and read it, just as he had read the names of earlier herders. It is the stuff of history. The self-portrait—usually a rough sketch of a male figure—belongs to this category as well.

The dated evidence in Copper Basin begins around the turn of the century because no older aspens survive. As an example of the ephemeral nature of these trees, Ramon Garat's 1901 carving that I recorded in 1990 was no longer there in 1997. The name is probably connected to the YP Ranch, owned by Garat-Yndart, and the carving might indicate that in 1901 YP had sheep or cattle in Copper Basin. The other possibility is that Ramon Garat worked for Williams.

Besides Garat, we recorded no fewer than nine Basque names (one was an initial) dated from 1900 to 1911. One, Michel Kopentipy, was still herding in 1934. According to Joe Zubillaga, of the Santa Fe Hotel and Restaurant in Reno, Kopentipy was almost surely from Aldude, Benafarroa, where Zubillaga himself was born. He knew some Kopentipy herders who probably came in Michel's wake. In fact, one of the priests the diocese of Baiona regularly sends to the United States was named Kopentipy.[132]

There is no denying that the researcher is the prisoner of the data. That is why people like Gregorio Barruetabeña and Manuel Aramburu are so much appreciated. In spite of the length of their names, these two men took the time to inscribe them prominently over much of the range.

The story of the former is intriguing. He first came to the Jarbidge area in 1911, and my guess is that he did not leave until 1915. Then he either gave up the job, moved to another area, or perhaps went back to Europe. Incredibly, he returned to Copper Basin in 1948, when he must have been more than sixty years old, and continued working until 1954. Several of Barruetabeña's carvings say that he is a Bizkaian, born in Ybarruri, a little hamlet at the foot of Oiz Mountain, overlooking two green and forested valleys. Sometimes his carvings are adorned with a fancy cross on a pedestal.

As for the Aramburus, there were three of them: Teles, father of Julian and Manuel, all from the coastal Bizkaian hamlet of Bedaru. Although Bedaru looks down on the misty Atlantic Ocean at the Bay of Biscay, it is a farming rather than a fishing town. Possibly the brothers preferred herding sheep in Nevada to working on a tuna boat or a merchant ship. Manuel carved most of the inscriptions. Their dates range from 1934 to 1936, a very short period for the three of them, but many more arborglyphs remain to be recorded.

In the summer of 1915 J. Z. was on the southern hillside of Copper Mountain peak enjoying the shade of the aspens. From where he stood he had an awesome view of the basin and the surrounding mountains. He—along with Barruetabeña—was almost certainly herding for George Williams. I would love to find J. Z.'s name fully carved. So far only one of his carvings has been found, but what he recorded on that aspen deserves repetition because it encapsulates the sheepherder's life so masterfully that even a Neolithic sheepherder eight thousand years ago would have agreed with him: "Before I used to chase girls, now I chase sheep, always dreaming."

The way the herders carved their names speaks a great deal about their individuality. Tall, highly visible letters suggest authors with big egos. Calligraphic carving indicates the work of a somewhat educated or artistic man, like Pete Forua (dated 1938), who inscribed capital letters with almost medieval Celtic detail. Arno Arretche (dated 1933) was even neater. He had better control of the knife, and his letters and the crosses, even after sixty-four years, are still sharp and striking.[133]

A herder who really wanted to make a statement would carve not only his name and the year but would also provide information regarding his origin. Barruetabeña and Manuel Aramburu did just that. But Zaracondegui, who was from the same hamlet as Aramburu, stated it in the Basque language: "bedarutarra" (native of Bedaru), as did Arrate, "navarniztarra" (native of Nabarniz). Other herders from Baigorry (Iparralde), Irurita (Nafarroa), and Bolibar (Bizkaia) hailed their respective hometowns as well.

Inscribing the name of one's hamlet instead of the region or nation indicates a somewhat medieval mentality. An American observer might be mystified by such rationalization, unaware, for example, that Bedaru is a hamlet somewhere in Europe. Perhaps, but the carver's intended audience—other sheepherders—knew where Bedaru was. Furthermore, the practice is intricately tied to Basque culture. As soon as Basques meet someone, the first thing they want to know is where the person is from. Robert Laxalt wrote that his father was from Zuberoa, and when people met him they would ask: "You are a zhibero, eh?"[134]

A Basque wants to nail every individual down to a location. After one finds out the stranger's hometown, one turns around and introduces him or her to others not by surname but with the announcement: "Hemen Gernikar bat" (Here is one [person] from Gernika). This denotes a cultural adaptation to life in the United States. In the Basque Country the point of reference is not the town but the *baserri* or *borda* (farmstead). Peru Munioz residing at the Urtzaga farmstead was simply known as Peru Urtzaga (shortened to "Urtza" in normal conversation), but in the United States farmsteads no longer worked as references. On the sheep ranges, town names were used instead. Culture, once again, explains this. Most Basque surnames are toponyms that refer to a certain place in Basque geography. In the United States, the first thing we want to know about a person is what he or she does for a living. Thus, in English there are many last names ending in "-er," like "Baker," which give information—or did originally—about the bearer's occupation. There was a lot of pride involved when a herder provided information on his hometown. At the end of the season, when the lambs were shipped, the scales would tell whose lambs were the biggest, thus establishing the champion herder and elevating the status of his hometown.

In everyday conversation, most herders were referred to by their nicknames. Often, the herder's hometown became his nickname as well, for example, Lekeitto and Markina (pronounced Lékeitto and Márkiña).[135] One Copper Basin herder was called "Akerra" (billy goat), another "Kantinflas." When inscribing for posterity on the aspens, however, the overwhelming majority rendered their given names in the official mode, in Spanish or French, and sometimes English, the nicknames being in the minority. One Bizkaian herder gave his whole name in Basque, "Arrate tar Edorta" (something like Edward von Arrate), a good indication of Basque nationalistic sentiments. Celestino Garamendi—great-uncle of John Garamendi, insurance commissioner for the State of California in the 1990s—preferred to identify himself with his trade, which was prospecting and mining. He always carved three symbols below his name: a spike or chisel, a hammer, and pincers, or sometimes two hammers instead of pincers. Many herders also carved self-portraits next to their names, often exaggerating physical attributes that made them look better, manlier. Bragging about one's sexual prowess was also acceptable because no one took it seriously.

The most prolific herder in Copper Basin was Castillo. Altogether in his long career as a sheepherder Castillo must have carved several hundred trees with his name. PIT volunteers recorded forty-five, all of which were dated. Sometimes he carved several in one day, and we were able to retrace his footsteps up and down the range. But Castillo never wavered from that norm and never carved personal messages, not even his first name, so we do not know what else he was thinking in the summer range from 1949 (the earliest date) to 1975 (the

20. This arborglyph combines many themes: religious symbol (cross), self-portrait with a Basque beret and pipe, name and date (Alberto Viteri, 30 August, 1967), hometown and region (Marquina, Vizcaya), and sheepherding (1250 ewes).

21. "Celestino Garamendi July 19 1943," Elko County, Nevada. Garamendi, who liked prospecting, always added figures of hammers and chisels to his name.

last). Elko's Star Hotel was Castillo's home in Nevada, and he died there.

Some, like Castillo, broke most of their ties with the Old Country and never returned. Others, like Dionisio R., were forced to return because the family was too much to give up. A fellow sheepherder left a message informing readers of his tragedy: "Dionisio R. is leaving on the 15th of August, but he is going against his will, [because of] family problems." Interestingly, I found a number of similar carvings in which a confidant informed the rest of the herders about the true story behind rumors of a sudden departure.

The Yearning for Women

The tree carvings in Copper Basin overwhelmingly support Aristophanes' argument, made in his play *Lysistrata*, that without female favors men feel miserable. An occasional visitor who has seen the aspen groves might go home convinced that the Basque herders were not only chauvinistic but also oversexed.[136] Both perceptions are at least partly wrong.

During the week's research in Copper Basin we recorded fifty-seven glyphs that refer to women, of which twenty-seven were erotic figures of females in various positions and states of nudity. That is a low percentage (less than 5 percent) of the total number of drawings, much lower than in other arborglyph areas, one possible reason being that several hundred carvings by non-Basques are included in the Copper Basin count.

The arborglyphs reveal the herders' fond memories of their girlfriends in the Basque Country. One poetically compares her to a *churro*, a sweet fritter: "Una nobia fina churrin mio" (A fine fiancée, my sweetie). Another carver, with misgivings, gave the worst possible news about his girlfriend: "Today my fiancée got married, and she will enjoy the night more than I." This carver included accents, a rarity.

A couple of herders hailed the whores and the girls back home. Jakes Oyharcabal carved: "Long live the girls of Lasaka, France."[137] Another tree reads: "The whores of Spain are no more." Someone else did not find this amusing and underneath added "Putero" (whoremonger). The word *puta* ("whore" in both Basque and Spanish) is one of the most common on trees. One Spanish herder even entertained the fantasy of a whorehouse run by a "Madre Superiora" (Mother Superior). In another erotic scenario, someone carved the figure of a burly man with the caption in English: "Big fellow, I do Basc-AR." Who is talking here? Is it a woman, or is it just another fantasy?

Not surprisingly, most of the erotic/sexual messages we recorded refer to whores in Elko, who are hailed as the world's best. In 1860, after the discovery of the Comstock, California miners affected by "rush-to-Washo" fever could not wait for the snow to melt on the Sierra. Reading the trees in Copper Basin, one gets the impression that the herders were struck with a similar "rush-to-Elko" frenzy, which affected them very badly in the few days preceding the long-awaited visit to town. Here are a couple of messages by different people: "How I long to see a whore." "The whores are calling us to Elko pronto and with lots of money." The original of the latter is hilarious: "Las putas nos llaman a Elko pronto kon i mucho dinero." For a non-Basque it might be unclear. The herder is not saying that the whores are enticing them to

Elko with lots of money, not for a minute. It was the other way around.

One tree standing in the middle of a verdant grove vouches for the fame of Elko whores: "The whores at Susi's [*sic*] screw very well, one can come three times before pulling out."[138] One would expect to find a corollary, a reaction, to such a statement, but there is none. Close by there is a different carving, the figure of a man in the opposite predicament: he is holding his phallus, but the accompanying caption says: "Mamicho cannot maneuver his penis."

In general, non-Basque carvers were more imaginative, more verbal in expressing their feelings, and Copper Basin is no exception. A Peruvian herder carved a number of inscriptions in which he detailed his erotic anxieties, or rather frustrations. One gets the idea:

> "Long hair, short skirt."
>
> "Crazy desires and solitary life."[139]
>
> "Fleeting love and wicked money."[140]
>
> "Perverse love, why do I seek out torment-giving creatures?"[141]

In order to understand these and similar carvings left by South American and Mexican herders one must realize that many of them were married and had left wives and families back home.[142] The pattern among the Basques was the opposite. The same Peruvian herder also carved "I feel very sad" and "I am very bored." Basques generally abstained from carving such statements, which were regarded as a sign of weakness.

Copper Basin has many more erotic carvings than Peavine, and I am inclined to believe that geography has something to do with this. The former is much more isolated; therefore the sheepherders' anxieties were considerably heightened.

22. "Hoy se casó mi novia y pasará la noche mejor que yo." (Today my fiancée got married, and she will enjoy the night more than I.)

Other Topics

The Copper Basin arborglyphs recorded thus far contain fewer references to sheepherding and its vicissitudes than are generally found elsewhere, but future research in the area might change that. In every grove one can find quasi-bitter messages criticizing camptenders. One we recorded in Copper Basin says: "All the camptenders are vagabonds." Perhaps the carver envied his camptender's freedom to run around; perhaps he went to town too often and neglected his range duties. We recorded only one other carving making a reference to a camptender. It says: "Heguy Dendary camptender."

A number of carvings refer to dogs and burros, some by name. We also recorded more than half a dozen good-byes, which are never lacking on any sheep range. Of course, not everyone was returning home. As Paulo G. put it on at least two different trees: "No way going back to life in Spain." Indeed, he married and moved to Winnemucca, where he later died.

One visually arresting glyph shows a herder standing on his horse and pointing with his *makila*. The caption says "Trel" (trail). This sort of arborglyph was useful and welcomed by herders who were new to the range. Nowadays U.S. Forest Service people welcome them, too, as many old stock trails have become difficult to identify.

Messages chronicling quarrels among herders, and especially among ethnic groups, are found in a number of groves, and in Copper Basin there is at least one glyph indicative of such problems: "I shit on all the Basques." This type of carving is usually the work of Spanish herders from Spain and commonly dates from the late 1960s. It is remarkable that I have yet to find an inscription describing the opposite scenario, that is, a Basque herder cursing a non-Basque. Furthermore, it appears that Basques and South Americans quarreled less than Basques and Spaniards, which is logical. An even stronger message, apparently aimed at Anglo-Americans by one Latin American herder, says: "Death to the red heads."[143] Possibly this could be interpreted as a complaint against the American sheep owner(s).

References to the United States

No carving is neutral on this issue. For those who did not take the isolation well, life in the United States was awful. Most Basques, but not all, disagreed with them. Copper Basin offers varied reactions to America and its people: "No American has balls as big as a Spaniard's." This could be taken as a joke but may not be. On a different note, a retired Basque herder who lives in Elko carved this in the early 1970s: "Long live the United States of America," a clear indication that he was grateful to his adopted country. And indeed, this was the overwhelming sentiment among the Basques. Mexican herders were the least content. I will refer to them again later.

The Copper Basin aspens have preserved some interesting views of American politics during the time of Presidents Kennedy and Nixon. The following were all carved by the same herder in different parts of the range:

> "Do not ask what the country can do for you."[144]
>
> "The Yankis [*sic*] have been screwed by the Chinese."
>
> "Mao screwed Nixon in the rear."

It may seem odd, but here was a herder with strong opinions about U.S. politics. His sources were probably Mexican news agencies. He believed that Nixon's overture to China was definitely flawed. This type of carving was nonexistent before the era of the transistor radio, which ended the herders' isolation from world events. Even World War II and the 1936–39 Civil War in Spain received minimum coverage on trees—and none at all in Copper Basin.

chapter two

Culture in the Carvings

Euskara, Homeland, Politics, and English

> Do you remember Don Marcelo Ruíz our teacher? He was very cruel . . . he used to give us beatings. . . . If we spoke one word in Basque he would say, "Take the wallet!" and he would give us the "wallet" and then whoever had it at the end of the week had to bring a *peseta* to school . . . we kids would spy on each other . . . it used to be frightening![1]

I grew up in the Basque Country after General Franco's 1939 takeover, when he instituted dictatorship in Spain. Franco's persecution and the Falange's[2] hatred for all Basques reached absurd proportions, supposedly because the Basques fought on the side of the democratically elected republican government and wanted to run their own affairs. The repression of Basque culture was so brutal that many Basques were led to believe that speaking Euskara was a sure sign of backwardness. Comic strips and newspaper clips routinely made fun of *baserritarrak* (Basque peasants) who spoke Castilian Spanish badly.[3]

The animosity against everything Basque was nowhere as evident as in the school system, where the Basque language was especially targeted. For years the use of Euskara in public places was forbidden. Not only was all official business conducted in Spanish, but priests in remote hamlets of the Basque Country, where no one spoke Spanish on an everyday basis, were instructed to deliver their sermons in that language. Here in the United States we try to be sensitive toward Native American gravesites, but under Franco, graveyards were violated simply because they contained tombstones carved in Basque, which were destroyed.

At age five, when I began attending school, I spoke only Euskara. During instruction I never heard the word *Basque* mentioned, except to punish us for speaking it. The teacher—who was always non-Basque—never mentioned the Basques as a people. Castilians were portrayed as model Spaniards; they had conquered and colonized America, so it was glorious to be Castilian, but everyone else in Spain had to be just a "Spaniard." I grew up with the idea that the Basques had accomplished nothing in history. At the end of each day we all raised our right hands in Fascist fashion and sang Fascist songs. Later on, I realized that Catalans and Galicians, the other two groups targeted by the Franco regime, were treated a lot better than the Basques, but their cultures, too, were persecuted.

It was not simply that all instruction was conducted in Spanish—or French, in France; it went much further than that. No one could utter one word of Euskara during class time without punishment. Children as young as five were encouraged to spy on each other and denounce any classmate who spoke Basque. Students would follow other students home hoping to catch them speaking Basque, if not among themselves, then with their parents.[4]

In any case, under the Franco dictatorship the Basques in Spain grew up without seeing one word of Euskara written in public. True, even before Franco, Basques, whether in Europe or abroad, did not normally write in their language. Imagine, therefore, my surprise and excitement when I discovered that a long time ago, sheepherders in the remote mountains of the western United States, totally removed from the political situation in Europe, had actually used the Basque language when carving tree trunks.

Iparralde Basques have not suffered such blatant cultural repression by the government of France, but ever since the French Revolution they have been subjected to a strong and steady "homogenizing" pressure to become "French."[5] The French Basques did not benefit much from "liberté, egalité, fraternité." They were stripped of their ancient laws and system of government, and some three thousand were forced into exile, where more than half died.[6] The invading Germans during

World War I may have completed the French government's task. Basque veterans returning from the fighting brought with them not only the French language but also a patriotic outlook.[7] Perhaps that explains the many "Vive la France" and "Biba Frantzia" messages incised on aspens in Nevada and California by the sheepherders from Iparralde.[8]

The Significance of Euskara Carvings

The Basque clergy, both in France and in Spain, were Euskara's only partners because the Church, which had some freedom to act, claimed that Euskara was necessary to minister to the needs of near-monolingual Basques in the rural areas. The Basque clergy in France took a more active role in using written Euskara to reach their parishioners, an effort that started in the sixteenth century when rulers such as Jeanne d'Albret, queen of Benafarroa and Bearn, supported translation of the Bible and publication of the earliest books in Euskara.[9] The first book in Basque, published in 1545, and the first translation of the New Testament, in 1571, were both written in northern dialects. The thriving fishing and whaling seaports of Lapurdi may have furthered the use of the Basque language in France because business there was transacted entirely in Euskara, whereas in Hegoalde the language of the bourgeoisie was Spanish.[10] This may be one reason that Iparralde Basque sheepherders carved most of their messages in Euskara and few in French—if we exclude the monotonous "Vive la France."

Yet, Euskara carvings are uncommon. But let me qualify that statement. Most of the carvings are simple names and dates, and the herders used the official version of these; therefore, a Basque from Iparralde would carve Pierre Mendiboure, while one from Hegoalde would render it Pedro Mendiburu. For a Basque speaker, both are the same. In their everyday communication and interactions in Euskara, these two individuals were probably identified by a nickname. If a surname was used in the conversation, both would say "Mendiburu," because the Spanish spelling conforms better to Euskara phonetics than the French one.[11] But I doubt that, in the herder's own mind, carving "Pedro Mendiburu, 1934" on a tree constituted a conscious act of using the Spanish language. Therefore, if we disregard the great majority of the arborglyphs that fall into the names and dates category, the percentage of Euskara carvings becomes much larger.

Euskara on the Range

The policies of Spain and France led to a curious situation in Euskal Herria. The peasants spoke mostly in Basque but were schooled to write only in Spanish or French.[12] The majority of the Basques, then, were illiterate or nearly so in their own language.[13] Euskara has a reputation for being impossible to learn—the devil himself could not learn it, according to a legend. But it was mostly the lack of literature and the fact that government, the media, and businesses did not use it that handicapped Euskara, and because the Basques themselves had been brainwashed against the practicality of writing in it.

The Basques who came to Nevada after World War II had probably received a primary school education, but the immigrants who preceded them had less education. The further back we go in time, the greater the number of illiterates. On the eve of the French Revolution, only 50–79 percent of the bridegrooms in Iparralde were able to sign their names on the marriage register. In fact, this was quite an accomplishment because compulsory public education did not materialize until a century later.[14] As for Spain, in 1868 more than 75 percent of the people were illiterate; in 1934 only about 50 percent received some basic education.[15] In 1998, on average, 7.6 percent of Spaniards were still illiterate, versus 1.4 percent in Euskal Herria.[16]

The situation was not so dismal in the Basque Country. Esteban Garibay Zamalloa, a sixteenth-century author, said that men there "were given to matters of the pen," which means that public schools operated in the 1500s and even before.[17] But not in Basque, and not everyone attended them, least of all peasants. The Church may have been Euskara's best ally, especially in Iparralde, but children were taught mainly to memorize catechism in it rather than to write it. In 1693 Zuberoan ordinances required that schools be established in every parish. Gipuzkoa's general assembly ordered the same thing in 1721, and similar rules were passed in other Basque regions.[18] But Agustin Cardaberaz wrote in 1761 that it was shameful to see Basque gentlemen unable to read one line in Basque without stumbling.[19]

Among the sheepherders who came to Nevada and California were some Spaniards and Mexicans, as well as some Basques from Nafarroa who did not speak Basque. Whenever an outfit included one or several such herders, the rest of the Basques usually accommodated and spoke Spanish as best as they could. On the other hand, I also know of herders who learned to speak Basque in the United States. American sheep owners often tried to learn the language of their herders, but that seldom included Basque.

There is no doubt that most Basques learned new languages in the United States, including knowledge of other Basque dialects (in the Basque Country, the chances of a Bizkaian learning the Benafarroan dialect was practically zero, and vice versa). Those who spoke mostly Euskara learned some Spanish in Nevada and California, and they all acquired English. The two main languages of the sheepherders were Euskara and Spanish; French came a distant fourth, behind English. Iparralde Basques were first and foremost culturally Basque rather than French. They spoke only Euskara among themselves, and the majority had a poor knowledge of the French language. Hegoalde Basques were likely to speak both Spanish and Euskara, but when dealing with this topic the most important factor to consider is the time period. Before the 1940s, peasants in many Basque-speaking rural areas spoke Spanish only when they were forced to—say, by the presence of an outsider. During Franco's dictatorship Spanish for the first time made inroads deep into the heart of Euskal Herria. The difference is evident in the lower quality of the Euskara carved on trees by the herders who came during that period.

Generally speaking, the bulk of the Bizkaians and mountain Nafarroans who came to America knew more Spanish than the Iparralde Basques knew French. Bizkaian and Gipuzkoans wielded plenty of political rhetoric in defense of the Basque Country and its culture, but for centuries most educated people wrote in Spanish; thus they earned the following poem, written by Joanes Etxeberri in 1636:

> I make fun of Garibay
> as well as of Etxabe
> because they have spoken
> in foreign tongue about Basque people.
> Since the two of them
> were Basque-speakers
> they should have written
> their histories in Euskara.[20]

If Etxeberri had a chance to visit the high Sierra in Nevada he might have taken pride in the fact that the uneducated herders tried to carve in Euskara. Linguistically and culturally, the little Euskara found on trees must be considered significant.[21] Consider, for example, the following inscription on a tree in Humboldt County: "Arima baduca nic ez ikusia dut 15 aout 1925 zer astoa den Jakes Marchanta arima saldut" (If he has a soul, I have not seen it, 15 August 1925, what a donkey Jakes Marchanta is! I sold my soul). It would appear that "saldu du" (he sold his soul) would be more appropriate than "saldut" (I sold it), so this part of the carving is not altogether clear, but its theme and the vocabulary render it very unusual, and the gist is clear.

American Basque

No one could have expected the sheepherders, with just a few years of primary education, to start writing/carving in Basque in America, although they did. In fact, they were instrumental in fashioning what we might loosely call a "Basque American" dialect.[22] A linguist would argue that the type of Euskara heard in the United States does not qualify as a dialect. That may be true, but I still think Basques in Nevada and California speak a Euskara that is not heard in any of the Basque regions in Europe. Some differences are subtle, but others are glaring, like the Basquesized English words.

In the American West, where Basques lived free from the officially imposed pressure of Spanish and French cultures, their own language became the official one. In many cases Euskara enjoyed a dynamic rebound. English replaced Spanish and French in the official capacity, but in the sheep camps the herder had little chance of learning or using English. Euskara gained in exposure by default. The herders spoke different Basque dialects, and at first some of them could not understand each other. Herders from Nafarroa and Benafarroa had no difficulty conversing with each other,[23] but the Bizkaians and the Benafarroans stumbled immediately beyond the initial greeting.

His first linguistic encounter in America was quite frustrating to a young man from Mungia, Bizkaia. After he arrived in Bakersfield in the late 1960s he went straight to a Basque hotel to wait for the sheep boss. While sitting at the bar of the *ostatu* (boardinghouse), he noticed two men at the other end speaking a strange language. After a while he realized that they were speaking Basque! This was his first encounter with Iparralde Basques, and he found out that he could barely make out a few words. Shortly afterward his employer, an American-born Basque, showed up and greeted him in Basque. But the

Bizkaian could not understand him either. The boss switched to Spanish, which he had learned from the Mexicans in California. The Bizkaian spoke Spanish, too, but was equally lost; the boss's Mexican Spanish was strange to him. Finally, the sheepman lost his patience and rebuked him: "You don't know anything, neither Basque nor Spanish." The young Bizkaian was mortified, but he said to himself, "Since I came all the way to America, I guess I will have to put up with it."[24]

Today, this Bizkaian has no problem holding a conversation with any Basque in America, in any dialect. But he no longer speaks his native Bizkaian dialect like he used to; now he mixes in a number of Benafarroan words and even a few grammatical constructions, such as "eiten ahal" (able to do), which the majority of the Basques in Nevada and California have adopted.

The further back we go in time, the more Euskaldun the herders generally were. When Bizkaians and Nafarroans met for the first time, they were not comfortable speaking Basque, and the herders who came in the 1960s and 1970s preferred to resort to Spanish, a situation that was strengthened on account of the Nafarroans who spoke only Spanish. I myself have been fooled by Nafarroans who spoke Spanish to me but conversed fluently in Euskara with fellow Nafarroans.

Due to chain migration, if the sheep company was owned by a Nafarroan, the herders were usually from Nafarroa. If among them there happened to be a Bizkaian, he had no recourse but to learn the Nafarroan dialect, and in a hurry. The same was true in the reverse situation, when the Bizkaians were in the majority, as was the case with Benafarroans who went to work in Idaho.

Within a month of interaction, as the herders assimilated each other's key vocabulary, many of the linguistic barriers came down and ordinary communication flowed as smoothly as it could be expected to flow. Once they achieved a certain level of understanding, there was no problem with each one speaking in his own dialect. The only difference was that the dialects were increasingly sprinkled with words borrowed from other dialects spoken in the camp. Each dialect had terms and expressions that seemed more appropriate than others and were adopted and used rather than similar words from other dialects. For example, *erran* (to say) appears to be winning ground against the Bizkaian *esan*, but *biga* (two) is losing against the Bizkaian *bi*.

I personally know Bizkaians who speak Bizkaian among themselves and Benafarroan with those from Benafarroa, or at least they try to. But when they meet people from Nafarroa they revert more to their own Bizkaian dialect because Bizkaian is closer to Nafarroan than to Benafarroan. Ambrose Arla, a native of Donastiri, Benafarroa, who worked in the Austin area with Bizkaian herders for sixteen years, explained the linguistic phenomenon that took place on the western range: "I learned Bizkaian, but they did not learn my Euskara."[25] He was actually saying that they did not speak his dialect, because the Bizkaians were in the majority, but that does not mean they did not learn some of Arla's dialect.[26]

How did the spoken Basque affect the carved Basque? The mixture of dialects is less apparent in the Basque carvings because written Basque is more established than spoken Basque. At times Bizkaians copied Benafarroan herders—for example, by switching from *gora* to *biba* (hail), which is not employed in Bizkaia. The influences ran both ways, however, as observed in the following example found in the Toiyabe Range in central Nevada.

Jesus Ibarluzea, a Bizkaian, worked in the area from 1963 to 1968, and in a small grove he left no fewer than twenty-one inscriptions in Basque. His message is repetitive and, with some alterations, goes something like this: "Ni nas Jesus viscainoa. Ni nas arsain pobre bat" (I am Jesus, a Bizkaian. I am a poor sheepherder). In spite of the brevity of his carvings, the messages carried such force that even sheepherders from the highlands of Nafarroa and Benafarroa were inspired to copy him.

One fellow from Hazparne, Benafarroa, carved: "Ni nas Bernardo Tocoua arsain pobre bat 1963" (I am Bernardo Tocoua, a poor sheepherder, 1963). In the same grove two other herders from northern Nafarroa copied Ibarluzea. One, whose name was Fidel, made a joke of it: "Ni nas Fidel Castro 1971" (I am Fidel Castro, 1971). Normally, Tocoua and Fidel would have carved *ni naiz* (I am), not *ni nas*.

Why did Ibarluzea use the word *vizcainoa*, a term never used in his homeland of Bizkaia but widespread in the American West? *Bizkainoa*, spelled in various forms, was simply a concession to the dominant dialect spoken by herders from Nafarroa and Benafarroa. But this begs the question: How was *bizkainoa* introduced into America in the first place? Certainly not by Bizkaian sheepherders, who would normally say *bizkaitar* (Bizkaian). This appears to indicate that non-Bizkaian herders spread the term, which might be of American origin. In the Americas during the three hundred years of Spanish colonial period, *vizcaíno* meant "Basque."

One day in Winnemucca I had the following conversation in Basque with an old ex-sheepherder from Benafarroa:

"Are you a *bizkaino?*" he greeted me.

"I am *bizkaitarra*," I responded. "And you?"

"*Arnegiarra*." (From Arnegi).

"You mean *arnegianoa*," I retorted, emphasizing the word.

"No, no, *arnegiarra*," he answered.

"If I am *bizkainoa*, you must be *arnegianoa*," I said.

For an instant he remained silent. Euskara was the primary language of this man, and he quickly realized the logic of what I was saying. He smiled and answered: "I understand, I understand."[27] Then I asked him where he had learned to say *bizkainoa*, and he said in the United States. In fact, it is so pervasive that even Bizkaians normally use *bizkainoa* today.

It became clear to the herders that they needed a common linguistic ground to communicate. Thus a new type of Basque was used in the range. As a man from Elko said: "In America I learned a new Euskara that I did not speak before and now I can talk with any Basque, it does not matter if they are from Nafarroa or Zuberoa."[28] This was no attempt to achieve linguistic unity in the manner of the academics who created Batua, the unified Euskara taught in Basque schools today, not by any stretch of the imagination; it was a compromise. The herders learned to use vocabulary from a common pool. What really made the "American Batua" work was the fact that the sheepherders learned the basics of each other's dialects and vocabulary. After that everyone could continue speaking their own dialect, but with certain concessions. I have a Nafarroan friend who speaks a mixture of Nafarroan and Benafarroan. Every so often in conversation, whenever he feels that I may not understand some words, he will remark, "You Bizkaians say this other word," and he will insert the Bizkaian variant for my benefit. Other Basques use similar tactics in conversation, but without any fanfare.

Many of the early Basques in Nevada lived in such cultural isolation that even their children were monolingual until they started school. I have had many conversations with American Basques who remember being made fun of because of their "peculiarity." Some of those who lived in northern Humboldt County do not recall any harassment or discrimination because Indians, Basques, and other minorities were the norm, not an oddity.[29] This was also the case with urban Basques who had Basque neighbors. The children played together and spoke the familiar language.[30] Without these "ethnic pockets" few American-born urban Basques would have retained Euskara. Many parents, in fact, encouraged their children to speak English and discouraged the use of Euskara.[31]

In families in which the father and mother spoke different dialects the children were exposed to both but generally favored one over the other. This was the case for Mari Irueta, who was born and raised in Winnemucca and lives there still. Her father was from Baigorri and her mother was from Lekeitio. Mari recalled, "At first my father spoke a very thick Benafarroan dialect, until he learned Mother's Basque. We children spoke Bizkaian."[32]

The Language of the Range

Louise Shadduck, who wrote about Basque herders in a Scottish outfit in Idaho, said that some people "say that sheepdogs and horses understand Basque as readily as English."[33] Bilingual animals? Not impossible, but it is more likely that they understand commands given in one primary language.

In the summer of 1993 I was visiting Biskay Land and Livestock's sheep camp in Carey, Idaho, owned by Pete Cenarrusa and family.[34] It was once a solidly Bizkaian company, but in the 1990s all the herders were Peruvian or South American. Still, they had several sheepdogs with Basque names, and Mike Cenarruza, the manager, said to me: "It's funny; we used to speak Euskara to them, but now they understand Spanish only."

Today the sheepherders from South America pick up Euskara from their Basque bosses, just as Basques once picked up English. In the summer of 1991 in Bridgeport, California, a Peruvian herder explaining the ways of the sheep to me, said: "When the ewes go takataka . . ." Surprised, I interrupted him:

"What do you mean by 'takataka'? Do you use that word in Peru?"

"No, but my boss uses it all the time."

When I asked him if he knew what it meant, he answered: "Yes, it means that the sheep are going on the trail at a steady pace." Indeed, that is what *takataka* means in Basque.[35]

Carved Basque Literature

I must warn the reader at the outset that the word *literature* may be a misnomer, for there is little of it here. Possibly 5 percent of the carvings are written in Basque—the percentage varies depending on the period and the sheep range. The style and gist of the Basque language that some herders carved is extremely authentic, considerably more genuine than most of the literature produced by writers educated in standard schools, I would say. The principal reason rests in the fact that these messages are actually thought out in Basque, not in Latin, French, or Spanish and then converted into Basque. All educated Basques used Romance languages throughout their school years and careers, and the grammatical and mental idiosyncrasies of these languages creep unnoticed into their writings in Euskara.[36]

Immediately one can see the "advantage" for sheepherders who were not educated: they were not dependent on non-Basque erudition. They were educated at the oral "university" of Euskara, a living language that for thousands of years has been a vehicle of information for Euskaldunak as well as a repository of their accumulated common knowledge. You can tell that the Euskara in Nevada's aspen groves was not exposed to Spanish or French literary traditions because it is sharply reminiscent of the Basque spoken decades ago, before foreign media flooded all of the Basque Country.

One thing stands out in these carvings: Basque conciseness. On paper you can write at length, you can elaborate; however, when carving trees it is necessary to encapsulate your thoughts. This was fine with the Basques, whom centuries ago a Castilian characterized as being "short in words."[37] Tree carvings are succinct, not unlike the technique of the *bertsolari* (improvisational Basque poet).[38] Both forms are products of common people, and both are intrinsically associated with Euskaldun (speaker of Basque) culture.[39] Therefore the tree-carving activity was already adapted to the Basque character. It will not come as a surprise that several of the longest carved messages I recorded were by Spanish herders.

It was with great satisfaction that I read and recorded the first *bertsoak* (rhymed verses) on aspens. Unfortunately, few herders carved them because they require quite a bit of space. Further, it is difficult to carve them without losing your train of thought; it is totally against the grain of *bertsolaritza* to have to wait to finish your verse. Most of its merit comes precisely from the fact that it is quickly improvised. *Bertso* are more commonly found on media other than aspens, such as water tanks, and written mostly in pencil.

In the South Warner Mountains are two *bertso* carved on neighboring trees by "Banka Txarra" (Little [Guy from] Banka), the nickname of Domingo Louisena.[40] Neither of the verses is dated, but Louisena made sure to leave a record on a nearby tree of the date on which he was inspired: 19 August 1915. It is extraordinary that both trees—they are not large—have survived.[41]

Huna ene izena	Here is my name
Domingo Louisena	[which is] Domingo Louisena
bromera iten dena	who loves to have some fun
ala dik almena	[and] has ability for it
dediela	can go to hell[42]
sinesten ez duena	whoever does not believe it.
Gark[43] emanik	After he/she had done it
ingelesa	in English
terzera franzesa	the third one in French
hori vehar degu	we need that
guzion juge zat	as judge of us all
alua puta	fuckin' ["cunt"] whore
nork erdi[44] baduke	someone could give birth
crande[45] zabala	large and wide [broad]
niorekin	with no one
denekin	with everyone
dena beza.	he/she can have it all.

The words of the last verse are mostly clear, yet I fail to understand exactly what Louisena was singing about. It must be understood that he sang these verses before and after he carved them on the trees, probably more than once. Singing is an integral part of *bertsogintza* (the art of improvising). And what a place to sing! From that position Louisena could see green meadows, forested mountains, and the glistening summit of Mount Shasta in the far distance. If we had been there in 1915 to listen to him, the verses would have made a lot more sense. The problem appears to be the rendition of the verses in written form.

It may not be a coincidence that at least two other *bertso* are carved in a canyon about two miles away. I mentioned that the herders took inspiration from earlier arborglyphs they happened to find in the forest. So it was with G. L. in 1929. Next to the verse he carved the figure of his farmstead, a huge, square, three-story house, and beside it he wrote "aldudarra" (native of Aldude). G. L. was related to Jean Battitta Laxague of Surprise Valley, California, one of the herders who was killed by Indians in 1911. His verse contains some of the typical ingredients a sheepherder might sing about:

Baska dugu	We have pasture
oray yana	already eaten
hortako ibilik	that is why we work
baduzula	and you have
eta nirezasko	toward me
gana	desire
Mari nagusiek	[my] boss Mary
betirana	forever
Biba ni eta	hurrah for me and
ene buztana.	my dick.

The *bertsolari* whose verse is quoted below, Jean Donamaria from Donibane Garazi, Benafarroa, was daydreaming under the aspens. Enjoying the cool shade, he surveyed scenery worthy of the Garden of Eden. Across the soft, verdant meadow he could watch the sheep enjoying the succulent grasses on the hillside. It was 1949, and he was sitting or kneeling because the verse is carved very low, almost to the ground, on the trunk.

Biba Donibane Garazi	Long live Donibane Garazi
Pariseri ere	to Paris even
da nauzi	is superior
han ez da	because there, only
yende onezi	honest people
eta nezka	and
donsella	unmarried girls
baizik bizi.	live.

This is a good representation of what was said earlier about pride in one's hometown. The population of Garazi in 1949 was a few thousand, but for Donamaria every one of them was upright, including the girls, who maintained their virginity; Paris could not boast of any of that. Large cities had a bad reputation, especially in the countryside. Henry H. Longfellow wrote: "There is not a virtuous woman in Madrid, in this whole city."[46] In regard to virginity—the idea implied by the word *donsella*—it could be a joke to be taken not literally but rather in the opposite sense. One must be aware of both the sheepherders' mind-set and the irony that is at the heart of Basque humor.

I am told that many herders whistled and sang during their idle time, and the more gifted ones composed verses. Singing old songs has always been a favorite pastime for Basques, especially during summer picnics and after a good meal with friends and relatives. In fact, this love of song has led the Basques to initiate the Euskal Kantari Eguna (Basques Singers' Day) celebrated in Gardnerville, Nevada, where several hundred ex-sheepherders gather every year.

In July 1951 Jean Biscay of Heleta, Benafarroa, transcribed the lyrics of a haunting popular Basque song onto a tree near Reno:

Iguzki denean	When the sun is out
zoin den eder itzala	how beautiful the shade is
maitia mintzo zira	[my] love, you speak
plazer duzun bezala	any way you wish
egiten duzula	as you provide
mila *disimula*	one thousand *excuses*
inorant zirela	*that you do not know anything*
erraiten duzula	*you say*
bertzetarik naizela	*that I am like the others [but]*
falzuki mintzo zira.	you speak falsely.

Biscay left out the part in italics (perhaps he could not remember the whole song). On the other hand, many popular songs are orally transmitted and have slight variations. He sat under the aspens in a lovely spot by a brook while singing his blues away with this beautiful melody.[47]

Biscay was obviously a lover of old songs and expressions, although it is not always easy to figure out what led him to inscribe particular messages. This one, for example:

Arno onak	The good wine
parerik ez du	has no equal
basombat baino hobe	better than one good glassful
biga	[are] two
ez dea egia	[but] isn't the truth
banaski hobe.	certainly better?[48]

The verse seems to be as much a question as it is a statement. Was he speaking of someone who brought him a good bottle of wine or did him a favor but at the same time lied to him? Or "arno ona" may be an allegory for a cheating woman.

Another herder, apparently from Zuberoa, who had problems with his girlfriend in the Old Country wrote all about it on a tree dated 1954 or 1955:

Gustura ny(n)tzan	I was contented

e(ne) sortean	with my destiny
hemen maiterik gabe	here without a lover
zahartuz	having to grow old
gero Ziberoan	later in Ziberoa

A Basque can certainly appreciate the directness of such a sorrowful statement. The story does not end there, however. A carving made the following year indicates that it was all over between them, but the sheepherder did not appear to be angry. With an alluring sense of poetry, he called the mountain his "castle" and recalled the good days he had with his woman friend:

Zer pina	What a fine
zer chato ederra	beautiful castle
hau lehen	this used to be
neure Marie Guartekin[49]	with my Marie Guarte

This is all in past tense, but I do not know whether the words must be taken literally or figuratively, for it is rather difficult, though not impossible, to imagine that Marie was once on the mountain with the herder.

In the Mono Lake area of California, sheepherders used large galvanized steel tanks to water the sheep and wrote on them with pencils. As everyone knows, pencils write much better than knives. The tanks are fairly recent, from the late 1950s, but these messages are the closest the herders came to "real" literature; most of it is in Euskara. Some incredible figures are found here as well; for example, a bearded, bushy-haired old man with a nose unlike any I have seen carved. A herder who saw it and apparently took exception wrote a *bertso* underneath:

Ene bizian ez dut ikusi	In my life I have not seen
olako gizon itxusirik	a man as ugly as this one
bizarra du	he has the beard
auditzen asia	beginning to grow
sudur tipia	small nose
eta belarriz gorria	and his ears are red
nai nuke jakin nor den	I would like to know who is
au egin duen gizon ikasia	the learned man who has done this.

This *bertso* is unique in that it contains not a single misspelling, is grammatically correct, and is readily understandable by any Basque reader. This was quite an achievement at the time it was written (probably the 1960s) because Basques still received all their education in Spanish or French. The word *ikasia* (learned) may be indicative of the author's education, and since he used no *h* before *olako* and *au*, he was very likely from northern Nafarroa.[50]

Nicknames

Until the 1970s Basque babies commonly received Spanish or French names at baptism, the only ones sanctioned by the Catholic Church. This explains why the first names of the herders were Juan, Pedro, Pierre, Jean, etc., and that is how they carved them on aspens. (Sometimes they used English equivalents.) Today the practice has been totally reversed, and few babies are named after saints.

That is only part of the story, however, because most herders had nicknames as well. Most nicknames, following the Basque preoccupation with a place on the land, made reference to the bearer's hometown. At times, they referred to a physical condition, like "Manes Tipy" (Little John), or a combination of physical attributes and the hometown, such as "Banka Txarra" (Little Banka), or some other particular attribute. I even found a French nickname "Noir Banca" (Banca the Dark). Some do not have a clear meaning or translation. Below are examples of nicknames found on trees:

Acherito (little fox)

Airako Belxa (dark one from Aira)

Aldude Debru (devil from Aldude)

Archtato (little bear)

Arro Zahaguia (puffed-up pigskin; this could be worse than a nickname)

Arzamenta (a place-name)

Asto Pitilia (donkey dong)

Asto Pito (like Asto Pitilia)

Baskito (little Basque—two such nicknames were found, one spelled "Basquito")

Basoch Kuku Donasti (Donasti is a town in Benafarroa)

Bilbo (from Bilbao)

Bolinero Chocolatero (miller, chocolate maker)

Dominico Abasolico Biscayico (wordplay)

Eri Motza (lacking a finger)

Etcheandia Asto Handia (big house, big jackass)

Etchelar Belxa (dark one from Etxalar, Nafarroa)

Ezkerra (left-handed)

- Franses Eskualdun Trampa (French Basque tramp)
- Franses Tripa Andia (big-belly French)
- Gazte Andueza (Andueza junior)
- Jean Tipy (little John)
- Kiki Miki Arrupe (wordplay)
- Lepabeltz (one with dark neck or skin)
- Lepo Luzia (long neck)
- Manes (John, nickname for many Basques from Iparralde)
- Palo Blanco (pimp [Spanish])
- Pedro Lapico (Pete the cauldron)
- Poteros (no translation)
- Quipucha (from Gipuzkoa)
- Tirrit Tarrat Truku (onomatopoeic)
- Tosia (could be a place-name)
- Traste (dumpy-looking man)
- Tximistie (lightning, a man who does everything fast)
- Zorrik Berri (someone whose last name was Zaldiberry)

For a while I was led astray by "Eri Motza," a nickname carved on several trees and dated in the early 1930s. There was never a name, just "Eri Motza." At first I thought that someone was disclosing a medical condition, an ugly sickness at that, because *eri* means "sickness or sick" and *motza* in my native dialect means "ugly." I felt sorry for the guy. Obviously, he was sharing his problem with the trees because there was no one else around. *Andra mina* (venereal disease) was not unheard of among sheepherders. I saw quite a few markings by "Eri Motza," and in one of them he had added "Vive France," so I thought, well, as bad he felt, the poor fellow still had some patriotic feelings left.

The next day, as I proceeded to the other side of the mountain range, I sighed with relief when I found the following carving: "Eri Motza kuet," which I understood to mean that the herder had finally gotten rid of the ugly disease, because *kuet* is Basque for "quit."

Although I thought that I had stumbled on a very personal story here—one with a happy ending—I was not completely satisfied with my interpretation of "Eri Motza." One day I asked a retired herder what *eri motza*

23. A message that may refer to one or all the Iparralde Basque herders: "Franses eskualdun trampa" (French Basque tramp) by an unknown carver.

meant in Benafarroan dialect. "Someone lacking a finger," he said. I sighed with relief. "Eri Motza kuet," therefore, meant that he was quitting the sheep, which made more sense.

Herders often derived their nicknames from some trivial occurrence. Here are a few collected from oral sources:

Bixkuter (small man, named after a tiny car "Biscooter")

Blaki (dark)

Burdinas (iron man, someone very strong)

Ixil Ixila (very quiet one)

Kafie (coffee, because someone liked it so much)

Kaioti (pronounced "káioti" [coyote], someone you cannot trust)

Kakati (shitty one)

Makuxe (slight man)

Markina Baltza (dark one from Markina)

Maxingan (machine gun, a stutterer)

Mediokilo (skinny fellow, half pint)

Pedro Kalbo (Pedro the bald)

Piper (pepper, someone with a bad temper)

Pistolero (fellow with two thumbs, eleven fingers)

Prakazar (old pants, raggedy)

Terrible (habitual liar)

Tony Kalbo (Tony the bald)

Traste (junk, old stuff)

Trumoie (thunder, someone predisposed to flatulence)

Txatarrero (a man who loved junk)

Txif (chief)

Txinpas (rascal)

Zigurre (someone who seldom took chances)

Nicknames are so much a part of Basque culture that as research progresses, more will no doubt be found.

Mother Tongue and Intimate Feelings

Years ago, when speaking with my priest friends Eusebio "Sakone" Osa Unamuno and Patxi Lasa, both from Gipuzkoa, the conversation revolved around the power of the mother tongue. They remarked that in highly emotional moments, people instinctively revert to their native tongue. For example, when the Basque missionaries, who for political reasons never spoke Basque in their own country, prepared to leave for the missions of Africa or America, they felt compelled to say their last words and good-byes in Euskara. It happened all the time. According to his helper, the Jesuit missionary Francis Xavier spoke in an unknown language, probably Basque, as he lay dying off the coast of China.[51]

Basques are said to be reluctant to confide their secrets or to talk about their emotions as openly as, for example, the Spanish. On the other hand, the isolation prompted many of them to break the cultural mold and talk to the trees more freely than they would have talked back home in the Pyrenees. Some of the herders could not keep secrets, and anything and everything went "out the door," into the woods, and onto the tree trunks. It is noteworthy that many of these intimate confessions are in Basque. The herder imagined that the deserts and mountains would forever protect the secrets of his carvings. After all, not even the devil could understand Basque. He knew, of course, that coworkers would read them, but he did not care because they were all members of the sheepherder community.

Many messages are intimate confessions. A few carved in Basque can be considered family secrets and are among the most interesting. For months I wrestled with two carvings found near Reno. I could read the individual words, but not the overall meaning. What did J. B. have in mind when he carved on two aspens: "J. B. wild dove trap gipsy lacotea philopina big teeth" and "J. B. fine wild dove in the district of Antxartei"?

Luckily, in the summer of 1989 I met some visitors from the Basque Country who knew J. B. and could shed light on the cryptic messages. According to their interpretation, he was reminiscing about wild dove hunts in the Pyrenean mountains, where the birds are caught with large nets.[52] J. B. must have had a powerful reason for wanting to reveal that his mother, Philopina, had big teeth and was a gypsy from Lacotea. She was still alive in 1989.

The isolation of the mountains, so conducive to introspection and leisure time, gave the herders opportunity to meditate and sort out their feelings. Many carvings consist of spontaneous emotions blurted out in a fit of anger, loneliness, resignation, or gloom. Whether they talk about a girlfriend in Europe, an illness or death in

24. "Gaizo Manez joan zinen mundu huntarik 1936. MANES 1932 BAIGORRY." (Poor Manez, you departed from this world. MANES 1932 BAIGORRY.) Authored by two herders—notice that "Manes" is spelled two ways. The part in large capital letters was carved in 1932 by Manes from Baigorry. Four years later Pierre Gortari, from the same town, carved the first part as an epitaph, because by then Manes had died.

the family, or a sickness the herder is suffering, nothing about the carvings seems contrived or insincere.

Some messages are so emotionally charged that the moment when the herder carved them is almost palpable. The scene can be re-created easily: It is a lazy summer afternoon, and a herder wanders through a silent grove in a distant canyon. He is so lonely that at times he doubts that he will ever see another human being again. In order to fight the stillness that surrounds him he has to remind himself again and again that he is *not* the only person on earth. Suddenly he is jolted from his ruminations by a familiar name carved on a large tree, a name inscribed by another herder from his hometown, Baigorry, Benafarroa. He eagerly approaches the aspen. The incisions are only a few years old, so they have not yet scarred over. He reads the message easily: "Manesa puta, Baigorry 1932."

Manesa is the name of a herder he knows well. Perhaps the two had emigrated together and worked in the same sheep outfit. But something terrible happened. Manesa died. How? The aspen is silent about details. Shivering in the summer heat, the herder runs his fingers slowly over the relief of the inscription, and his eyes begin to fill with tears. The sheepdogs beside him look up at him; one whimpers sympathetically. This scenario was basically lived by Pierre Gortari, who in 1936 stood by an aspen carved by Manes and above his countryman's name carved the following epitaph: "Gaizo Manez joan zinen mundu huntarik" (You poor Manes, you have departed from this world).

Few messages I have read throughout the forgotten canyons of Nevada moved me as powerfully as this one. What makes carved messages different is the immensity of the land that envelops you. With no one else around for miles, the message on the tree is the closest thing to a human voice, and sometimes it can scream at you. We may cut this aspen someday, transport it to town, and put in a museum, but it will not emit the same energy as when it stood in the grove, and it will not speak with the same voice.

Hundreds of miles from the grove where Gortari carved, in a much more hospitable and greener environment, but one just as remote, stands the testimonial of a sheepherder who became ill. I found several carvings that mention sickness or indisposition. For a sheepherder it was a tough situation. Lacking transportation, the nearest doctor might be days away. Besides, the herder could not, would not, leave the sheep alone. This particular carving in central Nevada is inscribed in three languages: Basque, Latin, and Spanish.

> Heart of Jesus,
> forgive me, oh Lord
> Oh, God, hurry
> to my assistance
> Lord, the one you love
> is ill
> G. O. 15 August 1965.

The Less Fortunate and Others

The literature on Basque sheepherders contains much more about successful people than about those who achieved neither riches nor notoriety. Men who started as herders but later became sheep owners who married,

25. A prayer in three languages—Basque, Latin, and Spanish—by G. O., who must have had some education: "Jesusen vioza parce Domini o Dios venid pronto en mi ayuda Senor el que amais esta enfermo GO 15 agosto 1965" (Heart of Jesus, forgive me, oh Lord, oh God hurry to my assistance, the one you love is ill, G.O. 15 August 1965), Toiyabe National Forest, central Nevada.

had children, and became part of the community have received greater coverage than all the rest of their countrymen combined, even those who remained sheepherders for many years or even throughout their lives—that is, the true professionals. What do we know about these sheepherders "par excellence"? Very little; and not much more than that has been written about them. Fortunately, a number of them carved some of their most intimate thoughts on trees. These nuggets of information displayed for the casual observer are invaluable in penetrating the outer shell of many sheepherders. The following carving nearly tops them all: "I am B. T., I have been in America many years, but I am still poor, 1963."[53]

A lot of herders could have signed that statement! B. T. and many others immigrated to America not to stay but to accumulate capital and return. But many never realized that dream. I have known some of them, withering away in the Basque hotels. They work the sheep from spring to fall and spend winters in the hotels. Their savings go to the bottle and cigarettes, when not to the casinos or the whores. In the spring, broke or in debt, they must return to the mountains until the fall, when the cycle starts once again.

We do not hear enough of the less successful type, the "black sheep" of the family who came to America *mundu ikusten* (to see the world),[54] a type described to me one day by a Basque veteran of the range. During his first two years as a herder he worked for three bosses. He left the sheep to become a miner, then a construction worker. All along, the Immigration Department was after him. He died in an accident, driving while intoxicated.[55] We hardly know these Basques, perhaps because they pose little attraction for writers. The curious thing is that in the Basque community some of these people who never "made it" can be rather popular, in the sense that they are often talked about, and not necessarily in derogatory terms.

As far as their carving activity, sheepherders can be divided into three major categories:

1. The sullen and the resigned, silent types and those who liked sheepherding and decided to make a career of it;
2. The perennially discontented ones, typified by Frank Rodriguez; and
3. Those who saved every penny because they could not wait to return to society to lead normal lives, marry, and raise a family.

The following statements by F. G. tell us which of the three types he was: "The life of the sheepherder is a sad life." The verb in the familiar (*hika*) mode makes this statement even more interesting. The herder is preaching to himself, trying to convince himself to make a decision. Indeed, F. G., unlike Charlie Paul, Martin Banca, Frank Rodriguez, and so many others, reached a decision: "Very soon I must leave this job in order to live a long life."[56] He did, and he lives in California still.

In 1926 Arnot Urruty entertained the same thoughts. He never quite gave up hope, and year after year repeated his words: "Here I am bored/tired, someday the time will come to leave this place."[57] Many of the arborglyphs belong in this category, carved by those who knew that they had better get out of the mountains but, for whatever reason, tarried making the decision.

For some, that day never came. What made them stay in a situation they so disliked? Once again, a philosopher herder in the mountains near Gardnerville, Nevada, put it succinctly: "Bakochak bere aldian penak" (Each one carries the suffering inside himself). That "suffering" was the enigma of some herders who could never break the yearly cycle forced on them by transhumance.

26. "Ni nas Bernardo Tocoua urte asko Américan eta beti pobre 1963" (I am Bernardo Tocoua, I have been in America many years but I am still poor, 1963). Tocoua was from Iparralde, but here he used a Bizkaian verb, proof of dialectal assimilation on the Nevada range.

The Importance of the Homeland

In addition to the ubiquitous names and dates, many of the carvings provide information on the herder's place of origin, a clear indication that the hometown was very important to the Basques. It would be interesting to know whether this preoccupation with national and regional origins is duplicated among other groups of immigrants to the United States.[58] At first, most immigrants feel "transplanted," and emotionally they continue living in the Old Country.[59]

The preoccupation with hometown remained even for herders who had been five, ten, or more years in Nevada. There is every indication that they were very proud of their hometowns, no matter how small. In fact, I recorded only one exception, carved by a Bizkaian working for the Goicoechea Sheep Company of Elko who apparently did not share the traditional feelings toward his hometown: "Juan Egaña was born in a pitiful hamlet, Ereño."

The sheepherder Isasi could not have disagreed more with his fellow Bizkaian Egaña. Using an exceptional identity marker and grandiose vocabulary to show his pride of origin, Isasi carved: "Ignacio Isasi, a national of Rigoitia" (a hamlet overlooking Gernika, Bizkaia).[60] Another carved: "My birthplace is Erasta in the Basque Country."[61]

This pride of birthplace is discernible in the details the herder provided, such as the following example: "Pete Ampo, 16 July 1929 native of Balkarlos, province of Nafarroa, Alonbro Arocor house."[62] Gratien Laxague carved in two aspens the longest and most detailed information about a birthplace that I have seen:

> juliet 1926 Gratien Laxague ne Gorba dans les basses pyrenes cantone baigorry commune desaldude maison Yrocho.
>
> Le 7 juillet l'anne 26 Gratien Laxague nele (?) 20 en 1901 en France dans les basses pyrenees on (baigo)rry andes maison Pritcha oila (a) on presse . . .[63]

These messages, in French, contain misspellings and some unreadable words. There appear to be at least two names of farmsteads, the one where Laxague was born in 1901 and the one where he lived at the time of his emigration to the United States. A few years later, on 25 August 1929, but hundreds of miles away, Martin Larzabal, no less proud of his roots, carved the following: "Urrunarra, Echeberrico semia" (from Urruna, son of the Echeberri farmstead).

All these messages indicate herders' nostalgia for their country. In a strange land, in the solitude of the mountains, it is not surprising that they often thought about their families and hometowns. But again, the strongest symbol of the Old Country, the hometown, the family, is the *baserri* (farmstead), which figures prominently in the groves. The *baserri* has a strong personality; each one has a name, which does not change even if it is sold to new owners. In America few would find a substitute for it. In the urban world of the United States, the home for most was a street address, a number.

The number-one hometown name in the glyphs is Aldude, a town in Benafarroa; and, indeed, more herders hailed from Aldude than from any other Basque town. They proudly carved "aldudarra" (native of Aldude) next to their names. A close second is Baigorri, followed by Arnegi, Garazi, Latsa, Bidarrai, Irisarri, Banka, Ortzaize, and so on, not necessarily in that order. All these towns are in Benafarroa. One herder, instead of using the "native of" form, carved "Iruritacoa," which means "from Irurita," a town of Nafarroa. But only a herder from a Benafarroan town showed solidarity with his fellows and carved: "Biba Bidarraitarrak" (Hurrah for the natives of Bidarray).

The herders from Nafarroa carved not only the name of their hometown but that of the valley as well, because each valley was an administrative unit. Sarratea, for example, carved: "Almandoz, Esteribar Valley, Nafarroa."[64] Bizkaia is not organized by valleys, but the peasants to this day cling to the ancient political unit of the *auzo* (hamlet). Several *auzo* make up each village. At least two herders, one of them from Abadiano, Bizkaia, used that identification system.

Baztan was probably carved as often as Aldude because the villages in this Nafarroan valley, taken together, sent more herders to the United States than Aldude did. Iruñea—also known as Pamplona—the ancient capital city of Nafarroa, was carved less often than Baztan, followed by Luzaide (sometimes written Valcarlos, Balcarlos, etc.).

Even when the place of origin is not stated, it can be deduced from the carver's surname. Sorhouet and Laxalt are Basque surnames written according to French phonetics and grammar rules, while Uriarte, Gabika, and Bermeosolo are clearly Bizkaian, and Urtasun, Goñi, and Elizondo are Nafarroan. Only a few surnames are general enough to make deducing the herder's native region difficult.

The most commonly carved place-names are those of the larger countries: France, Spain, Nafarroa, and Bizkaia, in that order, but written diversely. By now, Peru, Mexico, and Chile probably outnumber even France. America and Nevada are about even.

The sheepherders from Benafarroa often added "France" to the name of their hometown, sometimes written in Basque, "Frantzia." I have found only one "Bachenabarre" and one "Benafarroa." On the other hand, many herders included the initials B(asses) P(yrenees), without accents or punctuation marks. They tended to skip the longer names. This can clearly be seen in the names of Nevada towns. There are many "Elkos," "Renos," and "Elys," but not one "Winnemucca," which for a Basque ear was difficult (they say Binemuka or Minemuka—*u* sounds like *oo*). California, on the other hand, had a nice ring to it and was carved despite its length.

The names of towns were often carved with an audience in mind. The first loyalty was reserved for one's hometown, but sometimes to carve "Iragi" or "Ereño" was not enough. These names were not very well known, so "Nafarroa" and "Bizkaia" were added to them. Since the herder population always included some non-Basques, it was important to make clear to them that Nafarroa was in Spain or that Ortzaize was in the department of B.P. (France). In fact, few Bizkaians knew where Ortzaize or Iragi was located, just as few Nafarroans could identify Ea as a Bizkaian town (see appendix 4).

Ethnic Identity

The literature routinely divides American Basques into Spanish and French, to the point that it is almost mandatory to ask, "Are you French or Spanish Basque?" Only non-Basques would ask, however, because if the conversation is taking place in Euskara, the dialect reveals quite clearly on which side of the Pyrenees an individual was born.

Among Basque speakers, a person's name does not matter as much as where he was born. Most herders were peasants whose families had lived on the same farmstead for generations. Therefore, the idea that an individual was bound to a particular piece of land was of extreme importance to them. This is the underlying thrust that inspired the sheepherders to express their place of origin in tiers: first the farmstead (carved fairly often), then the hamlet, the hometown, the valley, the region or province, and, finally, the state. None of these—except, perhaps, the town and the nation—would have made sense to the Anglo public.

However, by dwelling on divisions we run the risk of downplaying cultural homogeneity. Over and above the national divisions of French Basque and Spanish Basque there exists a powerful concept of Basque identity that is based first of all on the language, Euskara. The difference between knowing and not knowing Euskara runs deep. The dialectal variations were significant in Euskal Herria but less so in the United States.

Although herders rendered their names either in Spanish or French, a few individuals identified themselves more clearly in Basque according to the rules established by Arana-Goiri'tar Sabin, founder of the Basque Nationalist Party. In the following example, the herder meant to follow the rules but did it backward, carving: "Candido Olanotar" instead of the more orthodox "Olano'tar Candido," which means something like "Candido of the Olano Clan." Other examples that I recorded are:

"Justo Sarria, Lekeitio, euskotarra" (Justo Sarria, [from] Lekeitio, racially Basque)

"Gilen Yrigoyen, 1939" (William Yrigoyen)[65]

"Guilen Borda, 1913" (William Borda)

"Inaki Plaza" (Ignatius Plaza)

"Battita, Bancar[r]a" (Battita from Banka)

"Erramun Borda, bidarraitarra, 1914" (Erramun [Ramon] Borda from Bidarray)

The most common ethnic identity marker was "eskualduna" or "euskalduna," which is what Basques call themselves. Those from Mexico called themselves "mexicanos," without variation; those from Spain called themselves "españoles," and the Peruvians were "perua-

27. An imposing Basque farmhouse.

28. Many herders unabashedly hailed their own egos. This arborglyph in Nevada, carved in big, bold, Basque-style letters, is dated in Basque and says: "Biba Battita Bancar[r]a ustaila 1957" (Hurrah for Battita, the one from Banca, July 1957).

nos." But the Basques, struggling under the shadow cast by stronger national cultures, could not present a united cultural front. It took a whole batch of terms to define Basque identity as displayed on trees: "euskalduna" (the Batua form today), "euscalduna," "ezkulduna," "escualduna," "basco," "bazco," "bazko," "vasco," "Bask," "Basc," "Basqo," "Basque," etc. All of these forms and more have been recorded, and they reoccur persistently throughout the sheep ranges.

In 1909 P. Laxalt made a strong statement of ethnicity when he carved "Bazco . . . Raza," indicating that he was of the Basque race. (Another word separates these two, but it is too distorted to read.) In the 1940s Fermin carved "euscalduna" under his name, and on 24 July 1971 Cisco Aguirre identified himself as "Euskalduna." In 1975 Joseba Guisasola from Eibar (Gipuzkoa) expressed the same sentiment when he carved a short rhyming verse: "Gora euskaldunak dio Guipuchak" (Long live the Basques, says the Gipuzkoan). But in 1950 Pierre Lanathoua clearly intended to overcome the political boundaries when he carved: "Long live all Basque sheepherders, French and Spanish."

Rarely, the herders took the unusual step of dating the carvings in Basque; for example: "Seigarren illea zen ogeita zortzie" (It was the sixth month, the twenty-eighth), "Arrasaldeko seyetan" (At six o'clock in the afternoon), "Uztailaren 26an" (On the 26th of July).

If the carvings are naked, elementary, incisive, and devoid of artifice, it follows that they can be used to explore the Basque character. But we should first distinguish between urban and Basque peasants—the latter are less sophisticated than the former—and between the Spanish character and the Basque. Some prominent writers believed the Spanish character to be rather simple and lacking in intelligence and imagination, like Don Quixote.[66] The Basques themselves, however, think the Spanish possess potent and fiery imaginations.

What do the carvings tell us in this regard? In general, they indicate that the Spanish speakers—including Spaniards, Peruvians, Mexicans, and even monolingual Spanish Nafarroans—are more outspoken, extroverted, and imaginative than the Basque speakers. Carving after carving indicates that the Euskaldunak are much more reserved.[67] But the carvings also demonstrate convincingly what others have noted as well: that the personality of the Basque develops more fully after he emigrates. Emigration does not have to be all the way to America; Castile will do just fine. E. Lafuente Ferrari, for example, noted that the Basque painter Ignacio Zuloaga's art took off in Segovia, Castile.[68] But this is probably true not only of Basques but also of most emigrants.

Regional Politics

Often, nostalgia, regionalism, and patriotism mix with politics in the carvings, and in that respect the herders did not imitate the flocking instinct of the sheep. Rather, each went his own way, or "zenbat buru, hainbat aburu" (there are as many opinions as there are heads), as the Basque proverb goes. Some hailed France, others Spain or Nafarroa, yet others "Gora Euskadi" (Independent Basque Country, variously spelled) or "Aupa Vizkaya" (Long live Bizkaia, variously spelled).

The herders from Nafarroa were generally more regionalistic than Bizkaians and carved "Viva Navarra" (Long live Nafarroa) often. But many of those from the northern valleys (Baztan, Erro, Bostiri, and the Bidasoa area) considered themselves simply Basques, or Basque Nafarroans, and spoke primarily Euskara; their loyalty was more to the idea of Basque in general than to just Nafarroa.

The political statements in the glyphs create a confusing picture. A few herders hailed the Basque Country, France, Spain, America, and, sometimes, the whole world. Below are some examples of sheepherders' loyalties:

"1915 Long live the Basques and Igantzi, Spain, Juan Manterola."[69]

"1930 Long live the Basque Country, Sorhuet, Arnegui, long live France."[70]

"1956 Long live Elantxobe Bizkaia, Jose Armaolea."[71]

"1958 Jose Lesaca, long live Spain."

"1958 Long live the Basques, Miguel Recarte, Erratzu, Nafarroa, Spain."[72]

"1959 Long live France and the Basque Country, long live Benafarroa, Marc Gortari, August 27."[73]

"1963 Luis Ojer, long live Nafarroa, always onward."[74]

"1963 1964 Long live Baztan, Lekaroz, Nafarroa."[75]

"1966 Long live independent Basque Country, B. Azkoitia."[76]

"1967 Long live France, Martin Tocoua."[77]

"Long live Bizkaia" (anonymous carving, no date).[78]

"Long live free Basque Country, praise to God and long live the Old Laws."[79]

"Biba Eskual Herria" (anonymous carving, no date).

"Long live the Basque Country."[80]

"Biba Escual Herri Maitea" (Long live beloved Basque Country).

The Hegoalde Basques, especially the Bizkaians, were more politicized; since 1833 they had fought three wars to defend their home rule against the centralizing forces in Madrid.[81] Curiously, Spanish Basques who carved in Euskara rarely carved anything more than political slogans hailing an independent Euskadi. French Basques, who cared nothing for an independent Basque Country, carved many things in Euskara. No one took offense at seeing "Vive la France," but it was a different story with "Viva España." Below are examples of reactions to three such carvings:

"Good-bye my beloved Spain, you are in my heart, H. T."

(Carved underneath): "Bastard."

"Viva Espana."

(First response): "Some Spain."[82]

(Second response): "It seems so."

(Third response): "And Franco."

"Nicolas Huarte, the year 1953, long live Spain."

(Response): "Shit for whoever wrote this."

French Basques were oblivious to politics on the Spanish side, and for that reason alone the message by Jean Azconaga of Iparralde is unusual and curious. Between

29. A carving in three languages: "Jean Goyhenetche 27 du July 1938 French Basko" (Jean Goyhenetche, 27 July 1938, French Basque), central Nevada.

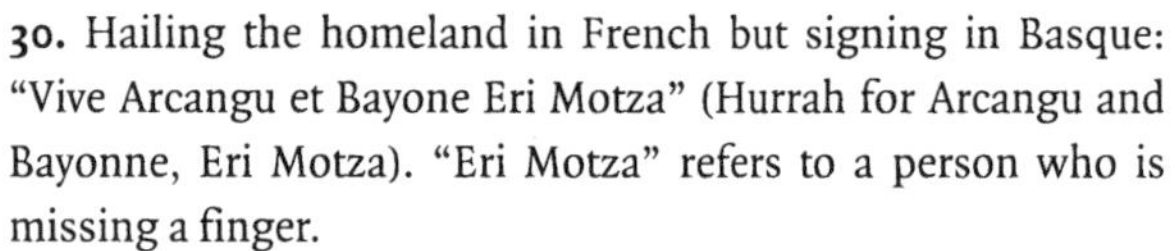

30. Hailing the homeland in French but signing in Basque: "Vive Arcangu et Bayone Eri Motza" (Hurrah for Arcangu and Bayonne, Eri Motza). "Eri Motza" refers to a person who is missing a finger.

31. Government in Nafarroa is still largely based on ancient valley units, and this carving is an example of it. Several symbols are incorporated into the message, which says "Valle Aezcoa" (Aezcoa Valley). For this herder the valley was his hometown.

1936 and 1940 he carved in Spanish: "Biba Francisco Franco palo blanco cabron maricon" (Hurrah for Francisco Franco, pimp, bastard, fag).

Some carvings are mutely intentional. A herder with the initials R. F. carved: "Long live Francia Ezpaña Amerika." On 19 August 1930, when Paul Sorhuet saw it, he thought something was missing in this patriotic equation; on the nearest tree he carved: "Biba Ezcual Herria" (Long live the Basque Country).

In general the Basque herders who arrived in the 1960s were the most politicized. This was the initial period of ETA (Euskadi ta Askatasuna, or Independent Basque Country and Freedom), the movement that galvanized and united the Basques in Spain against Franco's regime. Political slogans became much more common on the sheep range in those years. The following, riddled with misspellings, was carved by J. G. and is one of the most elaborate as it paraphrases a militaristic song:

Eusco gudariak guera	We are Basque soldiers
euscadi escatxeco	who will free Euskadi
eta belari motxak	and the short-ears [the Spaniards]
orcoes sapaltxeco.[83]	will trample underfoot.

Six years earlier, not far from where this tree stands, R. B. A. had carved a pompous political statement that was as long as it was misspelled:

Señor y Dios	Lord and God
nuestro Lose [*sic*] Antonio	our Jose Antonio[84]
este contigo	be with you
nosotros kemos lograr	we want to achieve
aki la España dificil	here the difficult Spain
y erekta	and erect (strong)
ke es nuetro anvicion	which is our ambition
Espana una	one Spain
España grande	great Spain
España livre	free Spain
arriva Espana	up with Spain
arriva	up
viva Franco	hail to Franco
viva 1961.	hail, 1961.

32. Entwined flags with the caption "Biba Franzia eta ni" (Long live France and I), carved by B. T.

No doubt J. G. saw this and his message is in part a response to it. R. B. A.'s message is either a joke or a very fascist political statement. I am suggesting the joke because the carver was none other than a Gernikan, a citizen of the holy city of the Basques, which was destroyed by Nazi airplanes allied with Franco on 26 April 1937. Taken at its face value, it is the carving with the most "vivas" in it.

Carving symbols and flags was another way to show political convictions. In the 1960s and 1970s herders, especially Bizkaians, eagerly carved the *ikurrina*, or Basque flag, which looks like the British flag but with different colors—red, white, and green. It was forbidden in the Old Country. "Gora Euskadi" (Long live Independent Basque Country) was the typical political cry of the Basque patriots.[85]

There are also several "Gora ETA" arborglyphs hailing the Basque guerrilla group fighting to liberate the Basque Country from the clutches of General Franco. In the late 1960s an ETA member who had been tortured by the Spanish police ran away and came to northern Nevada to herd sheep for a short stint. He was told that he would be safe from the Spanish police there because, as sheepherders like to say, even God sometimes gets lost in the wilds of Nevada. But clearly the young man was not cut out to be a sheepherder. He was from the city of Bilbao, college educated and from a well-to-do family, and when he bought a brand-new car even before he banked his first paycheck, the Basques in town knew he would never make it as a sheepherder. Within a few months he accidentally shot himself in the hand and left the range.[86]

He probably was not the only ETA member to seek temporary refuge in the sheep ranges of the West. An earlier message from someone else who apparently had skirmished with the Spanish police marks an aspen in a Humboldt County grove. He first carved an unmistakable *txapeloker* ("crooked hat," as the Basques called Franco's Guardia Civil, with their tricornered hats), and the caption accompanying it illustrates the point: "Watch out, you donibarra dog, or I will get you."[87]

In the 1970s Jesus Mallea[88] carved many "Gora Euskadi Askatuta" (Long live free Basque Country) messages in Plumas National Forest and even more figures of rifles and handguns. He identified himself as "the best gunman" and inscribed fiery political slogans. In one he used the word *euskoak*, an unusual term that academics might use to describe Basques in general.

33. Basque flag, or *ikurrina*, with the caption "Gora Euzkadi Azkatuta" (Long live Free Basque Homeland). The tree was cut down in White Pine County and was retrieved by Dave Goicoechea of Reno.

Even older political symbols have been found. On 13 July 1933, Antonio Garcia, native of Arrieta, Artze Valley, in Nafarroa, carved a fairly detailed coat of arms of the kings of Nafarroa consisting of a golden crown and chains arranged in an octagonal shape. Interestingly, the chained rectangle is wider than it is tall and really looks more like the *ikurrina* than the official Nafarroan coat of arms (which is taller than it is wide). Perhaps he was trying to combine both symbols into one, something that, as recently as 1999, Basques politicians have not been able to agree on. In this remote corner of northwestern Humboldt County, Garcia's mind boiled with precocious political ideals, which, thanks to this tree, have not been totally extinguished.[89]

World News and Politics

It seems strange that lonely herders would think of hailing the world as a whole, an idea that squarely contradicted their otherwise narrow view of fatherland, but we cannot argue with the trees. It makes a certain sense, for the herders lived in the biggest house of all, the outdoors. The arborglyphs hail a variety of causes:

Gora dena (Hurrah for everything)
Viva Rusia
Biba Paris (the city)
Viva el comunismo (Long live communism)
Viva Fidel Castro
Viva Cuba
Viva el Rey Juan Carlos (Long live the king Juan Carlos)
Viva Hitler
Biba Amerika
Viva USS ([*sic*] for USA)
Viva Peru
Viva Mexico
Viva Republika
Long live the workers
Viva Axturax (a region in northwest Spain)

One would expect the 1936–39 Spanish Civil War to have generated a great deal of interest among the herders, but I found references by only two individuals. In the Inyo National Forest, south of Lee Vining, California, one herder followed the war with great interest and carved several trees with messages condemning the Fascists, the rich, the clergy, and the military. He advocated socialism, jobs for the working class, and freedom of the press. What is even stranger, no one contradicted him or added comments nearby.[90]

The Iparralde Basques were not affected by the Spanish Civil War, but they were involved in the French Revolution, the Napoleonic wars, and World Wars I and II. Messages hailing France were carved more often during the wars than during any other period. Arnaud Joseph Ezponde from Gamarta, Benafarroa, carved a number of variations on this theme in Tahoe National Forest, California, where he worked from 1942 to 1950. Several trees read "Vive la France"; on others he added "Paris" or "America." Finally, he decided to go for "Amerika eta Eskual Herria" (the United States and the Basque Country).

I have found half a dozen comments on the last world war. One pointed arborglyph marks the beginning of World War II succinctly: "Guerla diglaratia 1939 gaicho Franzia" (The war has been declared, poor France, 1939).[91] Another carving provides more details about the conflagration. It appears that the reputation of Marshal Henri Pétain of France had spread even to the sheep ranges of Nevada. The following message in the Marlette Lake area appears to echo that sentiment: "Six to one Petein" [*sic*]. This fellow was betting on Pétain.

Saralegui read it and carved this comment below: "This [guy] too is German." Saralegui must have been ignorant of Pétain's identity, or perhaps he was referring to his collaboration with the Germans. Someone else was in favor of Germany, as shown by an unsigned carving on a nearby tree: "Long live Hitler, champion of France."[92] "Champion" here means "victor." But Saralegui was pro-American: "Long live America, boss of all," he carved on another aspen.[93] Finally, in the summer of 1946, a fellow named Fermin announced that Japan had been defeated by the Allies. Apparently it took a full year for the news to reach the sheep range in the high country. In contrast, the message "Kennedy presidente" was prescient. It was carved in 1961 in Columbia Basin, Elko County, by J. Lopategui—three months before the election took place. In general, however, herders preferred to discuss women, sheepherding, and humorous matters to politics. One carving put it quite bluntly: "I shit on politics."

There were always a few individuals whose self-appointed job it was to inform their fellow herders of world affairs. In July 1938 on Glass Mountain, Inyo National Forest, someone carved: "Howod went around the world in 91 hours." This is Howard Hughes, of course, and the exact duration of his flight was 91 hours and 14 minutes. And in Plumas National Forest in 1962: "This year Marilin [*sic*] Monroe killed herself. Poor Marilin." This carver probably heard the news on his transistor radio, but the former had to depend on newspapers or word of mouth.

34. Most herders received letters from home and were generally aware of the major affairs that affected their countries. This aspen in eastern Nevada testifies to the reaction of a herder from Iparralde at the beginning of World War II: "Guerla diglaratia 1939 gaicho Franzia." (The war has been declared, 1939, poor France.)

Euskara and English

Some of the terms needed in everyday American society did not exist in Euskara, so the Basques liberally borrowed English words. But since the two languages are so dissimilar, Euskara structures were not affected by the loanwords; further, the loanwords were adapted to Basque phonetic canons. For example, Thanksgiving turned into "Saniskibi" and "Saniskibil," Eureka into "Irurika," Duck Valley into "Dokabale," and Jordan Valley (Oregon) into "Jornabale." Three different sheepherders called themselves "siper" in tree carvings (see the glossary for additional examples).

The Romans had triumphal arches under which captive enemies were forced to pass as a sign of defeat and surrender. Immigrants of many nationalities went through similar experiences at the port of entry, where they often surrendered the spelling of their surnames to the Immigration officer. The Basques were no different, as the spelling of their names in the U.S. censuses attests. Many Basque surnames—I would guess more than half—are misspelled. Furthermore, some individuals turn into two people by magic, simply because their names are spelled differently in different places.[94]

As a rule, Basque composite surnames were chopped off; thus Guerricabeitia turned into Guerry and Marcuerquiaga emerged as McErquiaga. A Bizkaian fellow with a long name was told that he would have to change it. "Why?" he asked. The judge told him that he could not read his surname. The Bizkaian answered: "That is your problem, Your Honor, not mine." And he kept his last name intact.[95]

Broken English and Other Sheepherder Jokes

Anglo-Americans tell many jokes about sheepherders trying to speak or write in English, but I have probably heard just as many told by the Basques themselves. Of course, nowadays it is easy to laugh at bygone sheepherders trying to express themselves in an alien language, but at the time, everyone cringed under the linguistic barrier. Their attempts at expression in a foreign language were appreciated and even celebrated by fellow herders, who, through the years, retold the funnier incidents they remembered. In the end, these became part of the oral culture of the range.

Some Anglo jokes told about sheepherders reflect stereotyped notions of Basques held by cowboys or town dwellers, or are simply stories that poke fun at an identifiable group of people who also happen to be immigrants. Many of these stories underscore the lack of education. The sheepherders' own jokes have to do mostly with language problems and are framed against the historical backdrop of life on the range.

The following happened in the 1940s in southeastern Oregon, just north of the Nevada border. One day S. A., a herder from Bizkaia, was herding on Steens Mountain when an American cowboy came up to him and asked him about range boundaries and why he kept his sheep restricted to one side of the creek. S. A. could understand English better than he could speak it, and he answered: "El de oder side de crik gobernamente pey" (this should be read according to Basque phonetics). It is doubtful that the American understood, but S. A. was trying to say that the creek was the boundary of government reserve land, where the sheep could not graze without payment of a fee.

The next incident showcases the herder's resourcefulness. He used three languages to explain things to a monolingual cowboy. It happened in 1949 in Yakima, Washington. The herder, from Busturi (Bizkaia), was known as "Káioti" by his fellow herders. As he was herding, an American approached him and inquired why he kept a goat among the sheep. Káioti must have racked his brains before he came up with this answer: "Dis fella auntz maneja ship."[96] It is mystery whether the American understood the herder's "Koiné," but when other Basques heard about it they had a good laugh, even though they themselves could not have explained it any better in English. Káioti used English, Basque, Spanish, and English again. "This fellow" (is) "auntz" (she-goat in Basque), "maneja" (controls or handles, in Spanish), "ship" (sheep). He meant to say that the goat was the sheep's leader.[97] A lot of the Basques in Winnemucca know this story, to the point that when conversation turns to a topic that seems to have no clear answer, or a politician is speaking on TV, someone will say: "Dis fella auntz," and everyone has a laugh.

A number of Basques who have been thirty or forty years in the United States still speak little English. That poses no problem for fellow Basques, but some Anglo-Americans have difficulty understanding them, among them a Reno judge.

In the 1970s a Basque herder went to court to request citizenship papers. Since most of his knowledge of English consisted of half a dozen swear words, his uncle went along as an interpreter. Right away, the judge wanted to know why the herder, after eight years in the country, had not learned the language. The man from Nafarroa replied: "Your Honor, the sheep have been in America longer than I and they still do not speak any English." The magistrate, recognizing the truth as well as the humor of the argument, granted him the papers.[98]

In another true incident, a Basque herder was put in jail for trespassing. Frustrated because the man didn't speak a word of English, the judge sentenced him to four years in jail. After serving the full term, he was led into the sentencing judge's chambers once again, and this time the herder spoke fluent English. The judge, surprised, asked where he had learned it, and the herder replied: "Your Honor, I learned it at the four-year college you sent me to."[99]

chapter three

Sheepherding According to the Sheepherders

Biba artzainak eta hemen egoiten ahal direnak. (Long live the sheepherders and those who have the guts to stay here.)

In them days, we no sooner got off the train . . . we found ourselves in the desert. We had our provisions, or bedroll, a carbine, strong walking shoes, an American hat, a burro, and a dog. And, oh yes, three thousand sheep. The boss would take a stick, and looking at that miserable desert stretching out there forever, he would scratch a map on the ground. To show where the water was, where the good feed was. Then you just moved out. . . . In a year we would walk thousands of miles.[1]

The arborglyphs document the sheepherders' summer months in the high country, typically from late June to early October, and shed new light on their day-by-day experiences when they were alone and charged with considerable responsibilities. In addition, however, the carvings reveal the sheepherders' overall impressions regarding life in general and life in the United States.

Many people in the American West have two great misconceptions when it comes to Basques: they think that the Basque Country is predominantly rural and that most Basques are sheepherders. While the former may have been fairly accurate a century ago, the reality today is quite the opposite. Less than 5 percent of the population in Euskal Herria is employed in agriculture and fishing.[2] The figure was 28 percent in 1977 for Nafarroa, and probably a bit higher for Iparralde.[3] Most farmers today keep only small flocks of sheep that graze in fenced pastures and require no herding.

Many of the herders who came to California and Nevada, especially those who came from Bizkaia, had no previous experience with sheep whatsoever; those from northern Nafarroa and Iparralde were more likely to have worked with them. During the 1960s and 1970s all sheepherder candidates from Hegoalde were required to take an exam at the American consulate in Bilbao before they were admitted into the United States. Every herder I talked to agreed that it was more a formality than a test. For example, in 1960 a prospective herder from Sollube, Bizkaia, was asked if he had sheep at home. He answered: "Yes, I have forty, and each one has new shoes."[4]

He obviously had no idea what he was talking about, but he passed and came to America anyway. Perhaps it is to the likes of him that the following carving refers: "Asensio B. is going to Ameriga. Poor Ameriga."[5]

Many arborglyphs were carved in reaction to the weather (cold, rain, snow) or the quality of the pasture; the vicissitudes of sheepherding, for example, predators (e.g., coyotes or bears), moving the sheep, and shipping lambs; or personal problems with fellow herders or the camptender. Loneliness, homesickness, and related topics are discussed later in this chapter.

Some of the de rigueur themes found in almost all treatises on sheepherding, such as the wars between cattlemen and sheepmen, are totally absent from the tree-carving literature.[6] Of the twenty thousand or so carvings I have seen, not one mentions such problems. A puzzling discovery? Perhaps—or perhaps an accurate view of the past.

To Be Man Enough

It is said that farming and animal husbandry are in the blood of Basques because the majority of Basque surnames refer to ancestral farmhouses. At the same time, most sheepherders were aware that they held jobs no one else wanted. They also knew that their friends employed in town earned more money and had cars, girlfriends, and wives, lived in normal houses, and could socialize on weekends . . . all of which made them jealous. What they may not have known was that few uneducated blue-collar workers in town could afford to deposit almost their entire paycheck in a savings account, as the herders did. Saving money was usually the reason a Basque man became a sheepherder in the first place, but it was tough going. Every day, a herder had to reassure himself that after a few years of sacrifice he would have something to show.

Those who were looking forward to their last day with the sheep—the majority—sometimes felt that the wages were not worth the discomforts, as in the following carving: "Fuckin' life, there is none worse, but without America there are [no] dollars, *foken*, 1965."[7] One herder who saw this message did not agree; underneath it he added: "Whiner." Most Basques, in fact, decided that deploring the situation did not benefit them and tried to psychologically boost their spirits. This is how the numerous "hurrahs" and "long live" messages must be understood. From "long live I" to "long live my country," they were ultimately designed to shake off loneliness.

Sheepherding built character and self-reliance.[8] The herders knew that being entrusted with fifteen hundred ewes and as many lambs was quite a responsibility, but throughout history Basques have shown that they are very capable of shouldering responsibility and discharging their duties with competence. The Basques call this trait *gizon izatea* (to be a man), or, less euphemistically, *potruak eukitzea* (to have balls), to accomplish a tough job. This strength is not *indar* (physical strength) at all, as has sometimes been written. It has to do with Old World society, where it was very important for a male—or a female—Basque to be considered hardworking and dependable by his peers. In the same vein, I recorded many carvings by herders who took a dim view of the "whiners" among their ranks.

In the rural world the first farmer to plow the ground in the spring or the first one to start cultivating the corn always earns praise from his neighbors. The last farmer to do so earns a dubious distinction. To be lazy is the worst disgrace, beneath even that of being a drunk. When someone died in my hometown, I remember my mother saying, "He certainly was a hard worker," which in her mind constituted the greatest possible praise. If it were a woman, she would say, "She was all work."[9]

To be labeled *ganorabako* (lacking basic skills, not dependable) is almost as bad as being lazy. Dependability goes hand in hand with trust. Among Basque peasants, business is transacted by verbal agreement; papers do not exist. Written documents are a concession to the outside world.[10] For a Basque, a verbal agreement is as good as a promise, and someone who breaks his word (*hitza jan*) seriously tarnishes his or her reputation. Such were the qualities that sustained the Basques on the lonely range. To go back on one's word or to be proved incompetent were disasters to be avoided at all costs.

Their dependability was the characteristic that, according to some observers, differentiated Basque herders from others. An English sheepman in Nevada said that the Scots were "the most skillful" herders, but "they and the Irish had a breaking point. . . . When they got fed up, they would walk away from their sheep and go to town for a tear. Not the Basques. They would stay with those sheep until they dropped in their tracks, or went nuts from being alone."[11]

That is what M. O. must have had in mind when on the desolate Cherry Creek Range he proudly cried out for all the world to hear: "Biba artzainak eta hemen egoiten ahal direnak M&O 1950" (Long live the sheepherders and those who have the guts to stay here).[12] He was saying that in the mountains you could tell the men from the boys. The message contains an ironic aspect as well, for the Basques did not think highly of Nevada's barren sheep ranges. M. O. was also saying that tending sheep in Nevada was better suited to monks than to gregarious young men. William Smallwood of Idaho told me that the Basques are the least suited culturally for sheepherding.[13] Yet they will endure the pain stoically until the job is done. As M. O. did and thousands like him.

In 1950 near Gardnerville, Nevada, Pierre Lanathoua carved that same thought in awful Spanish that translates as: "Long live the Basques, in the whole world there are none like them for work."[14]

Yet, many herders felt that society did not appreciate their sacrifices and hard work. No wonder they were having second thoughts about coming to Nevada. On a cold and rainy spring day in 1993 a Basque American sheepman said to me reproachfully: "We sheep operators provide jobs, and produce meat and wool out of desert grasses that would otherwise go to waste."[15] "Yet no one seems to value our work," was the obvious corollary, but, being Basque, he did not state it.

Some sheepherders believed themselves to be lucky in

35. A cry in the wilderness by a defiant herder: "Biba artzainak eta hemen egoiten ahal direnak M. O. 1950" (Long live the sheepherders and those who have the guts to live here, M. O., 1950). The Basques were well aware that their jobs were not the most desirable in the United States, but many nevertheless felt proud of their performance.

comparison with others. Factory workers, for example, had to endure smoke, noise, and the slavery imposed by the time clock. Basques routinely cited poor economic conditions at home as their reason for emigrating to the United States to herd sheep, but they could have found work in the industrial cities of Bilbao or Eibar. But *baserritarrak* like C. O. preferred to herd sheep in the open spaces of the American West. He carved: "C. O. artsaya nere borondati" (C. O. sheepherder by my own choice). Many herders probably would have left the mountains if given the opportunity, but certainly there were some who enjoyed the lifestyle and led fairly contented lives.[16] Few carved about it, however, because although it is acceptable in Basque society to have a good time, publicizing how much fun you are having isn't.

View of Self

How did the sheepherders see themselves? The many self-portraits indicate that they had a definite self-image of their capacities and strengths, and that they were proud of their work. The cowboy would not be without his horse, but the herder relied on his own two feet. The typical sheepherder portrait shows him alone, standing with feet firmly planted on the ground, and holding his *makila*. Often he is smoking and wears a hat and a Levi's jacket or coat. Some carvings depict the herder's dog, donkey, or other pack animal next to him. Others show him riding a horse. More rarely, the portraits include the tent, the rifle, and sheep.

I found one tree carving in which the herder is surrounded by a donkey and three dogs. At first I thought they were sheep or lambs, because the carving is severely distorted. But another herder who had been there when the figures were newer clarified it for me. The first carver wrote: "This is Vicente with Pagola's donkey." An unidentified herder carved on the side: "Why do you need so many dogs?"

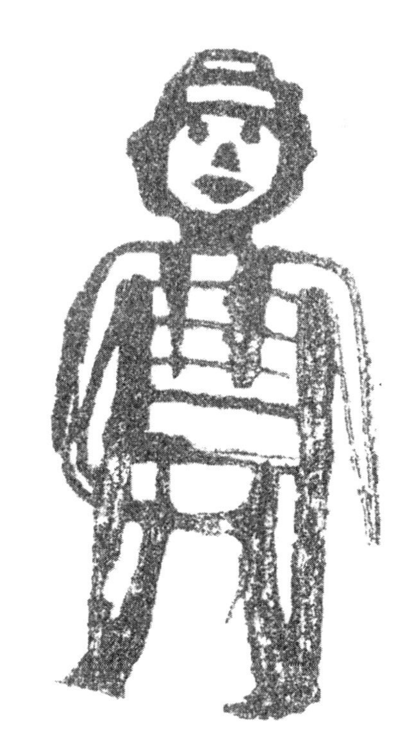

36. If it is not a self-portrait, this carving from the Lake Tahoe area, Nevada, must be an Old Country dancer, or an alien.

38. Sheepherder's self-portrait.

37. Those for whom carving the name alone was not enough added a self-portrait—which might be more or less realistic.

39. Cowboys were lost without their horses, and herders without their *makilak* (walking sticks). Here the herder, accompanied by a sketchy female figure, holds a rifle.

It is significant that sheep depictions are very rare. During one field trip a student asked me why there were no carved sheep. Were not sheep at the center of the herder's life? It is a good question, one that puts a new spin on the way we sometimes interpret historical data, and art. Arborglyphs reflect the psychological environment as much as the physical one. They reflect the presence of individual herders on the land, but they also reflect what was missing from it. The thousands of references to women do not mean that they were part of sheepherders' lives anymore than the lack of carved sheep means that sheep were absent. If we put the student's question to a herder he would surely respond: "Why would I want to carve the figure of a sheep when I was sick and tired of the thousands of real ones? I carved figures of what I wished to have, women and friends, of which I had none around."[17]

B. E. M., sometime in the mid-1950s, must have spent a long time carving an elaborate representation of his immediate physical surroundings and emotional state. First he carved his name on top: "Bruno E. M.[18] from Erro" (a town of Nafarroa). The *o* in his first name is a large and elaborate circle inside of which he added some unrecognizable symbols or figures, perhaps his farmstead. Below he carved a large *zato*, or wine bag, including its brand name, "zzz Bota," which is manufactured in Nafarroa. Underneath he wrote: "Hurrah for the wine and the naked women."[19] At the bottom is carved a large, triangular tent secured on both sides by ropes tied to stakes driven into the ground. A smoking chimney protrudes from the roof of the tent, and a long, crooked stick with a couple of appendages—it may represent a tree—towers above. The flap opening is closed, and beneath the image are the words "Mi carpa" (my tent). It took the entire circumference of a large tree in Humboldt County to carve all of this.

The herders carved almost as many figures of men as they did women, but the rules for representing the two sexes were different. For example, unlike women, there are no armless men. To the contrary, men are generally shown performing some activity: holding objects, shaking hands, holding a woman or a pipe, or just as often urinating or holding their penises. Why was the act of urination so important to them? (No men are shown defecating.) It might have been an excuse to show off their manhood.

40. A self-portrait by an unknown herder best described as weird. The arms are missing, unusual in carvings of men, and the genitals dangle from his belly.

41. Burly herder with hat, White Pine County, Nevada. Like most such figures, this fellow has puny arms.

42. Herder with *makila* raising his left arm in salute. Below is a lamb and the caption "Aupa" (Up with you).

43. "Viva el vino y las mujeres las dos cosas son buenas 1959 FB" (Hurrah for the wine and the women both are good, 1959, F. B.). This type of message seldom went unanswered by other herders who visited the area.

Daily Routine

Frank Aramburu of Paradise Valley, Nevada, described a sheepherder's life during the summer:

> We set camp close to the creek or water. We put food and drink in the water, so they would keep fresh. In the morning we were up with the dawn, and we drank coffee. Then we took the sheep to pasture. Sheep want to eat while the grass is fresh and cool. When it starts warming, they lay down. Approximately at 8:30 A.M. we returned to camp, and at 10:00 A.M. I enjoyed eating the beans cooked the previous day. Until 4:00 P.M. the herder is free to cook, carve trees, or take a nap. If within three hours there was a trout stream, that is where I

usually went, to catch them and eat them. At about 3:00 P.M. we would eat supper and one to two hours later we went back to the sheep. We took them to water and, after they grazed some more, they bedded down for the night near the creek. After that our day was done.[20]

Day after day the sheepherder's routine changed but little, conditioned by nature's clock and the sheep's demands. Each herder had his own favorite camping spot; some preferred to be beneath the trees, others near them but in the open. Some liked to cook; others spent more time fishing or napping than making bread or washing clothes.[21]

A concentration of carvings meant that the place had been used as a temporary campsite by previous herders. Sometimes the trees told the herder where to camp, or at least the pattern of carvings suggested a good site. Raymond Mosho thought he had found a suitable place on the eastern shore of Lake Tahoe and inscribed an aspen with news of his discovery. He was from Iparralde but tried to communicate in awful Spanish: "Notice, good for sheepherders, good here camp today, 17 August, 1909." In case future herders missed the message, Mosho carved another nearby aspen with the same information. Perhaps the Laxalt brothers heeded this notice when they picked the area as a campsite.[22] Mosho's choice of language for this message is significant. Herders in America behaved like many writers in the Basque Country who, in order to reach a larger audience, opted to write in Spanish.

In northwestern Humboldt County in 1954, Arnaud Etchamendi expressed the same idea, carving in a Basque that would be the envy of any writer: "au leki on da" (this is a good place), for camping as well as for sheep. The pasture and weather must have been good. Every herder camped near water, either a creek or a spring, if possible. Springs were often protected from sheep and other animals with a fence or other means. One such place, identified by name, says in Basque: "Spring, sheepherders' cool water."[23]

Most of the hours of the day were the herder's to dispose of as he wished, but there were a few crucial chores that he had to attend to meticulously day in and day out. Every morning he had to get up before the sheep had a chance to stir; that was about 4:00 A.M. The consequences of a late start were painful, and more than one herder may have experienced them. Rarely was such information written on trees, but sometimes we are fortunate. For example, on 24 July 1967, S. A. from Nafarroa overslept while working near Lee Vining, California. He left the following *bertso* penciled on a galvanized steel water tank:

Aurten lenbiziko aldia	This has been my first time
tanke untan	[working] by this tank
egibat badaukat argatik	incidentally, I have a story
mingañaren puntan	[to tell] at the tip of the tongue
goizeko seyek arte	until six o'clock in the morning
egon naiz ofian lotan	I stayed in bed sleeping
ta guero ardiak	and afterward the sheep
ezin sartu urian	I could not get them to the water
amar orenetan.	for ten hours.

We can be sure that was the first and last time S. A., a mountain man from northern Nafarroa, slept until 6:00 A.M. The *bertso* is written fairly well, with only a few minor misspellings.

On 12 August 1996, I drove by chance into an isolated sheep camp in the South Warner Mountains in northeastern California. It was getting dark, and two herders had obviously arrived at the camp just a few minutes ahead of me, because I found them in the process of building a fire for their supper. Unlike the typical sheep camp by a creek, this one was perched on the very mountaintop, under a huge, lonely pine that had been cut in half by lightning. It had been thundering and lightning for a good portion of several days, and I thought it was plain crazy to camp there, but I did not say anything. I parked my van away from the tree.

One of the herders was an old Bizkaian, the other a young Peruvian. The latter did all the work while we talked. Imagine cooking supper in semidarkness in the light of the fickle flames. Forget about a balanced diet, counting calories, and especially cleanliness. I was asked to share the meal and accepted a couple of lamb chops that had been so overcooked in the high heat that, even while swimming in oil, they tasted like wood. Anyway, we ate and we chatted, and no one did the dishes that night. We went to bed and I had "one sleep."[24]

Morning arrived quickly. Both herders were up with the stars and started building the fire for the coffee in the near-freezing cold. After drinking it I grabbed my camcorder and followed the old herder to the place where the sheep had bedded down for the night. I engaged him in a little conversation and learned something that might explain much of the bickering in the carvings. The old Basque, swinging his *makila*, said: "In here you need this, otherwise the young ones get ahead of you into the pasture."[25] In other words, some herders do not respect each other's boundaries, especially when the grass is greener on the other side. This is a serious offense against the herder's code, and one such incident is enough to upset a man and influence his relationship with his fellow herders for the whole summer.

I felt that I now understood some of the countless carvings by two people, the second a profane reaction to the first. One example that comes to mind is "M. Mariano," and below, added by another, "Shit for him."

Dirt and Dust

Basque women may be a better source than men for learning about certain aspects of the sheepherders' lifestyle because the women pay attention to details that the men deem less important. I asked Eugenia Mendiola, a ninety-six-year-old resident of Humboldt County, many questions, one of which was about cleanliness. How often did the herders take baths, wash their clothes, brush their teeth, shave, etc.? She said that in the winter the sheepherder probably did not take baths, but kept his clothes cleaner than in the summer (a statement corroborated by the herders themselves). In the summer they used creek water for personal hygiene, but all the dust made it difficult to keep clothes clean. Some herders took baths often; others didn't. As in everything else, different herders had different habits.

Mendiola said that she knew a notoriously unkempt Bizkaian from Natxitu. This fellow changed underwear once a year, but the word *change* had a particular meaning for him. He would remove his undershirt and dust it off against a tree trunk, then reverse it and put it back on. "Aaaah, fellows, this feels gooood," the man used to say.[26] Probably every sheep outfit had a sheepherder whom others regarded as peculiar when it came to personal hygiene, for I heard other similar stories. There were two untidy herders in a sheep company in Elko, for example. One day, on a remote summer range, their shirts looked so black that a fellow herder could not resist lampooning them: "Hey guys, have you heard? They are going to open up a laundromat here soon."[27] The joke did not wash.

I was speaking about these matters and discussing the habits of a well-known and popular sheep owner with a matronly married lady. In typical Basque fashion—bluntly, that is—she said, "Yes, he is older and single, he is nice, he has money, and he could retire comfortably, but he does not keep himself clean. What woman would want to go to bed with him?"[28]

Food and Drink

This aspect of the sheepherder's life rarely appears in the carvings, even though eating and drinking were always a highly anticipated part of the monotonous day. I learned the details of the herders' eating habits through conversations with ex-sheepherders and the women involved in the sheep business. Most herders told me that they had not suffered from hunger back home. There was usually plenty of beans, potatoes, and vegetables to eat, as well as fruit—mostly apples—and cornbread.[29] They did, however, enjoy being able to eat meat, which was available but expensive in the rural Basque Country.[30]

In fact, Basques ate better than neighboring Spanish and French peasants, who ate bread once a day and lard once a week.[31] According to a late-eighteenth-century Basque author, peasants ate a lot of corn as well as milk, cabbage, turnips, goat meat, old beef, salt pork, chestnuts, apples, beans, and wine. Rich people ate wheat bread, several types of fowl, and young fattened oxen.[32] This must be only a partial list because no potatoes are mentioned, nor apple cider, important staples.

A Nafarroan logger described his diet in the 1950s: "[In the forest] we ate lima beans, bread and wine every day, for nine years (in my case) . . . at night we had one old sardine for the four of us . . . no meat . . . if we saw a dog or a cat, it would soon end up in the pot, you bet! With the wine we drank, we suffered from heartburn constantly. To get rid of it we used to eat the tender shoots of oak leaves."[33] In my own home region of Bizkaia I rarely heard of dogs and cats being eaten, although in Spanish they say "meter gato por liebre" (to serve cat meat instead of hare; in other words, to deceive).

Some sheep companies were more generous than others with food, but I have seldom heard actual complaints about provisions. On the other hand, I was told that food—and wine—was the reason most herders changed outfits. All the sheep owners I interviewed told me that they fed their herders "better than others."

In the old days, herders generally butchered their own meat. For example, in 1924 Isidro S. carved: "I have killed a good lamb for meat." He would have been in trouble if his boss had read this. They were supposed to butcher older mutton—old or dry ewes and the like—not good lambs. In the United States today many people turn up their noses at lamb, but, according to Byrd Wall Sawyer, "sheep and their products were an important and essential factor in the development of the state [of Nevada]. Mutton and beef were the chief support of mining camp life." People in general ate more mutton than beef then.[34]

Butchering was an event and an excuse for the sheepherders to have a little get-together with conversation, drink, and song. That was the case for Nicolas Elissalde in 1932: "Butchers, God will pay you back," he carved and then added a little detail: "Adios a los piches," that is, "Good-bye to the peaches," apparently a treat they enjoyed during the butchering.[35]

On 15 August 1957 Bautista Barrenetche left a record of a similar operation: "We have done the butchering today." August fifteenth is a holiday in the Old Country, and the herders never forgot such days. It is conceivable that the choice of this day for the butchering was deliberate; it was a good excuse to get together and celebrate.

There was little variety in sheep camp meals. Soup, lamb, beef stew with potatoes, and the ever-present beans were the basic fare. Today sheepherders get bread from the store, but in the old days, and until the 1960s, they baked big, round loaves in Dutch ovens buried under hot coals. As a rule, all the *kanpo handiak*, or main summer camps, had an oven for baking bread. Recently I was lucky to witness the traditional bread-baking operation by an old-time sheepherder, Candido Olano. This was not a show—he had no advance knowledge of my visit—but rather a long tradition kept alive in the sheep camps of northern Nevada.

When the hot, golden loaf is pulled from the oven, the herder traces a cross on its bottom with a knife before slicing it. Emily (Zatica) Miller of Paradise Valley, Nevada, told me that sheepherders did not eat the bread fresh, but waited at least one day. While this may be generally true, I do not think there was any taboo against eating fresh-baked bread, and I doubt that every sheepherder followed the practice, especially if he happened to be hungry. I do recall that on one occasion when I was eating a sheep camp meal, fresh bread, though available, was not served. Fresh bread does not slice well, but an Old Country philosophy might also help explain this behavior. One never wastes food; for example, if a piece of bread is accidentally dropped, children are taught to pick it up, kiss it, go outside, and throw it skyward while saying the formula "*zeruko txorientzat*" (for the birds of heaven). Herders would not start eating the fresh bread until the old bread was first consumed.

If the sheep owner was Basque and had a ranch, then summer fruits and fresh vegetables were likely to reach the herder's pantry. During the Christmas holidays the herders, tired of mutton, feasted on chicken and homemade pies. The Fourth of July was another occasion on which herders tried to get together for a celebration and a change of diet.

Sugar was fairly expensive in Europe but cheaper in the United States, as were rice and canned milk. These are the ingredients of rice pudding, a popular dessert in the Old Country. Many a herder made a big pot of this dessert, only to be sorry later. One fine summer day in the late 1940s, a Bizkaian herding in Yakima, Washington, was looking forward to having his fill of rice pudding. Not only had it been made with undiluted canned milk, he poured still more milk on it before eating it. He ate so much that for days his bowels would not move and his belly hardened to the consistency of a rock. The man thought for sure that he was dying, but luckily a trip to the doctor cured him.[36]

South of Gardnerville, in 1932, a lucky herder named Cea caught a duck. He carved simply: "Eating duck." Near Reno someone carved: "Shit, here I am eating chicken drumstick for breakfast." Doubtless he was unhappy because it was the seventh of July and his heart was in his homeland, Nafarroa, running after—or before—the bulls in the streets of Pamplona (Iruñea).

Each tree carving, even the most prosaic, can be meaningful. The mere fact that the carvings were inscribed indicates their relevance to the author, but sometimes their significance eludes us. For example, in 1914 near Bridgeport, California, Nick Sario carved the following: "I won't eat it either."

Fresh-baked bread, a pot of Dutch-oven beans, and wine could keep any engine running strong. Batita, in Humboldt County in 1956, agreed: "Hurrah for my wine bag when it is full." The wine bag is the handiest container ever devised. It is easy to carry and drink from, and besides, it gives the wine a special flavor.

No carvings of bread, steak, or fruits have been found, but there are fancy whiskey bottles carved in one grove in White Pine County, Nevada. I recall only two groves in Nevada where isolated figures of bottles were recorded.

According to some sources, the practice of furnishing wine to herders was not common until the 1950s. Ordinarily the herder was allocated about a gallon a week. In Copper Basin near Jarbidge, Nevada, under the figure of a gallon jug, someone wrote: "Hurrah for the gallon of wine." It is not dated, but it looks to be older than the 1950s. That wine was an important issue is attested by the following carving in the Sierra Nevada, probably dated in the 1950s or 1960s: "Gluttons, hurrah for Dember, although he does not provide us with wine, he is a good sheep boss."[37] Obviously, at least one man in this outfit could survive without wine.

In Basque society, drinking, usually wine, enhances a good time among friends. People seldom drink alone and without taking food, and their tolerance of alcoholic beverages is thus rather high. Drinking songs, such as "Drinking, drinking is a pleasure, water [is] for the frogs that swim well," are a tradition. But because inebriation is not the goal, a drunken person is looked down on.

South of Gardnerville near the Nevada-California border, half covered by a young pine tree, is the self-portrait of a man with a prominent girth. Leaning dangerously backward, but still standing, he is holding a bottle in his mouth. The caption says: "You folks already knew that I liked to drink, R. B., August 1915." This carving is on a dead tree and will soon disintegrate beyond recognition. The same topic is covered by a charming arborglyph in the Lake Tahoe Basin, but this man was more prudent than R. B. He portrayed himself sitting down while drinking from a bottle. The accompanying caption is partly written in phonetic English: "Gut vizki salut ankaru" (Good whiskey, to your health, Ankaru). The last word is not totally clear, but it may be someone's nickname.

One late summer day in 1950 in Cherry Creek, White Pine County, M. G. was daydreaming of food and drink. But instead of the same old sheep camp fare, he was obviously thinking of more refined surroundings in town, of the few days he would soon enjoy among friends. His message in Basque can be translated as anticipated pleasure and reads: "Hurrah for the sheepherders and for the vacation in the wine cellar."[38]

Artzain, Borreguero, Siper

These three words mean "sheepherder" in Basque, Spanish, and English, respectively. That is what the herders were, and the words would not appear carved on dozens of tree trunks—with different spellings—unless they were proud of it. The word *sheepherder* itself doesn't appear in the carvings for two reasons: it is rather long, and the sheepherders could not figure out how to spell it. In three different mountain ranges, however, three different herders—all of them from Nafarroa—tried and carved "siper" (pronounced *shee*-per), which is an approximation of what the Basque hears. As far as carving frequency, "borreguero" and "artzain" are about equal. Iparralde Basques who had a slight command of Spanish carved "buregero," "burreguero" (perhaps they thought it had a connection with *burro*, or donkey), and so on.

Few herders carved the name of their employer. Had they done so, historians today would have a much clearer picture of the sheep industry. Those who did provide the information may have been happy or proud to work for their bosses. That seems to be the case with Bautista Amestoy, a fellow with a flair for carving in beautiful cursive letters. In 1924–28 he left at least two carvings with a repeated message: "Bautista Amestoy, native of Urdats, Spanish Nafarroan, sheepherder of Ramon Borda, Gardnerville."[39]

In the 1930s in Inyo National Forest, an unidentified herder wrote this about his sheep boss: "Solt Pork Hery [*sic*] M is shrewd, foxy, and deceitful." Henry, known by his nickname "Salt Pork," was from Bishop, California, and the herder was accusing him of skimping on the provisions.[40]

In White Pine County, a man named McDermitt was a sheep owner from 1881 to 1915. In 1901 he hired Bixenti Astorkia, a Bizkaian, and paid him well. In 1915 in Cherry Creek, also in White Pine County, Juan Manterola claimed to be a herder for Margullegy, a Basque. In 1930 Paul Sorhuet stated that he worked for Swin from Currie, Nevada. Steptoe is also mentioned on a tree as being a sheep operator during the 1920s and 1930s. On Steens Mountain in southeastern Oregon, a carving dated 1937 identifies Felix Urizar as the "main boss" of the "45 Sheep Company." In Tahoe National Forest a Mexican herder carved that his "patron" was Abel—that is, Mendeguia. Another herder in the Sierra carved that he was working for "TC," and in the Jarbidge area of northern Elko County a Bizkaian herder informed his audience that his boss was Goicoechea. A carving near Gardnerville mentions Loy Libras as a sheep owner, and one near Reno names Pradera.

A few herders gave an address, either their own or that

of the outfit they worked for. Pete Gamio provided two addresses, one for his boss, E. Wings of Red Bluff, and a post office box in Alturas, both in California. Martin Aguerre carved "Cedarrele Calif," for Cedarville, California. Pierre Gortari in 1936 inscribed "Gerlach Nevada" under his name. Others spelled it "Nebada." The list of owners' names carved on trees is a very short one, partly because many herders were working for themselves—they were so-called tramp sheepmen—and there was no extraneous "owner" to identify.

The Good, the Bad, and the Rest

Two things were asked of the herder: to take care of the ewes and to grow fat lambs. The trees speak loudly about several kinds of herders—good ones, bad ones, and those in between. Sheepherding was not a matter of skill so much as of attitude. A good sheepherder had to be diligent and hardworking. The lazy or the lethargic made bad herders. Old herders always thought that newcomers were spoiled and that they did not know what the business was all about. They tended to equate hard work, or more precisely "hardship," with skills. Old herders automatically believed that the younger ones were no match for them.

Old sheepherders often say that one problem that frequently assailed younger herders was anxiety. First-year herders worried too much about losing sheep and tended to overuse the dog to keep them rounded up. They were happy only when sitting on a vantage point from which they could watch and monitor the movements of the whole herd at all times. The more experienced herders knew that was not necessary.

One of the least-coveted jobs was that of herding yearlings (*primaler* or *primalerua* in Basque), which are not only full of energy but have not yet learned to respect dog or man. They are far more difficult to herd than the older ewes, which, according to one source, know the range better than the herder. Several herders told me that when a boss dislikes a particular herder, he puts him in charge of the yearlings.

I recorded several carvings that refer to that predicament. One individual, who may have been reprimanded by his boss more than once, sought to boost his low esteem by carving "Viva el fok dis up primalero 1947 año miseria" (Hurrah for the fuck this up yearling herder 1947 the year of misery). But perhaps it could be better interpreted as one of the many "I-am-fed-up-with-this" messages.[41] He was not alone. Another individual carved in desperation: "The yearlings do not even eat tobacco flower [which sheep usually considered a delicacy], therefore throw the dog at them."[42]

Sheep owners forbid the herders to let their dogs chase the sheep without good reason. It frustrates and stresses the band, and as a consequence the sheep do not gain weight. Lower in the same canyon where the previous arborglyph was found, another herder had already had problems with the yearlings; his inscription simply says: "Screw you, yearling." I asked an ex-sheepherder who was with me how this message should be understood. He looked at the tree and the surrounding narrow canyon and said that the herder had probably killed a particularly troublesome yearling there.[43]

One Basque told me that old sheep can actually be more difficult than yearlings because they can spot a new herder right away. Dozens of anguished messages I recorded certainly could illustrate such scenarios. "Sheep, you are finishing me off" is one example. (I have already mentioned that this type of carving was viewed as a sign of weakness.) Older sheep know the range well and are fond of wandering at will, regardless of what the herder wants. He must show them who is the boss, the same man told me. He must "dog" them constantly for four or five days; and he must not let them eat as much as they want either. That way he will earn their respect, and they will not forget it.[44] I presume few sheep owners would agree with this method, but it occurred to me that sheep and men at times behave similarly.

Some herders worried too little, and that could be as bad as worrying too much. A herder who napped too long or sat in the shade rather than walking in the hot sun and jumping over the brush in order to keep an eye on his charges might lose some of his sheep. The two individuals quoted below—the first Nafarroan, the second Mexican—were candid enough to write about their dire predicaments. The Nafarroan (ca. 1968) inscribed several messages in the same grove: "I lost the rams," he said in one. He walked around the canyon looking for them and carved again: "I am a sheepherder but a worthless dick. My rams are missing and I cannot look for them. Thank goodness. . . ."[45] He found at least some of them, because a third tree carries the good news: "Here I am counting the rams."[46]

The Mexican's plight was even less enviable, and he knew it: "Jose Hernandez was here on 4 July 1980. I lost my sheep in the mountainside. I have two dogs with me, one called Chispas and the other Chicote. I am in big trouble." This confession must have been difficult for him because sooner or later other herders would read it. Even the layman can understand losing a herd of rams, which in an average outfit numbered around twenty to forty. But how does one lose one thousand ewes? And Hernandez had two dogs with him. In the winter with foul weather and fog it is easy to lose the sheep or miss the camp. But this was the fourth of July! Subsequent herders who read these carvings no doubt had a good laugh at the expense of these two easygoing sheepherders. Yet we must be grateful that instead of hiding their misfortunes, they decided to record them.

Who were the best herders in the West? The trees bear various claims of being the best and taking pride in it. Ethnic, regional, and national factors undoubtedly came into play here. One Bizkaian called himself a "famous sheepherder," but Paul Ynchauspe proclaimed himself the "champion." In 1965 a non-Basque, Alberto Alonso, from Santander, claimed: "The people from Santander are good herders. I raised lambs of 111 pounds. I am unique, never before in the Sierra."

Many variables affect the weight of the lambs at shipping time. I heard a story of an Idaho sheepman who watered his lambs just before shipping so they would weigh more, but they made such a mess that he could hardly find a trucker who was willing to clean up after them. Proper feed and a week's delay in shipping can easily add several pounds to the lambs. A lamb of 90 pounds was considered good in Nevada; therefore Alonso's success was, as he claimed, excellent. But that was in the 1960s. Today lambs weigh much more. In 1993 I witnessed *txepena*, or lamb shipping, in Idaho, and the sheep owner instead of being happy about the weight of the lambs was sad because too many of them weighed more than 130 pounds, for which the buyer penalized him by a percentage.

A Dry Country

"This is a bad sierra, no water, no grass" (Humboldt County, Nevada). This typical message is not dated, but it looks pre-1930. There is evidence to suggest that many herders tried to take good care of their charges because personal pride was at stake. When circumstances beyond their control hindered performance and the sheep suffered, they recorded it. Naturally, climate was the primary variable in a herder's success or failure, much more so than predators, another chief concern. The carvings provide localized information regarding pasture conditions as early as the 1910s, some of which may not be available anywhere else. State and federal agencies collect data from some of the ranges but not from others, and the tree carvings fill the void.

In Nevada, rain was never a concern; drought was. On 4 July 1914 Jose Vizcay wrote succinctly: "No pasture." He had two more hot and dry months ahead; September is cooler, but it can be as dry as August. How was Vizcay to cope? On 11 August 1918 Vicentez Antonio carved a similar message: "There is no pasture, dry year."[47] Both of these carvings are in the Sierra south of Gardnerville. In Inyo National Forest, an unnamed sheepherder left this verse in Basque:

Viva gu eta pinalia	Hurrah for us and the pine forest
guztian dabile animalia	animals live all over it
bazka anitz eta gutigi yalia	lots of pasture and too little feed
aurten aragien partez	this year instead of flesh
aundiko zaie ilia.	they will grow wool.

The carving is not dated. This is an excellent example of a *bertso* that is also informative, although I do not understand what the author meant by "a lot of pasture and too little feed." Perhaps the grass was too dry or unpalatable. This verse addresses a major concern of sheepherders: the sheep were not putting on weight, just wool. The sheep owner could not make money that way; worse yet, the sheep boss could not pay the herder. Prices, of course, were always subject to market forces. Wool today is not the cash crop it used to be. Sheepman Reginald Meaker wrote that in 1923 the price for wool was 42 cents a pound, "a remarkable price,"[48] but in 1991 it was just 14.5 cents.[49]

Nevada suffers from periodic drought as well as its opposite, localized flooding. According to Meaker, 1924 was a very dry year. In the early 1930s Nevada suffered a five-year drought; Washoe and Honey Lakes dried up, and water had to be pumped from Lake Tahoe. The years from 1928 to 1931 were drought years in northern Nevada. Finally, in 1937, it snowed heavily with abundant spring rains and the drought ended.[50]

The following are additional meteorological observations gleaned from the aspens regarding grazing conditions in Nevada and the Sierra:

"1922 June 18 bad day"

"There is little pasture, Arnot Urruty, 24 Sept. 1926" (Lake Tahoe Basin)

"Poor pasture, August 5, 1931, Pete Etcheberri" (White Pine County)

"1931 bad year"

"1944 bad year a lot of wind" (Plumas National Forest)

"August 2, 1949 very dry, Michel Urrels, Yvonne" (Humboldt County)

"No rain in 105 days, 1957, Pedro Fernandez" (Humboldt County)[51]

"This year little pasture, 27 June 1957" (Monitor Pass, California)

"1959 24 July bad year, Beorlegui" (Humboldt County)

"Adios pasture" (Humboldt County)

"1960 bad year"

"Good weather for dry fishing, 2 June 1962"[52] (south Ruby Mountains, Elko County)

"1-9-64 bad weather"

"4 Sept. 1966, drought year"

"Attention, attention, this year 1966 has been a miserable one, Emiliano" (Topaz Lake area, California)

"Dry year, but the sheep do not die, Emiliano Alonso, 1966" (carved by the same fellow as above)

"86 bad year"

Observe that several of the carvings are dated in the latter part of the summer. Usually, by then a dry year would have been felt even in the higher mountain ranges. Until then pasture would have been barely adequate, but now the sheep were running out of feed, making the sheepherder's job that much more difficult. Was a dry year the reason why Joanes Saroiberri carved: "1920, year of the devils"?

Not all the carvings are negative, for even in Nevada it rains sometimes. The following two messages were recorded in Elko and Humboldt Counties: "6 August 1961, 4 P.M. and raining, Juan Urdaniz." "Michel Urrels 1949 July 3 rainy day." The next two messages were located in Humboldt County and are interesting because the herders seemed to be happy for the boss: "E. Irazoki, 1967, July 30 good weather, good profits" (perhaps Irazoki received a raise when profits were up); "J Iturriaga, Larrauri, Mungia, Bizkaia, 1969 good year for the lambs."

Beltran Bidart, a sheepherder who also ran cattle for Fred Dressler of Minden-Gardnerville,[53] carved in the summer of 1936: "Much, much pasture." This was very good news after several years of drought, and, of course, sheepherding also suffered from the depressed economy of the period.

Barren Land

This subject deserves a special emphasis because the carvings give it one. The herders' negative comments go beyond the weather to the environment itself. The barren country, the harsh climate, and the rugged terrain had powerful impacts on sheepherding. These messages must not be confused with the carvings that refer to loneliness and emotional deprivation—discussed later—although sometimes one may have influenced the other.

In a fashion consistent with Basque idiosyncrasies, the messages often contain contrasting images: the herders cheered green Euskal Herria when Nevada seemed too harsh and dry. The carvers resorted to the same tactic when they felt sad and lonely. When a foray into town did not turn out as well as he had hoped, a herder might hail his fiancée or himself. An overwhelming majority of such carvings refer to herders' profound disappointment with the land. Was the herder having a bad day when he carved it? If so, a lot of herders had a lot of bad days. "I shit on this goddamn sierra" is typical. Another reads: "This sierra is not worth a donkey's penis. Long live France." He contrasted Nevada and France, and it is not difficult to predict which was better. Another said: "Nevada is not worth a dick." One Bizkaian made a curious comparison in a sentence he did not totally finish; but there was no need to: "vai mutillak au sierra vaino neska gaien ipurdia . . ." (Yeah, boys, [I would rather have] a young girl's ass than this sierra). A disappointed immigrant carved in 1914: "I shit on this false land, good-bye."

The carvings cited in this section are all in the Sierra, the high-country summer range, where nature in Nevada is most generous. What would the herders have carved in the bleak creosote-bush deserts, had trees been available there to carve? We may be better off not knowing, but I remember what retired sheepherder Joe Madariaga of Elko told me one day. We were in Panpanika (Pumpernickel Valley, Nevada) and, pointing at the interminable alkali flat shimmering in the summer heat, he said: "You see this? Even jackrabbits pack lunch before crossing this flat."

Reaching the End of the Rope

It was not that the land was barren. It wasn't. Even the best herders had bad days, though, and when their frustration reached a certain point, the carvings reflect it. "They can go to hell with all the sheep," carved one.[54] This is the kind of thing that sheepmen, even non-Basque ones, believed the Basques were incapable of doing—abandoning their charges—but the author of the carving was a Bizkaian.

Many people who do not know sheep will wonder about the last carving. Are sheep not docile followers, you say, easy to handle with a dog and a few commands? In fact, sheep made martyrs out of some herders who took their jobs too seriously and constantly tried to confine the herd. A couple of Spaniards left clear evidence of their frustration in several strongly worded messages. A man from Cantabria carved: "Bad year, but no death yet, Antonio." Some ten years later and three hundred miles away another inscribed the following: "Let the sheep die, screw the sheep owners from the rear, and love their daughters, J. V., Asturias, 2 Sept. 1973." These two fellows used a similar tone in other messages they carved.

Non-Basques complained the most. Either they had less patience or they could not get used to the lifestyle as well as the Basques, or perhaps they just voiced their feelings more freely. Alberto Alonso from Cantabria carved statements railing against the sheepherders' contract with the Western Range Association.[55] The most strongly worded complaint found so far is by a Peruvian who apparently had reached the end of his rope: "Sheepherder, it is best that you shoot yourself in the head."[56]

Living Close to the Land

In Nevada's high country Mother Nature has a short temper and can change her mind without warning. The herders had no choice but to take whatever she dispensed. If filmmakers in Hollywood were looking for a strong, self-reliant westerner, they should have picked a sheepherder. Any old herder, highly trained in the school of nature, possessed those qualities.

Some individuals were tougher than others, but all of them had to be extremely strong, both mentally and physically, in order to survive alone in the primitive conditions. One fellow in Idaho was so tough that his coworkers called him "Burdiñas" (iron man).[57] But brute strength was not always enough. Sometimes *maina* (skill) was more useful.

Living entirely at the mercy of the elements was a constant liability for the sheepherder, as the following carvings indicate:

"1922 June 10, cloudy weather" (Humboldt County)

"Manuel Iturregi 1924 today is very hot" (Humboldt County)

"Very cold, Martin Larzabal, Oct. 20, 1923" (Monitor Pass, California)

"Sept. 24, 1926 cold wind, Arnot Urruty" (Lake Tahoe Basin)

"Gratien Oçafrain, it is very hot" (1930s, Humboldt County)

"1957 July 10, it was colder than hell, B. Fernandez" (Humboldt County, Nevada)

"Snow, 10-10-1973, Jesús"

When a sheepherder says it is cold and windy, that usually means that it is much colder and windier than normal. That is why they left a record of it. In northwestern Humboldt County I found this statement, which is neither dated nor signed: "November 31 [*sic*] it is snowing it is warm."[58]

Many sheep outfits summered in the Sierra and wintered in California, but the sheepherder, no matter how anxious, did not start moving until the camptender showed up with orders to start down the trail. There is no date on the following unsigned carving located well above Lake Tahoe: "A lot of snow, let us go to California."

Who has not heard tales of sheepherders caught in a blizzard? Old herders love to talk of days when it was "too cold to urinate." All of them say that winters used to be much colder. Whether or not that is true, they certainly felt the cold more in the old days because the clothes they wore as protection against the elements were not as sophisticated as today's coats, gloves, and footwear. More important, people today spend less time outside at the mercy of weather—riding a horse, for example—and a lot more time driving in heated vehicles.

The carvings cited above are all brief, and this sometimes may cause us to miss the point. The following refers to an incident that could easily have ended in tragedy: "7 October 1961 bad weather with snow, I am all fucked up, Florencio Sarratea." Mr. Sarratea now lives in Reno, and I asked him to comment on the carving. Eagerly he answered that he remembered it very well: "It was a very wet snow that came down hard above Gray Creek. I only had a miserable tent that became soaked, along with everything inside, clothes, food, etc. For two days I just tried to stay warm. I did not have to worry about the sheep [getting lost], because they were trapped like me."[59] It is noteworthy that the herd's welfare was paramount in the sheepherder's mind even in this predicament.

Another account of a real-life drama is carved near Hirshdale, California. We must try to fill in the details as best we can: "If it had not been for Valerio, Antonio would have been eaten by worms in some other place. He would have died of hunger." The carving is drafted in poor Spanish, and the carver's name and the date are missing. In a nutshell, Valerio is credited with having saved Antonio's life. If not for him, by now Antonio would be six feet under and eaten by worms. I do not know who Antonio was, but Valerio's last name might have been Zubiri, from Nafarroa. For a time he was the camptender for the Kuhn and Miller outfit, which ran sheep in the Whiskey Creek area. Valerio carved several trees, on one of which he called himself "El Cascabel" (The Rattlesnake). Sometime later in Squaw Valley, California, a mule kicked him in the knee and left him crippled. Valerio retired and died in California, according to Severino Ibarra, who worked with him.[60]

At least half of the time, though, Mother Nature was kind to the herders, who also carved messages about enjoying good weather. The majority of the herders I talked to admit that theirs was a very peaceful and healthy life. They ate well, most had wine, and they fished in some of the best trout streams in the country. Some herders were equipped for serious fishing, as this carving dated 1965 indicates: "Hurrah for the creek and the netting."[61] Frank Aramburu of Paradise Valley, Nevada, was not quite as well prepared, as the following story indicates:

> One time in 1923, I remember well, it was a dry year and at night the sheep ran away to the creek, which was within the Indian reservation. In the morning when I got there I saw trout in the water. I tied shoelaces to the end of the stick and I began fishing. Well, what would you know! Two wardens came and caught me. One of them, P. T., asked me: "Did you catch those fish with *that* fishing pole?" Yes, I said apprehensively. "Then, continue fishing and catch some more," he said. Wow! I got off easy that time.[62]

In 1970, in a bar at Midas, Nevada, I saw hanging on the wall a crooked stick about two and a half feet long to which a battered reel was clumsily tied. The sign said: "Basque fishing pole." At that, it was a more expensive rig than Aramburu's.

At night the herders slept under a blanket of stars, serenaded by the yelping of the coyotes and the hooting of the owls. One herder carved: "These are beautiful mountains to be in." And then he added: "But with a good woman."

Living Close to the Animals

It would be odd indeed if the herders' carvings were silent about wildlife, because no other occupation in Nevada required people to live closer to the "critters." The stories the sheepherders tell about animals would fill volumes; every herder has more than a few. Each thinks his tale is unique, but a good number of them are similar.

Take, for example, Rufino Berroetabena (he shortened his last name to Bena) of Elko. This Bizkaia-born trapper worked for the U.S. government for some thirty years until he retired in the 1960s. During that period he trapped or killed sixteen thousand coyotes, twelve thousand bobcats, and 116 cougars. "If I had their skins now I would be rich," he used to say. Today we think not of skins but of the slaughter those numbers represent. However, it is difficult to argue this matter with sheepmen. They contend that the killing went both ways, and that in order to protect the lambs they had to kill the predators. Mr. Bena told me about an elusive female cougar that in one night alone killed more than one hundred sheep.[63]

A herder from Nafarroa told me another cougar story. One day in White Pine County his dogs cornered a cougar. From a safe distance he aimed his rifle and emptied it, but he succeeded only in wounding the big cat. Grabbing the gun by the barrel he advanced toward the cougar. For an instant, overpowered by the cat's blazing eyes, he almost faltered, but he forced himself to take two more steps and strike the cougar over the head, killing it. The following nights the herder, unable to sleep, broke into a cold sweat and shuddered at the memory of the fire in the cougar's eyes.[64]

44 and 45. Carving of a cougar lurking in the Sierra Nevada forest.

It was a good thing that Dan Sanchez was a fit herder with a strong heart. One day around the year 1941 near Cherry Creek, White Pine County, he entered his tent and came face-to-face with a mountain lion. Apparently both were so startled that neither had time to react, and the lion casually slipped out. For days and weeks Sanchez did not say a word about the incident, afraid that people might think he was going crazy.[65] On Mount Rose, near Reno, I found an excellent carving of a cougar with an impressive head that appears to be lurking at the edge of a clearing.

Sierra Nevada herders talk about bears, too. Ambrose Arla used to see them in the Little Valley area of Washoe County, but they did not bother him or the sheep. That was not the case in Plumas National Forest in northeastern California, however, where two carvings refer to bear predation. One says: "Sheepherder, keep alert, bears in the area . . . 1935." And the other: "The bears are eating all of my lambs."[66] The same herder may have carved both messages.

The figures of snakes found in many aspen groves in Nevada are a different story. Some are small, others large; some are elaborately drawn with conspicuous rattlesnake markings, tongues, and rattles, but most are simpler and with fewer details. Every sheepherder has stories of snakes—that is, rattlesnakes, the only dangerous snakes they encountered. Rattlers are not abundant in the high country, and in the wintertime on the desert ranges they hibernate, so the danger they posed was relative. But they were known to kill sheepdogs, scare horses, and even bite an occasional herder.

In the late 1940s a Bizkaian camptender traveling through deep snow in the Yakima area of Washington noticed a large, flat rock where the snow had melted. Intrigued, he drew closer until a strong smell stopped him. Putting his handkerchief over he nose, he continued forward until he saw a huge ball of rattlesnakes. He retreated in a hurry.[67]

The herders' abhorrence of snakes sometimes borders on fascination. I heard stories of snakes crawling into a herder's sleeping bag for the night. It is said that when snakes are cold they seek warmth and do not strike. One cold fall afternoon in northwestern Nevada, several herders were leaning against some rocks enjoying the last warm rays of the sun when one of them noticed a slow-moving rattlesnake crawling inside another's jacket. The snake had slithered from the rocks where it, too, apparently, was enjoying the warmth of the sun, and crawled up Candido Olano's backside and down his neck. Olano did not see or feel anything, he said, until the herder in front of him grabbed the snake by the tail and pulled it out.[68]

Birds frequently figure in the carvings as well. J. P. Laxalt carved many fine quail and sage hens. One such carving south of Lake Tahoe appears to be a grouse taking flight. In 1918, near the Black Rock Desert, Francisco Aldunate carved a long-beaked aquatic bird that would have been more at home at Stillwater, Nevada, than in an aspen grove. East of Gardnerville, many of the represented birds appear to be blue jays or woodpeckers, which abound in aspen forests.

There are no representations of mosquitoes, but several carvers complained about them. Anyone who has been to an aspen grove in the summer knows that you must be ready to battle the mosquitoes. Leave your shorts at home, wear long pants, and bring plenty of bug spray. You will never see a sheepherder in shorts.

Two Basque women *bertsolariak* who ventured into a damp aspen grove one morning in 1999 and encountered an army of hungry mosquitoes wondered how the herders ever survived. In between swats, Estitxu Arozena and Oihane Enbeita composed the following verses about a seldom-mentioned battle that the sheepherders must have faced every day during the summer in the high country.

AROZENA:
Etxetik urrun zeuden artzaiak
hemen bete zituzten heuren nahiak.

The sheepherders who were far from home
here tried to satisfy their needs.

ENBEITA:
Bakarrik bertan gazte zebiltzan
aukeratu beharra bizitzan.

Alone, the young men lived here
because of the life choices they made.

AROZENA:
Etzen beharrik
etzegon Jainko ez Lege Zaharrik
eltxoen menpe zeuden bakarrik.

There was nothing they could do
there was no God and no Ancient Basque Law
but they were subject to the mosquitoes.

46. Cute sheep head by Nikolas, who titled it "Seep," a common misspelling, dated 1971.

47. A donkey nurses a gigantic snake while the foal looks on. There are several of these in the Sierra Nevada, carved by M. L.

A few of the arborglyphs depict fish.[69] Sheepherders knew the best fishing holes and were experts at catching trout, even with their bare hands. Near Reno one herder carved a portrait of himself catching fish on 7 July, the first day of the festival of San Fermin in Pamplona. "Here I am fishing," it says.

In almost every grove there are unintelligible carvings. Almost always this is because they are simply difficult to read, but at times, as in the next case, the words are clear but the meaning is not. In a carving made on 20 September 1924 near Ely, Nevada, Isidro Sala noted that he saw a bird, and he included an expletive. What was the point? Was it an unusual bird? Did it scare him?

Coyotes

Coyote, the Trickster, is an important figure in Native American folklore.[70] But coyotes were no less important to Basque sheepherders, who had never seen one until they came to the United States. The coyote was the sneakiest of the predators, and its reputation is secure as far as the carvings are concerned.

"El mero cabron" (The very bastard), the caption on an arborglyph in the Columbia Basin of Elko County that shows the figure of a coyote,[71] is the perfect description of the herder's view of this creature. In 1990 I went to the Columbia Basin and saw half a dozen dead coyotes dangling on the fence that lined the road leading to the sheep camp. Anyone can kill a coyote anytime; it is always open season on them.

Being a champion sheepherder meant caring for individual sheep, and especially for the weakest—the lambs. To protect them from predators the herders had to be ever alert, which was no easy thing. Sometimes the crafty coyote caught the sheepherder literally napping and nabbed a lamb. The average herder took such incidents philosophically, but the good herders, like Faustin Murkuillat, were grieved. Near Bartlett Butte, Humboldt County, he carved his reaction in his alluring Zuberoan dialect: "Collote poutak bildoxa chiestan hila 20 sept 1938 Faustin" (Fuckin' coyotes, a lamb [was] killed during siesta time, 20 Sept. 1938, Faustin).

The coyote is without a doubt the most celebrated animal in the sheepherding world. It is both loathed and admired—despised for killing the sheep, but respected for its ability to survive. There are several carvings of howling coyotes, "Auu auu auuuu," in northern Nevada. One tree I read in Elko County indicates how dreadful that sound must have been for the herders: "The coyotes have started their howling."[72] The herders killed every coyote they could. A 1924 carving by Iturregi proudly announces: "Today I skinned a coyote."

A Nafarroan I interviewed vividly recalled some of his encounters with coyotes. Once he spied one in a flat in the desert and, jumping on his horse, started after it. After a chase, the coyote stopped. The herder dismounted

48. "Collote poutak bildoxa chiestan hila 20 sept 1938 Faustin" (Fuckin' coyotes, a lamb [was] killed during siesta time, 20 Sept. 1938, Faustin), Humboldt County, Nevada. Most sheepherders lamented over such affairs but rarely carved about them because it was embarrassing to admit that such a loss happened in broad daylight.

and advanced toward it carrying his *makila*, but it took off again. The herder jumped back on his horse and continued the chase. The coyote repeated this maneuver five or six times, until both it and the horse were soaked in sweat and too tired to continue the game. The herder jumped to the ground one last time and approached the crafty animal. Taking aim, he threw the staff at it and struck it in the head. The coyote faltered, and the herder ran up and tied its snout with his handkerchief. Then he threw it on the horse and brought it to camp. He tied it to a post and told it: "You rascal, you will starve right there." The coyote looked and listened in silence (the herder assured me that the coyote understood him). Later the same day, however, a U.S. Immigration officer showed up in camp and, seeing the coyote, advised the herder to shoot it, which he did, reluctantly.[73]

Transhumance News

The herder's life was a never-ending journey. Even in the summertime, when he camped in the same spot for a few weeks, he had to follow the sheep every day. The carvings constitute an invaluable tool for the historian in that they mark the sheepherder's whereabouts as nothing else could. In fact, if aspens were long-lived and grew wherever Basque herders went, we would possess something like a diary of their migrations, as well as a map. When a particular herder carved a lot, we can trace his whereabouts throughout the canyon.

Some fellows left messages telling where they were going or coming from. The Bizkaian Jose Maguregui carved the day he left the summer range of Timber Creek near Ely: "Good-bye until next year, 20 August 1967." It is an early date, indicative of the range restrictions that the Forest Service was beginning to impose on the sheepmen during the 1960s. In earlier days the herder would normally have stayed in the forest reserve until the middle of September or the first snowfall.

Halfway up Mount Rose, south of Reno, is a carving that tells of the movements of another herder: "Jesus is going toward the top, hurrah for the meadows, 1966-6-25." The message was probably carved by Jess Arriaga. The herders went up Thomas Canyon and over the flanks of Mount Rose to the high meadows, where they had it pretty easy. But first they had to pass through Astozulo (donkey hole), a spot in the trail where several donkeys lost their footing and rolled down the embankment.[74]

On 9 September 1954, J. Urreaga arrived on Peavine with his sheep band and carved a message saying that he had come from "Kacepik" (Castle Peak, California) in the Sierra Nevada. Another message he carved hints that Urreaga was probably returning to Peavine after the pasture there had been allowed to recover. Sometimes, too, the lambs were shipped to market from the foot of the mountain. On 10 July 1956 Urreaga carved in the adjacent grove: "This same day I am going to the Sierra." Castle Peak and Peavine appear to have been Urreaga's summer range. Normally, Peavine received the assault of the first sheep band of the year by early to mid-June, but the mountain is too small to sustain a herd for the whole summer, so when pasture there dwindled, the herder moved on to higher country.

The sheep were moved along established trails, and in the mountains transhumance was predictable and relatively safe. But driving the animals across roads and train tracks—i.e., civilization—could be an adventure, if not downright hazardous. The following laconic carving suggests that scenario: "Shit, [it happened] when I was just crossing the tracks, Juan." It is not dated.

Other carvings also give news about herd movements. South of Reno on 11 September 1956, Jose Almirante left word that he was leaving the mountain for the lowlands and was going to the winter quarters in California in October. From a carving in the same general area we know that the lambs were shipped on 9 July in 1965. In contrast, in northern Elko County, lambs were shipped in late August or early September. We also know when the herders of the Espil Sheep Company left Clover Valley in Plumas National Forest in 1940: "October 8."

The carvings are nuggets of information, too brief to mean much to outsiders but enough for the fellow herders to whom they were addressed. The herder who carved about being caught on the train tracks was aiming to alert others of that danger. And those who read about the date of departure from a given range appreciated the information because it gave them a basis to determine how long the pasture would likely last and allowed them to plan ahead.

The Wilderness "Community"

It has been said that "loneliness is caused not by being alone but by being without some definite needed relationship or set of relationships."[75] That definition may not apply to the herders' situation. The physical isolation itself was chiefly responsible for their loneliness.

Accounts of the sheepherder's life always emphasize the isolation and loneliness, which were particularly acute during the summer months. But it is possible to overstate—rather than exaggerate—the effects of loneliness on the herders, and I say this only after surveying thousands of carvings. For example, the herders in the Columbia Basin of Elko County knew a number of other herders working around Mountain City, Gold Creek, and Jarbidge. All these summer reserves are rather isolated from each other, but in the 1950s and afterward, when pickup trucks became commonplace, it was not impossible for herders to drive over to visit their "neighbors" for a few hours. I have even heard stories of sheepherders going into town for an evening—that is, leaving the sheep alone in the mountains—but I cannot say how many did it. Such outings were necessarily furtive, and for obvious reasons the herders do not make a habit of talking about them.

Humans need the society of other humans to survive, and the herders created one in the wilderness. The arborglyphs offer insight into this sheepherder "society." The carvings provided a bond between the old-timers and the newly arrived. Sometimes the current "occupant" of the grove knew the carvers or had heard of them, and he took their warnings or advice seriously. Thus the communion extended itself through several decades to future generations of herders.

The men depended primarily on the grapevine and letters for news, or on the camptender, but the arborglyphs played a part as well. Herders sometimes combated loneliness by leaving messages for one another, as when two herders grazed the same area, one in late June and the other a month or two later. Some carved questions that were meant to be answered by subsequent herders, whomever they might be. Often the exercise was for fun, but serious matters were sometimes discussed as well. For example, adultery, rape, and incest are described in carvings on three different mountain ranges.

One of the most common messages is "Biba artzainak" or "Aupa artzainak" (Long live the sheepherders), which undoubtedly reflects a sense of camaraderie. Often the carver inserted the word *eskualdun* (Basque speaker). This message is usually carved in Basque, but Spanish versions are not lacking, "Viva los borregueros" being the most common form.[76]

Occasionally even the Nafarroans, who spoke only Spanish and normally viewed themselves as Navarros rather than Basques, in the strict sense of the word, carved "Viva los pastores vascos" (Hurrah for the Basque sheepherders) or similar statements. In rare cases Anglo-Americans joined in hailing the sheepherders, as in the carving in Tahoe National Forest that reads: "1952 Meaker Reno Nevada Biba ezkuldun [*sic*] artzainak." The message appears to have been carved by the Englishman Reginald Meaker, a sheep owner from Reno who left carvings in several Nevada sheep ranges. The undercurrent of these messages is a common solidarity. But carvings by Mexicans or Spanish herders sometimes convey a patriotic sentiment as well.

It is significant that many of the carvings hailing the sheepherders are in the Basque language. Most messages are supportive, carved to encourage new herders or those who would follow. Some examples, all in Basque, are: "Cheer up boys, don't be discouraged, onward." "Hurrah, sheepherders, don't worry." "Onward boys, don't ever be afraid." It was easy to be afraid, especially if one was new to the range and under twenty years of age (which was not unusual), alone in the night and hearing strange noises in the darkness. The messages in the ancestral language carried an additional dose of comfort for these people. A herder named Etchemendy understood the feeling and expressed his message of courage in a classic manner worthy of the great Pedro Axular[77] himself: "Kuraya har gaiso artzayna" (Take courage, you poor sheepherder). A number of short carvings, two or three words long, begin with "gaiso" (poor, pitiful), as in "Gaiso Mariano." Obviously those referred to were fellow herders struck with some kind of misfortune, perhaps mental illness, but the glyphs' meaning remains a mystery.

Although the sheepherder's trade was not conducive to making friends, the men found ways to get together. Those who became friends in the mountains were likely to remain close in later life. Most herders whom I asked to comment on their fellow workers had grateful memories regarding particular individuals who helped them in various circumstances. A 1938 carving by Joaquin Pagola is right on target. It says simply, in Basque: "Faustin my first sheepherder companion."[78] Faustin was Zuberoan and Pagola was Nafarroan, and the carving is dated 7 July, the feast day of San Fermin.

Not everyone was friendly, however. Many outfits had one or two grouchy old sheepherders, fellows who had spent too much time alone with the sheep and tended to take advantage of newcomers or laugh at their ineptitude. Such jokes were not necessarily made out of

49. A seasoned herder having some fun with the newly arrived recruits: "Alo gazteak zer diozue" (Hello there, young ones, what do you say?).

50. A highly unusual message reveals a fellowship between two sheepherders: "Faustin ene lehen arsain laguna, Joaquin Pagola, Julio 7, 1932" (Faustin, my first sheepherder companion, Joaquin Pagola, 7 July 1932). Pagola also spelled his first name Joakin. He was from Nafarroa, and Faustin Murkuillat was from Zuberoa. They must have gotten together on this day to celebrate the running of the bulls in Iruñea/Pamplona.

cruelty, but were simply part of the Basque culture, which believes in mocking someone's behavior as the quickest way to teach a lesson. The old herders also believed that the method would allow the newcomer to cope with reality quicker.

Every old-timer will tell you that newcomers were too soft to herd sheep in Nevada and needed to be taught the hard lessons as soon as possible, for their own good.[79] One message appears to have been carved by an old veteran with a smirk on his face: "Alo gazteak zer diozue?" (Hello there, young ones, what do you say?). The young herders in turn consoled themselves by laughing at the old-timers, whose odd behavior was the result of too many years in the mountains. They were *txamizuak jota* (sagebrushed) beyond repair.

In such an environment it was important for a herder to keep his head above the sagebrush and look at the big picture, and one way of accomplishing it was by singing his own praises. The message "Biba ni" or "Viva ni" (long live I) occurs in almost every sizable grove. It was the carver's reminder to himself and others that the lonely existence was temporary, that better days were coming.

Inscriptions that were clearly composed to elicit a response seldom went unanswered. Some herders responded to even the most prosaic statements. On 21 August 1907 Francisco Ezcurdia was upset about something and carved "caca" (shit).[80] Twenty-four years later, Joe Esain came around, saw it, and on a nearby tree carved next to his name the response "chisa" (piss).

Some herders were not satisfied with carving a question; they had to provide the answer as well. A message carved on a tree near Reno dated 1957 says: "Someone might think that whoever wrote this was dumb." It was

51. A frequently carved idea: "Viva ni" (Hurrah for me).

the perfect message for someone else to respond to, but the carver could not resist and continued: "but whoever reads it is a bigger fool."[81] Years later, someone added "Bastard" below it. An arborglyph dated 1909 says: "Everyone here that reads these letters [words] is a fag." Another, written in Basque and sprinkled with expletives, warns all not to read the intimate confessions. One fellow even threatened anyone reading his writings with spiritual damnation and excommunication. All in good humor, I presume.

The grove became a conference room, but without a moderator sitting at the head of a long table. It might be five or ten years before a question was answered or a comment added, but it made no difference; everybody had plenty of time. Communication was not as fast as today's electronic "chat rooms," but the groves served a similar purpose.

Rivalry

Rivalry was an integral part of the sheepherders' "communitas." Whenever several herders met on the range, they soon started playing *mus* and betting as they tried to outdo each other at lying, bluffing, and even cheating. When not playing cards, they would invent contests in marksmanship, skill at knife or ax sharpening, or doing tricks with a sheepdog. But the greatest competition of all was the job itself: who would grow the fattest lamb and lose the fewest sheep to predators? The arborglyphs reflect this competitiveness. One in Basque boasts: (I am a) "heck of a sheepherder, a champion sheepherder, one from [the town of] Latsa, Paul Ynchauspe, 1948, 1949).[82]

Just as in Euskal Herria the popular sports of hay cutting, wood chopping, and weight lifting are derived from everyday tasks, in the American West, sport derived from work.[83] Once a year the Basque sheepherders would leave the mountains and come to town to relax, meet friends, and celebrate their heritage, but even then they could not escape their competitive nature. There was and is a lot of teasing between the Iparralde (French side) and Hegoalde (Spanish side) Basques, much of it expressed in competition. A common carving might say: "The French Basques are tramps." But the latter were likely to return the same or a similar accusation. A tree in the Mahogany Creek area of Humboldt County bears three carvings, one above the other, probably carved by two, and possibly three, carvers: "All the Frenchmen are tramps (sneaky)." "You Spaniards are not worth a dick." "And the Frenchmen [are] cuckolds." One herder carved his name and the date, "B. B., 1915, 1917," and someone else added underneath, "Sneaky."[84] Interestingly, the word he used was *tranpa*, which is very similar to *tramp*—a term settled ranchers often used in reference to itinerant sheepherders.

During the 1910s, two or more sheepherders from Iparralde—one from Urepel and the other(s) from Bidarray—carved several mutually mocking inscriptions in Euskara. One inscribed by Batita Borda and dated 1915 and 1917 depicts two men shaking hands, but other nearby carvings indicate that the handshake may not have been totally friendly. G. B. depicted a burly man with a huge belly. The caption is difficult to translate faithfully, but in spite of the strong language its style is basically humorous: "Here is Yuanes Aldude. What he says there is shit. He is a devil from Aldude and a sneaky Frenchman with a big belly." This suggests that the term *Franses* (Frenchman) was somehow derogatory and was used even by the French Basques themselves. Perhaps it meant that one was a better Basque than the other. Aldude is probably a nickname in this case.

Two more carvings with the same general characteristics have been found in the vicinity. One shows the sketchy figure of a man with an exaggerated penis hanging down between his parted legs. The caption says: "Aldude. His dick." Another undated carving is similar:

"John donkey dong and S., 17 August." On the side, by another carver: "Shit to you, too."

It is safe to say that none of these men became too upset over such pranks. The inscriptions are not always signed, but there are ways to figure out who was responsible. On a heavily inscribed tree in the same area is "JS" and, underneath it, two statements in Basque that must have been carved by two different people: "You are a shit, too. You toad, kiss my ass, it is pretty." The response reads, "Dirty swine." On the same mountain but in a different grove is an unsigned inscription in the same vein: "Tosia and Arzamenta are two pigs."[85]

Most of the bickering between herders in the same outfit was in jest. But, for example, whenever a Bizkaian and a Zuberoan had problems, national affiliations might become involved—not that nationality itself was paramount. Linguistic differences were a much more serious impediment. Here are two examples collected on widely distant ranges: "Frenchmen, male whores, suckers, J. Y., 11 July 1918." "Frenchmen, pimps, wart suckers, s.o.b. suckers" (1960s). The last carver was definitely upset, but being aware that humor was essential in the mountains, on a nearby tree he changed his tone and left a rather funny message: "De Francia salen los gabachos unos a filar tigeras [sic] y otros a capar los machos" (From France come the *gabachos*, some to sharpen scissors and others to castrate he-mules).[86] Some carvings are just plain silly: "There goes S., he has to take a shit with his sheep."[87]

Nationality issues might have exacerbated the problems between J. C. and his herders in the 1960s. J. C. was from Iparralde, and his herders were Nafarroan and Bizkaian. One Bizkaian carved: "Screw the bosses from France. This year I am switching outfits. Long live Spain and the sheepherders, J. A., 16 June 1966." Curiously, J. A. normally did not hail Spain, but rather Bizkaia, his homeland. In this case he decided that invoking his nationality might benefit his case.[88] In other cases language had little bearing on the dispute. At one time J. C. owed his herder E. S. fourteen thousand dollars, even though both spoke similar Basque dialects. The following was carved by E. S., perhaps when J. C. owed him only two hundred dollars: "They [the bosses] have [it] pretty good but they work the sheepherder until he gets hernia, like myself . . . $200."[89]

J. C. seems to have been a controversial sheep boss. Someone carved his figure and underneath it the caption: "This guy is a real jackass." The carving is right on the road, and another carver—perhaps J. C. himself—saw it and replied underneath: "Up yours."

Difficult as it may have been for men separated by miles of wilderness, sheepherders argued quite a bit. There is no other way to interpret the dozens of isolated curses inscribed next to the names. The herder was like an absolute monarch on his range because there was no one around to contradict him. His word was law; his was one man's empire. So why the quarrels? Just that: the herder was so used to solitary life that the smallest things bothered him. These were lonely men who sought company, but at the same time they had a difficult time adjusting from their normal solitary lifestyle to human interaction.

Basques and Non-Basques

I have found no carving in which a Basque attacked a group of minority herders such as Mexicans or Spaniards, but the reverse is common. Spaniards, and a few Mexicans, carved a number of anti-Basque messages. I do not know why the carvings are one-sided; perhaps it was a way of bashing the Basque majority. Even Spanish-speaking Nafarroans carved an occasional anti-Basque message. I am convinced that their antagonism derived from the fact that they did not understand Euskara and thus felt excluded from the Basque brotherhood of herders. Their culture, to be sure, was no different. Four different carvers left unsigned inscriptions insulting Basques: "All Basques are bastards." "Son of bich [sic] Basque Borderre." "Bad shit Basque."[90] "The Basque are ass kissers, they think of themselves as very smart but we the Navarros [Nafarroans] screw them in the rear."

The latter was carved by a herder who in Nevada passed as Basque but in his own mind identified with the part of Nafarroa where only Spanish is spoken. He regarded fellow Nafarroans from the north as Basques—in the same category as Bizkaians, Benafarroans, Zuberoans, etc., who spoke Euskara.[91] These linguistic differentiating factors were less prevalent in America than in Europe, and such attitudes were either uncommon or offset by other southern Nafarroans who carved pro-Basque messages. Toribio Gallues from the town of Rocaforte, for example, carved the classic: "Long live the Basques." When they left the mountains and settled in town, these Nafarroans normally claimed to be Basque, joined the Basque clubs, and were viewed as such by Anglo-Americans. I know one Navarro who speaks little

Basque but nevertheless owns a truck with the license plate "Euskadi," a clear pro-Basque political statement.

For whatever reason, the Mexicans and Peruvians seem to have disliked life in America far more than the European sheepherders did. They carved some anti-Basque messages, but more often they vented their frustrations with anti-Yankee (read American) slogans such as: "Yankees, ass-fuckers." "The Yankees live under the eggs [testicles] of the Mexicans." Since most Basques (even those from Iparralde) spoke Spanish with Mexicans, Mexicans called all of them *españoles*, or Spaniards. Some Basques may have been included in the following inscription: ". . . let the Spaniards shit on themselves, long live the Mexican eggs [testicles], 1969, 1970."

Frank Rodriguez, a Galician, carved some of the most acrimonious diatribes against the Basques, but he seems to have been perennially discontented with everything and everyone about him. Rodriguez, who carved dozens of trees, was perhaps lucky to be a sheepherder. The mountains and smooth-barked aspens afforded him the freedom to say anything and insult anyone with impunity. Rodriguez loved needling the Basques, especially the Bizkaians. Once he carved a donkey head with the caption: "A Bizkaian before coming to America." Rodriguez did not sign these carvings, but the style of his letters is unmistakable. In another glyph he attacked all Basques: "The pig, the Basques, and the chicken are stingy animals, they eat shit." He carved two versions of it in case people missed the first one. Indeed, other herders saw the carvings; on the adjacent aspen an anonymous carver responded by hitting him below the belt, so to speak: "He is worse, because he masturbates in order to save money."

A similar message in the same canyon appears to have been carved by Rodriguez himself: "Antonio and Quipucha [Basque for Gipuzkoan] masturbate each other in order to avoid paying a whore. All Basques are like that." Since neither of these carvings is dated, I do not know which of the two is the original and which the plagiarism. Incidentally, these carvings and one other are the only references to homosexual behavior I saw.

The Camptender

The camptender, who normally stayed in the main summer camp, was responsible for baking bread for the herders and was generally in charge of the cabin and its environs. Some of them also carved trees in their free time. In the Lake Tahoe Basin I recorded a carving that says: "Hau da kanpo handia" (This is the main camp). Facing it, about one hundred feet away in a spot that could not be missed, is a response by a different carver: "Hau da kaka lekua" (This is a place to shit).[92]

The camptender was every sheepherder's whipping boy, yet in a way he was also like a medieval feudal lord with the herders as his peasants. The herders depended on the camptenders for their provisions, mail, and well-being. A herder's relationship with his camptender had an enormous effect on his attitude. This relationship is another aspect of the sheepherder's life that gets little coverage in the literature.[93]

When problems arose between the herder and his camptender, linguistic, regional, and ethnic divisions could easily aggravate matters. The camptender could make a herder's life miserable, and there was little the herder could do about it. After all, the boss seldom visited his herders in the remote summer camps. The camptender was the herder's lifeline, his only communication with the outside world. He could "accidentally" misplace a letter or a herder's note of complaint to the sheep boss. Herders who expected mail from home but were not getting any customarily accused the camptender of losing their letters.

The carvings indicate that herders' criticism of their camptenders was general and ongoing. In 1924 Martin Iturregui carved: "The camptenders of yesteryear were bandits, they did not bring firewood for the following year and I was fucked up." This comment, while interesting, is not particularly clear. Why, for instance, did Iturregui carve it near a *kanpo handia*?[94] This carving was the first I saw that indicated that gathering firewood during the summer was the camptender's job. The herder had plenty of time to procure his own wood, which was plentiful on the summer range. Perhaps Iturregui was referring to winter conditions in the desert, where firewood is scarce; indeed, in certain winter grazing areas it must be packed in by donkey or mule.[95]

During his sheepherding years in the late 1930s and early 1940s Jean Lekumberry read a carving in the mountains near Markleeville, California, that said: "Yesterday my burro ran away. Today it was my camp day and the camptender did not show up. Tomorrow, whichever comes first, the burro or the camptender . . . he's dead."[96] The next two examples of the adversarial relationship between herders and camptenders obviously involve Nafarroan herders because the carvers did not hesitate to use full names. Unfortunately, the first one, which is located in California, is undated: "M. is lazy, he is no good as a camptender, he abuses his herders, Antonio Yrigoyen." The following inscription refers to a fellow who is praised on another sheep range as being the longest-lasting camptender in the outfit, but at least one herder did not agree: "The camptender Pit [Pete] D. is an s.o.b. through and through, T. G., 1967–68."

Most Basques preferred to be camptenders rather than herders. The camptender had far more freedom—he could go to town to pick up provisions or have a drink. He also made slightly better wages. The herder, on the other hand, was likely to have more money saved. A herder in White Pine County was surely jealous when he carved: "A camptender's life is like that of a lazy sombrero-clad burro."

There is no question that camptenders got bad press, and since they lacked the herder's abundant leisure time to engage in carving, we may not have heard their side of the story. In favor of camptenders we must recall Valerio's rescue of Antonio, who was doomed to die of hunger. At least one Bizkaian thought well of his Nafarroan camptender: "Eugenio, you have been real good to me, but I do not want to return to this place, J. A., 1 August 1967." I did not finish my research in this particular canyon. A storm was approaching, and my truck was parked fifty minutes away, across two canyons and two swift creeks. I do not know whether Eugenio answered J. A.'s carving, but I do know that somehow he convinced him to stay another year with the sheep.

An Idahoan who was both a sheepherder and a camptender flatly stated in his memoirs that "the camptender ruins herders."[97]

52. Humor was very important if one was to survive the loneliness: "Nicolas Perez es un hombre honrrao porque nunca a parido y tampoco a estado preñao 21-7-68." (Nicolas Perez is an honest man because he has never given birth and has never been pregnant either, 21 July 1968.) Amazingly, I found no comment on this carving in Elko County, Nevada.

Humor

Humor was the cheapest and best antidote against loneliness. Many carvings begin with the herder lamenting his predicament but end on a humorous note. This strategy expressed the herder's determination to insulate himself from the harsh reality surrounding him.

One of the funniest and most original pieces I encountered is located high up in the Columbia Basin in Elko County, where a sign carved on a slender aspen says: "Hotel Derrepente" (Hotel Suddenly). I was not expecting a hotel sign there, so a few seconds elapsed before I began laughing so hard that I was rolling in the grass. A layman might not get the joke, but the two words were sufficient information for the sheepherders. However, the carver, wanting to be more specific, went ahead with the punch line: "still open three times a year." Below the date (1960) are several initials and names, but Jess Lopategui of Elko told me that Hotel Derrepente was his idea.[98] As a young herder newly arrived in Nevada, Lopategui was brought to this site by the camptender, who told him: "This is going to be your home for a few weeks." That is when Lopategui proceeded to carve "Hotel Suddenly," which became an instant hit; from then on that camp was known as "Hotel Derrepente" among the herders, not Badger Creek as before.

Typical Basque humor is dry, quick, and blunt, with contrasting or reversing extremes. The following, complete with Roman numerals, is a good illustration, "Jesus, on XIV-VII-LXXI, doing badly but happy." Many of the humorous carvings refer to ideas associated with banal expressions such as epithets or ethnic slurs.

Four-letter words and accusations abound in the humorous carvings, but there are also witty sayings and wordplays, like "Benito Camela." At first I took it as a normal arborglyph with the name of the carver, but the last name was suspicious and kept me wondering until it finally dawned on me. I had to separate the words in order to decipher the meaning, "Ben [ven] y tócamela" (come and touch me). Another herder borrowed the word *camela* and carved below it: "Que te camele el obispo" (Let the bishop screw you).[99]

Composing verses and singing old songs was another way to cope with isolation. And everyone talked to himself, not in the same way we all do, but aloud. A tree in Plumas National Forest addresses this issue: "Notice: No talking to yourself allowed." Almost all the rhymed verses I recorded are in Basque, but the following one is in Spanish:

Del mar salen	From the sea
las perlas	come pearls
de las perlas collares	from pearls, necklaces
y del corazón	and from the heart
de un amigo	of a friend
un millon de felicidades	one million happinesses.

Humorous Old Country stories were rehashed and retold countless times in Nevada sheep camps. In the early times, when literacy among the Basques was probably lower and there was no radio or television, storytelling was the only form of literature the herders knew, and it was much more important than it is today. Some of the Old Country stories I have heard in Nevada reveal aspects of Basque religiosity—or the lack of it—but one is not likely to hear them in Europe. In the United States, far from the parish priests who intimidated many peasants, the herders felt free to tell stories such as the following:

> One day a priest needed to cross a swollen river, but he did not know how to swim. Just then a *baserritarra* showed up with his little donkey and agreed to help the clergyman. Both men mounted the poor beast and they entered the river. When they reached the middle, the water was deep and the current swift and the priest began to fear for his life. He started to pray in earnest, until the old farmer told him: "Father, stop praying because if the donkey decides to kneel down we are both doomed."[100]

A Bizkaian I spoke to remembered walking to Gernika and passing well-fed priests enjoying their daily stroll. As he and his friends approached the priests, they respectfully took off their berets and greeted them: "Egun on, jaune" (Good morning, sir). The priests would reciprocate, but as soon as they walked away, the boys, sotto voce, added a rhyming verse, "Horretxek jan bigaittue!" (Those same guys are going to eat us up!).[101] In America the herders did not become exactly anticlerical, but looking back to life in their little Basque villages, they realized the great influence that the priests exercised over many aspects of society.

I think cursing can sometimes be regarded as humor. By now it must be obvious that many of the statements inscribed on trees contain expletives in several languages, as well as lots of slang. Swearing is not deeply ingrained in Basque peasant culture—some claim that Euskara contains no swear words—but it didn't take the herders long to acquire a considerable "dirty" vocabulary from English and Mexican Spanish in Nevada. The words most frequently carved were "cabron" (bastard), "puta"/"puto" ("whore," female and male), "fok" (fuck), "chingar" (Mexican for "fuck"), "joder" (to screw), and "kaka"/"caca" (shit). The everyday language of many herders is peppered with such words. Cursing acted as a release of pressure while deflecting feelings of loneliness.[102]

A few trees tell of individuals notorious for their cursing. According to Al Gallues of Reno, in his father's outfit there was a man from Nafarroa, C. O., whose colorful swearing was accompanied by theatrical gestures. He would raise and shake his fists while looking skyward and screaming: "Let lightning cut me in half if . . ." Gallues was right. I found a carving by C. O. complete with the formula "me cago en la hostia" (I shit on the Host). Curiously, only a couple of cases of the typical curse "me cago en" (its Basque rendition is *mekauen*) have been recorded.

Sheep versus Cattle

It is interesting that no carving found thus far mentions problems between cattlemen or other Anglo-Americans and itinerant sheepherders. The western literature is full of descriptions of disputes over water or pasture, anti-

Basque animosity in the local media, and even violence.[103] But the carvings do not even hint at any discrimination, injustice, or fear of violence. Is this a case of the "typical" Basque behavior of omitting the obvious?[104]

Herders, carving in isolation, were at liberty to express their sentiments about anyone. If they carved nasty statements against their own bosses and camptenders, they could have just as easily cursed the federal government or nearby cattlemen who ran them off the range or away from the water. Why such a complete silence if this was a vital issue for them?

Perhaps the importance of the topic has been overblown by the media. Stories about gunfights or fistfights between local cowboys and foreign sheepherders were fail-safe fodder. They were certainly more popular in town than on the range. The local media, naturally, almost always sided with the cattlemen and established ranchers, who were also the readers.

The data the urban media have provided to historians must be compared with the many reports from herders of their friendly relations with neighboring cowboys. The existence of tensions between sheepmen and cattlemen is undeniable and is acknowledged by older Basques in northern Nevada. But as Frank Aramburu reminded me, on the range the cowboys and sheepherders were mostly on good terms, and he gave a logical reason for it: "Cowboys don't pack lunch and often they go hungry and then they know where to go, to the sheep camp. I had cowboys come in, and their first step was toward the Dutch oven. They knew sheepherders always had food. A little bread and a swallow of wine and they left happy."[105] Herder J. Ithurralde concurred with Aramburu abut feeding hungry cowboys. I wondered aloud how many townspeople, including the newspaper editors, knew about this, but Aramburu simply shrugged.

In northern Humboldt County (and the Jordan Valley area in southeastern Oregon), where the Basque presence was strong, there was, as mentioned, little anti-Basque sentiment. In the 1910s, in Oregon Canyon, for example, a number of the ranches were owned by Basques such as Aboitiz, Achabal, Ansotegui, and Echabe; and in Coffee Canyon there were the Marcuerquiagas, Bengoecheas, Ugaldes, and Mentaberrys, among others. Pete Bengoechea, a native of that area and a Humboldt County commissioner for many years, does not remember any ethnic-related incidents, and rancher Frenchie Montero does not think the Basques in Humboldt County were discriminated against.[106]

Jordan Valley was overwhelmingly Basque.[107] In 1966 I photographed a sign written in English and Basque that stood by the highway west of town: "Entering Jordan Valley, Oregon. Home of the Basques. Hunting & Fishing. Kaishio etorri danok."[108] To this day tourist fliers label Jordan Valley the "Heaven of the Best in the West," and Basque hospitality is extended to all with the Bizkaian saying "Gaure echie da saure echie"(Our house is your house).[109]

Alone with the Mountain

When you ask a herder, "How's life?" his first response may be, "Lonely," but we should not assume that herders lived with loneliness every minute. They did not think about their isolation except during certain days and times. Many Basque herders made a deal with themselves about being alone. They saw it as temporary: "Torco dite urte onac" (Better years will come), as one carving says.

In the summer of 1999 two *bertsolariak*, Estitxu Arozena and Oihane Enbeita, walked through the aspens improvising and singing verses. In her last, masterly, verse Enbeita married the herder's loneliness with the art he created to combat it.

ENBEITA:
Zenbat marrazki, zenbat irudi
artzaia artista zela dirudi.

How many carvings, how many images!
it appears that the sheepherder was an artist.

AROZENA:
Mendietako bakardadea
isladatzeko zeukan fedea.

He believed that he could reflect
the loneliness of the mountains.

ENBEITA:
Artzai fededun
lagunarterik ez zeukan inun
ta artea egin zuen lagun.

Sheepherder, believer,
nowhere had he any company,
therefore, he made art his companion.

Not every herder reacted the same way to the loneliness, but all were affected by it. The theme of summer isolation in the high country is the third most common in the arborglyph literature, after single names and dates and comments on sheepherding. It is often linked to the theme of women. Two messages by different people set the general tone of such carvings: "Poor sheepherder" (Tahoe National Forest, 1925). "I am sadder than a pine forest at dusk" (Ruby Mountains).[110]

Some could not overcome their loneliness; others, like the herder who carved the following, took their jobs more philosophically: "Here lives Ampo, whether well or badly."[111] We should not be misled by the carvings of a few discontented herders. Often they were the most vociferous and prolific carvers—Frank Rodriguez, for example. Often, too, they were mediocre herders with more education than the others; and their minds were not with the sheep. The fattest lambs were grown by herders with an upbeat attitude.

53. "Estoy mas triste que un pinar cuando anochece" (I am sadder than a pine forest at dusk). Most herders took a dim view of those who carved such messages.

Manex looked down from his lofty post on the desolate Cherry Creek Range in eastern Nevada.[112] From the rocky outcrop where he sat he could see the rest of the world. The sky was the same old blue with a few scattered high clouds. The sheep and the lambs were grazing and bleating happily, and the dogs Beltxa and Pintto, his constant companions, were snoozing beside him. In the far distance Manex contemplated the north-south highway bisecting the lonely valley. It reminded him of a dead rattlesnake crawling with ants, but the ants were cars. "Is it possible that there are people inside those cars?" he wondered. "They do not even know that I am watching them," he said aloud. No one answered, but Pintto raised his head and pricked one of his ears as if to say, "Are you talking to me?"

Manex got up and walked toward the nearest aspen clump. He was lonely, but thinking about people hurt more than it helped. He knew a better way to deal with moments like this. He pulled out his knife and on a smooth aspen trunk he carved: "Biba artzainak" (Long live the sheepherders), and then he turned around, cupped his mouth, and loosed a piercing *irrintzi* (Basque war cry). His face red and his neck veins near bursting, Manex sustained the *irrintzi* for a long while. The dogs in chorus joined him, barking and howling, and even the normally passive donkey turned his head toward his master and started braying halfheartedly. Running out of breath, Manex let the *irrintzi* slide, finally ending it halfway between a coyote's yelp and a horse's neigh. The dogs quit howling, and the mountains reverted to silence. Manex listened. Beltxa and Pintto were already looking in the same direction. As if it were a lazy echo, a distant *irrintzi* rumbled through the canyon in answer to Manex's cry of triumph, desperation, and resignation. He felt better already. "I traveled six thousand miles to take on this job; I must have been crazy," he often told himself.[113]

According to Jean Lekumberry, not many Basque sheepherders thought they were insane, even though Anglo-Americans often said that a man must be nuts to be a herder.[114] An individual on the foothills of the Cherry Creek Range in White Pine County who signed himself ABV agreed: "Adios Kazeno, here I spent my winter and I went crazy."[115]

Sheepherding is an exercise in self-control. Day after day the herder must ignore his basic social instincts. He must tell himself that he can live differently, that he can stand alone, just for a while. This was exactly M. O.'s frame of mind on the Cherry Creek Range when he carved his proud message that a man needed plenty of guts to be a sheepherder.

Most felt like Arnot Urruty, who in 1926 inscribed his classic: "Here I am bored/tired, some day the time will come to leave this place." Where was "this place" where Urruty felt so bored? Was it some God-forsaken canyon or sagebrush-filled flat? Hardly. If he turned his head as he carved he could contemplate a spectacular view of Lake Tahoe. Perhaps Plato was right when he said that reality is in the mind of the beholder. Many people pay a lot of money to enjoy the crisp fresh air and the vistas that Urruty was willing to forgo. Many people would view it as a privilege to be standing in Urruty's place.

Sheepherders endured loneliness far more than cowboys, miners, and prospectors. They spent days, weeks, and months alone with the heat, the cold, the wet, and the dust. Alone and alert, like *harri mutila*. I received a taste of what the carvings describe when my research forced me to spend weeks alone in sheep country, although never more than five days at a stretch. The second day was the most difficult, mentally. By the third day I had almost accepted the situation and was actually starting to enjoy the mountains and the solitude.

Obviously there is no comparing my brief experience in the wild with that of the sheepherders. For one thing, I was completely alone—no sheep, donkey or dogs. In fact, with hundreds of fascinating arborglyphs to record all around me, I had little time to dwell on my isolation. What I hated most was fixing dinner in the dark or in front of the truck headlights. I envied the herders with their well-stocked Dutch ovens.[116]

Morning was a different story. I would get up before the sun and build a fire to boil water for coffee and to warm up the camcorder batteries. In the crisp morning air of the high country I walked around camp with a steaming cup of coffee in my hand, checking the frost on the creek bank, listening to the noises (or the silence), and watching the sun's red ball rise majestically from behind a distant peak—for me alone. The rest of the world—the cities, the people—didn't count at all, they were all too remote. I wondered if the sheepherders felt the same way.

Returning to civilization, I felt as if I was going to a new country, as if in five days I had forgotten what towns were. Some things in town made no sense; others annoyed me. I read newspapers avidly, and the news on TV and the radio was really new. I could easily understand why old herders often became men of few words, men who spent hours in silence or preferred talking with their dogs rather than with people.

In 1929 Ambrose Arla, sitting on a rock, cried on his first night as a sheepherder. He had come to Reno with his sister, but suddenly, it seemed, he found himself alone herding sheep at Lakeridge. Eugenio Sarratea cried too, as did his brothers Pedro, Fulgencio, and Ignacio, and Luis Ojer, Inazio Arrupe, J. L. Eguen, and others. The list is long, but the weeping stopped soon. It had to.

Those who came in the 1960s suffered more than previous sheepherders because they were used to a more comfortable life. At the time the economy was rapidly improving in the Basque Country, and these immigrants found living conditions in the sheep camp too harsh. As Ramon Zabalegui carved in Poco Valley, Plumas National Forest, in 1963: "Vida triste" (Pitiful life). Another carving in the same forest says: "What a miserable life we sheepherders lead here." A Castilian sheepherder I interviewed in Parker, Arizona, in 1971, was of the same opinion: "We live like pigs in America," he said.

When it comes to depicting the sheepherder's life, writer Robert Laxalt is a master. And with good reason; he was a part-time camptender for his father, who stayed with the sheep most of his life. Describing the worst enemy of the sheepherder, Laxalt wrote: "The hard part was loneliness. You would almost die from the loneliness, just to hear a human voice. Then a funny thing happened. You turned a corner in your mind, and you wouldn't walk over the next hill to see someone."[117]

Clearly, many arborglyphs were carved on the spur of the moment in response to loneliness. In the lush foothills of the Ruby Mountains, near Lee, Elko County, a lonely herder too homesick to appreciate the beauty of the place carved a very simple "Kaixo" (Hello). I wonder if he expected an answer from anyone. Tree carvings contain a number of hellos and *alos* (the French version). They are simple expressions of the herder's desire for companionship, powerful psychological weapons against the depressing isolation. Some of them are accompanied by single figures of a man or a woman. One shows both, a man greeting a woman with "hello." There are dozens of simple figures of a man and a woman just looking at each other. These are not erotic; the man stretches out his hand, seeking reciprocation. Carved in these remote wilderness areas, they are powerful affirmations of the social nature of humans.

Chin Chin: The Sound of Money

Why go through it all, the pain, the agony, the loneliness? For money, of course. Logically, one would expect to find a wealth of carved references to such an important topic. That is not the case. This omission might seem baffling, but it is consistent with the sheepherder's point of view. Money was a very private matter, more so than sex, and extremely few revealed how much or little they had.

A Basque song popular with emigrants tells of the material accomplishments of an *Amerikanu* (emigrant to America) returning home.[118] He sings it to himself more than to his sweetheart:

Amerikara joan nintzan	I went to America
xentimorik gabe	without a penny
handik etorri nintzan, maitia	I returned from there, my love,
bost milloien jabe	owning five million.
Txin txin, txin txin	Chin chin, chin chin
diruaren hotsa	the sound of money
haretxek ematen dit, maitia,	that is what gives me, my love,
bihotzian poza.	pleasure in the heart.

A number of Basques emigrated for other than economic reasons. When, in the nineteenth century, the Spanish state forced the Basques into military service, many preferred to leave the country rather than comply. My uncle was one of them. In the course of my years in Nevada I have met young men who were forced out of their country because the Spanish police were after them. Others came after impregnating their girlfriends or being labeled "troublemakers" in town. Still others simply wanted to see the world. But for the majority the paramount motivation was economic.

Many Basques learned about money for the first time in the United States. In other words, it was here that they began to have cash in their pockets and in the bank. In the rural areas of Euskal Herria cash was in short supply until the 1960s. People grew much of what they needed and bartered for the rest. Ironically, many left the country just as industrialization—and jobs—were coming to Bilbao and the larger towns in Hegoalde; but the Basque peasants preferred sheepherding to factory work, which was left for the thousands of Spanish emigrants who moved into the Basque Country.

Poverty in the Basque Country was never as extreme as, say, Ireland after the potato famine, but opportunities were limited. As an example, let us listen to a Nafarroan: "We were eleven of us at home . . . very poor. . . . My inheritance was a sow of 163 kilos. When I began working as a lumberjack I sustained myself a whole winter eating the pig. . . . One day, a friend that had worked in Ely, Nevada, for five years asked me if I wanted to go to America. 'I am ready to go right now,' I said. That is why I came, and then I brought my brother over."[119]

It is a misconception that herders had no place to spend their money and therefore saved most of their wages. The truth is that about 30 to 50 percent of them were unemployed during the winter months and without the benefit of unemployment compensation. These herders did not—could not—save much. A number of them were broke by the time spring came around and they were rehired for the lambing.

But everyone back in the Basque hometown expected the *Amerikanu* to return with money. How much? According to a story told by Frank Gallues of Reno, in Nafarroa they counted on a thousand dollars for every year with the sheep. Ten years as a herder meant ten thousand dollars in savings, and so on.[120]

For those who came to work as sheepherders for reasons other than economic ones, it was a rude awakening. I remember a herder from León, Spain, who was working by the Colorado River in Arizona in the winter of 1971. There was some kind of trouble in his outfit. The Basque herders, who were all from Nafarroa, declined to be interviewed, but he was talkative and agreed. He claimed that his family owned a business. "I did not have to come," he said. He hated sheepherding and could not wait to get out.[121]

I found only one message carved by this type of herder: "Lehen nintzen aberas" (I was rich before), it said, but it was not well received—perhaps his companions thought he felt superior to them—and someone carved a response next to it: "Gure poeta alue" (Our fuckin' poet).

Certainly many sheepherders in Nevada did make money. Their wages were low compared with those of city jobs, but they could save a lot more than the worker in town. As one sheepherder put it: (In the 1930s and in the previous decades) "they paid you thirty dollars a month and board . . . you made up your mind to suffer for that money."[122]

Although wages fluctuated, a dollar a day was the prevailing pay in rural America for quite a while. Sheepherders apparently were sometimes paid a bit better. In 1901, for example, Bixenti Astorkia was making very good wages working for McDermitt of White Pine County. He carved: "McDermit [*sic*] pays forty pesos a month." And as mentioned earlier, Ambrose Arla made fifty dollars a month during the Depression. When Abel Mendeguia came to Nevada in 1951, he worked for Pete Elia, a fellow Nafarroan. His wages were two hundred dollars a month (room and board was always included). Mendeguia worked for sixteen years and received a fifty-dollar raise at the end.

In the 1960s I helped a sheepherder in Elko with his bank account, and I still remember the engaging look the young female teller gave him, vocalized as: "Gawd, you are loaded." He had been seven years in America. Puerto Ricans came to New York City because they were told the streets were littered with money; one just picked it up as needed. Basques were told similar stories about America, as may be reflected in the following unsigned statement carved near Reno: "Long live America and money come."

The many carvings with "hurrah for America" hail the better life the herders hoped to find there. One herder got carried away and carved "Hail to the USS." On 7 September 1965, P.S. carved a grand *baserri* (farmhouse) complete with a coat of arms and two smoking chimneys, but in front he appended an American touch, a driveway and on it a car. The caption below reads: "Hurrah for the house and the car."

The herders who cheered the United States in their carvings were sold on its economy. One Zuberoan saluted the country and called the Columbia Basin "Amerika chokorik eijerena" (America, the most beautiful place). He would not have carved that if he thought that he was not doing well economically.

Generally speaking, Basques who emigrated before 1960 were more impressed with the United States. One sheepherder who returned to Munitibar, Bizkaia, around 1950 outraged everyone by saying that compared with American farmers, the farmers in Bizkaia lived in Adam's time.[123] Basque farmers did not appreciate hearing such things, and they usually counterattacked with the now classic, "Bai baina Ameriketan be txakurrek ortozik" (Yes, but the dogs go barefoot even in America).

Sometimes the problem was not accumulating capital but determining when enough was enough. Some old herders actually did not know how much money they had in the bank. The boss paid them annually and deposited the money in the herder's account year after year. When a herder needed cash, he would very likely borrow it from the boss and settle accounts with him later. The herder had almost no interaction with the bank. Juan Irazoki, whose carvings are dated as early as 1914, was still herding in the 1940s for Jess Goicoechea of Elko. One day Irazoki asked Jess to check his bank account because he thought he had only thirty-six dollars left. Goicoechea checked it and gave him the good news: Irazoki had forgotten to count the zeros following the 36.[124]

Thus they continued working, afraid they still did not have enough to retire on or to return home. To one such old man was the following carving directed. It took me a while to read it; in fact, I felt elated about the accomplishment because of the poor spelling and the advanced distortion of the bark: "Hello Manuel, leave these mountains, you have enough money." Like so many others, this message consisted of two very different parts. It ended: "Hurrah for me and the sheepherders."[125]

Sheepherders who were particularly tight with their money avoided coming to town. They knew money evaporated there. For comparison, a Peruvian herder took a two-week vacation in Elko in 1990 and spent one thousand dollars. A Basque in 1933 stayed in Ely for eight days and it cost him one hundred dollars. Both felt that they had "thrown" the money away. "How did you spend the money?" I asked them. Both had spent it purchasing similar things: room and board, drinks, several visits to the whorehouses, and a few new clothes.[126]

Many herders had very wrong ideas about America and life in town. They were envious of their urban Basque friends and decided that people in town lived much better and had more money. I recall in 1967 the words of a young Bizkaian who had quit the sheep a few months earlier for a higher-paying construction job in Reno: "Mallea, money, money; there is nothing else."

When making money became such an important goal in life, any setback was painful. Perhaps the biggest setback of all, for many sheep owners as well as sheepherders, was the Great Depression. I found a couple of carvings dated in the 1930s that must be understood within those general terms. In the following example, one herder talked out his complaints: "Damn Misfortune, you control the poor, 1930." Other statements are just as vague: "1922, 7th day of August, Juan P. Martinez is going crazy with his fixation with poverty." This obviously was not carved by Martinez, but by a friend who knew of his troubles.

After a fleeting visit to town a herder returned to his solitary camp and, feeling sorry for himself, carved: "Good morning for the rich." And: "This world belongs to the rich." The same herder carved the following as well: "Life is worth more than a few dollars." "Good morning [but] for the rich, one must suffer a lot in this world. Good-bye, I am leaving." All of these glyphs are in the Ruby Mountains of Elko County, carved apparently by a Spaniard.[127]

Basques had similar thoughts, but rather than dwell on them they were likely to "hurrah" themselves out of it. A typical example is this: "Young and poor, long live France."[128] The two ideas carved here have nothing in common but are brought together by the Basque principle of the contrasting equation. Sure enough, nearby I found: "Young and poor, but one must not give up hope," by the same carver. In the face of adversity the herders were buoyed by the knowledge that the situation was temporary. Some herders came to terms with the life while others struggled and suffered more than they should have.

54. This carver would never become president because he advertised something his companions wanted to forget: "Mas vale la vida que unos dolares" (Life is worth more than a few dollars).

The messages that comment on money are divided between those claiming that the sheepherder's life was not worth the compensation and those who disagreed. I think, overall, the balance tilted toward the latter. The Basques had come to make money, and they were making it. They were not suffering in vain. But I also heard people say that in the 1960s and 1970s the sheepherders could not save money "like in the old days."

Today the herders under contract to the Western Range Association make six hundred dollars a month and up. In the summer of 1989, near Truckee, California, I met an old Asturian man who earned eight hundred dollars. He said that in a few months he would go off contract and start working at the ranch for a thousand dollars a month. Ranch hands usually start at eight hundred to a thousand dollars, plus room and board. Another herder in the Ruby Mountains was not happy with his wages, but at least he had a sense of humor when he carved: "Antonio Hidalgo, Peruvian sheepherder. My friends, I have plenty cojones but little money, 1983."

I found only one carving in which the herder provided the dollar amount of his savings: "Next year [mid-1950s] I am returning to Nafarroa with $10,000, J. A." It was a small fortune, and J. A.'s carvings in Nevada and California indicate that it had taken him about ten years to amass it.

A non-Basque carved the following: "Next year I am returning to Spain and to the good life, Basilio Casares, 1966." Two years later, F. C. carved: "Two more years and with a car and a girlfriend I will go about the streets of Elizondo scratching my balls, 1968." Elizondo is the government seat of the beautiful and lush valley of Baztan in northern Nafarroa. As it happened, two years came and went, and this sheepherder was still carving trees in Nevada, but eventually he did return home.[129]

Few of those who made fortunes in the sheep industry returned to Europe; instead they settled down, bought ranches, and raised families. John Achabal of Idaho was an archetype, as was Pete Itçaina of Elko. Just about every Basque in northern Nevada has heard about the time Itçaina was denied a drink at a bar in Elko because he was dirty and looked like a bum. He called his banker/lawyer and ordered him to buy the place. Then he went back and again asked for a drink. The barkeep said something like, "I thought I told you we don't serve the likes of you," to which Itçaina responded: "I own this place now. Get out." Ana Hachquet, who knew Itçaina, said that he was a powerful man. She summed it up with the words "Sosak aiskidiak eiten'ttu, ba" (Money makes friends, you know).[130]

Most bankers were happy to accommodate the Basques, even those who could not write. The latter usually signed their checks with an X, and the banks accepted them.[131] Incidentally, halfway up the Toiyabe Arc Dome in central Nevada I found a check for twenty dollars *carved* on a thick aspen limb, but I couldn't cash it: it wasn't signed.

Many Basques returned home relatively wealthy but kept quiet about what their fortune had cost them in terms of human suffering and deprivation. They were also secretive about their money. It is said that farmers in the Old Country did not trust the banks, and that many kept their money *hormazuloan* (hidden inside stone walls). But Bolu (a nickname) was different, and when he returned from Idaho to Bizkaia, he went to put his money in the bank. When he was asked "for how long" —that is, the terms under which he wanted to deposit the money—Bolu answered: "Forever."[132]

All the sheepherders knew that their dollars were hard-earned, but the following carving indicates one's peculiar assessment of the situation: "The sheepherder masturbates a lot in exchange for many pesos." This individual clearly understood the sheepherder's life as a "quid pro quo" arrangement, as if he were saying: It is too bad that we have to masturbate so much, but at least we are making money.

If the arborglyphs are correct, a lot of the money the herders spent in town went to whorehouses and drinking. We have already noted some carvings that relate to drinking. The references to whores are innumerable. One carving says: "I am completely broke because of the whores in Carson Siri [City]." Another herder who was dismayed at seeing his dollars disappear so quickly admonished his fellow herders: "It is better to screw the jenny than to waste $110 on the whores of Mustan and Munlai [Mustang and Moonlight]."

Nevada towns, with dozens of tempting places to squander money, were dangerous to the pocketbook of the inexperienced herder. The casinos were always a temptation, but lack of language skills kept many Basques away from the blackjack tables. Most Basques probably gambled less than the average Nevadan. In the late 1960s, however, I met a Bizkaian who a few years earlier had won twenty-two thousand dollars in one night in Reno. For one day he was king. He got the best room in the casino, free food and drink, and whatever else he requested. By noon of the second day he had lost everything, including the brand-new car he had bought.

Temporary Jobs

In the 1960s a Bizkaian nicknamed Oxine went to America after telling his friends that he would never return. Two years later he appeared in the public square with a large suitcase in his hand. An old buddy greeted him and asked:

"Hey, Oxine, didn't you say that you were going to America for good?"

"Yes, I did, but for a mortal like myself that is no place to live. However, for someone like God, who lives forever, there are wonderful places in America."[133]

Many herders did not intend to stay in America. They left home after saying to worried parents, and perhaps a fiancée, "I will make some money and in a few years, before you know it, I will be back." That temporal concept is expressed well in a carving located near Bridgeport, California: "Julio Gorriz was here on 8-28-1968 for the first time. When will it be the last time?" He was already thinking of going home. Gorriz, and many others, did in fact return to Europe. Many Basques looked on sheepherding as a sort of novitiate, a transitional stage between two points. The anthropological concept of rites of passage can be applied to sheepherding as it was lived by Basques. In their rites of passage three main elements may be identified: (a) removal from normal society and yearning to return, (b) the passage or transition itself, and (c) reincorporation.[134] The carved messages attest to all three components, sometimes vividly.

Yearning for Home

Family ties are very strong in Euskal Herria, and herders alone in the mountains missed their families most of all. A number of the carvings refer to home and family. One man, perhaps a very young herder, simply carved: "Aita eta ama" (Father and mother) as a synthesis of the home life he had left behind. L. O. expressed the same nostalgic idea, but more poetically: "When I hear the *jota* [a type of song] being sung, it seems as if I were sleeping in my mother's bosom."

E. I., too, had his mother in mind, but for a different reason. A letter arrived from his family in Nafarroa announcing her death: "Today I have received the sad news about my mother, 8-28-66." By the time he received the letter she had long since been buried. He read the letter in a shady sheep camp not far from Reno. He must have reread it many times, refusing, perhaps, to believe that he would never see his *ama* again. Martin Goikoetxea hoped to see his father one more time, but it was not to be, and as a way to ease his grief he composed the following *bertso*:

Gure artean gelditua da	Among us, there remains
zure izatean gorantza	the greatness of your personality
ni horrekin arrotzen nintzan	I used to be so proud of it
iduri nula orantza	that I swelled like yeast
aita zure mesedetako	father, there appears that
eraman zaituan antza	your departing was propitious to you
mundu obera joan zerala	at least I hope that
nik hola det esperantza.	you have gone to a better world.

When the yearning for home struck, sheepherders were likely to carve crosses and figures of Basque houses, of which there are many. Some include the coat of arms; others are elaborately carved. In 1918 Frank Aldunate carved an elongated *baserri* with two chimneys. In the early 1930s, on the same mountain range, Joaquin Pagola loved to inscribe old-country farmsteads. On one tree he carved his own three-story home in Nafarroa.

Almost all of the houses found so far contain one important detail: the smoking chimney is always prominently displayed. The Mount Rose area contains a number of house carvings. P. Gamio carved a fancy one in 1951 and added stars and a tree next to it. A decade later, Pedro Sarratea carved a stately house nor far from Gamio's. An undated and unsigned arborglyph in Humboldt County represents a northern Basque farmstead, probably Zuberoan, with a tall chimney, complete with its name, "Olarite," inscribed below. In the Columbia Basin I recorded a most ornate house with three stories, several chimneys, and lots of windows, reminiscent of the houses overlooking the harbor of Lekeitio, Bizkaia (see figure 27).

The longing to be back in the village, surrounded by friends, was never as painful as in the summer when Basque villages were celebrating *herri jaiak*, the main annual festivities. Nafarroans seem to have felt more compelled to carve about these celebrations than other herders. No red-blooded son of Nafarroa forgot to recall San Fermin on 7 July. There are many inscriptions that hail "San Fermin," the patron saint of the old Pyrenean kingdom. A couple of herders could not believe they were missing the celebrations: "Today is San Fermin, and I am here!" "Today, 7 July, is San Fermin and here I am rubbing my balls without being able to touch a pussy, can you believe it!" In 1965 Pedro Sarratea celebrated the day by going fishing and then portrayed himself in the act of pulling a fish from the creek. Another carver, E. S., was still dreaming about the festival two months later. It was 4 September, well past San Fermin, but he carved himself and three of his friends filing through city streets with hats, staffs, and erect penises. The caption says: "Marching in Pamplona."

San Fermin's popularity among the herders is unmatched, but there are references to other towns and their fiestas as well. Felipe Errea on 10 July 1947 carved about his hometown: "Fiestas in Mezquiriz." In 1935 Bertrand Bidart announced: "The fiestas of Baigorry are next month," but he forgot to date it. On 26 July 1964, someone from the valley of Baztan, Nafarroa, carved: "Fiestas in Elizondo."

Some people worried when it was holiday time in their hometown, as was the case with Salvador from Jaurrieta, Nafarroa: "21 September, first day of festivities in my hometown. Some bastard will enjoy my fiancée and my sister, and what can I do about it?" At the opposite end of Elko County, F. Z. of Gernika, Bizkaia, worried about the same thing. After passing through the treeless Owyhee Desert, he carved on the first aspen he came to: "Today is Saint Peter's and I wonder who my fuckin' fiancée is running around with."

Thinking about the fiestas and recalling past good times was good therapy for the herders. They kept it up even after settling in town and raising families. They talked about the good memories—eating, drinking, and dancing back home—to the degree that American-born Basques had a mistaken idea regarding life in the Old Country. Emily (Zatica) Miller of Humboldt County was utterly amazed on her first visit to Bizkaia. From the

55. A Zuberoan-type *borda*, or farmstead, called Olarite.

56. A Basque farmhouse with two chimneys and a waterfowl, by Frank Aldunate, 1918.

many stories she had heard about the Old Country, she had assumed that people there did not work.[135]

The Transition

Before they knew it, many herders found themselves alone with three thousand sheep, a dog, and a donkey. For some, the decision to emigrate had been swift, with no preparation at all. One night J. A. told his parents: "Tomorrow I am going to America." And so he did, like thousands of others.[136] A Bizkaian's carving is amusingly candid: "I am not an American, I am not a sheepherder, but I am here."[137] Physically, most of them were in good shape, but to say that few were mentally prepared to deal with a big herd of bleating ewes would be an understatement.

The herders found themselves surrounded by unfamiliar mountains and plains. The grasses and trees were totally different. Their passage had begun. Not long ago a young man had been in Euskal Herria enjoying a farewell party with friends. Now he was on a mountain whose name he did not even know, and could not have pronounced if he did. The initial shock was particularly difficult on the herders who arrived after the 1950s because of their higher expectations. Martin Goikoetxea's first impression of the United States is captured in the following *bertso* he composed:

Honera etorrita	After arriving here
hasera zorrotza	a shocking beginning
hizketan ere ez jakin	could not even speak the language
hau leku arrotza	this was a foreign country
leku aundiak eta	the expanses were immense and
ibilera motza	wretched trekking
Amerikan bizi ta	living in America
Euskadin biotza	[but] my heart [was] in Euskadi
jaiotako lekuan	in the place where I was born
nahi det eriotza.	I want to die.
Etorrita gauza hau	After arriving, this
pentsa egon nintzan	is what I was thinking
desertura eraman	they have taken me out to the desert
basa piztin gisan	in the manner of a wild beast
gazte batentzat bizi	for a young man was that
modua hoi al zan	any way to live?
ez gendun denpoik pasa	we spent no time
herrian t'elizan	in town and in church
ordu tristeagorik	a more sad moment
ez det inoiz izan.	I have never had.[138]

They went for days without seeing a soul. Most camptenders took supplies to the outlying sheepherders every five to ten days, but some came only every twenty to thirty days. In the early days in the high country the herders were often provisioned once a month. In 1869 John Muir was herding in the Sierra from 3 June to 21 September, and during that time the sheep boss brought him supplies on 7 July and 14 August.[139] Emily Miller told me of her father's experience. "My father said that at first he was thirty days alone with the sheep. Camptender came every thirty days. He ate mostly meat and fish."[140]

Juan Kozkorroza of Ispazter, Bizkaia, arrived in north-central Nevada at the turn of the century and for thirty days did not see anyone.[141] Pete Itçaina's nephew Jean Ardans probably took top honors for endurance on the range. He came to Elko in September 1928 and herded sheep until 1952 without taking a vacation or coming to town—for almost twenty-five years, that is.[142] "Mañiz" Ithurralde of Lasa, Benafarroa, came to Ely in 1929 and stayed with the sheep for four years and ten days. When his boss, John Auzqui, suggested that he go to town, Ithurralde at first declined, but later he acquiesced.[143]

The newly arrived herders lived a borderline existence, with one foot in Europe and the other in the United States. "We knew absolutely nothing about America," Jesus Pedroarena told me.[144] The trees spoke to them about the tribulations and joys of former herders during the same transition, and the newcomers were bolstered by the carvings. They dreamt of the future but were emotionally caught between two continents. How did they make it through the transition? What did they think of their new lifestyle?

The literature has a definite tendency to empathize with them and perhaps to portray them as slightly larger than life. "Men of leather and bronze who had been rich as barons," Robert Laxalt described his father, Dominique, a lifelong sheepherder.[145] And indeed, they were tough as leather. But if we go straight to the source, to the sheepherders themselves, their stories are not always in agreement. Let us read the trees:

"I am a bastard, I am forty-nine, I cannot believe I came here." This was an individual who knew he had made a mistake, but he had company in singing the blues. In 1928 G. E. carved: "I am a poor sheepherder . . . I am a shithead and more . . . only a queer would be a sheepherder." G. E. obviously had both low self-esteem and a gloomy view of his occupation. He blamed himself for his predicament and castigated himself mercilessly for being a "bastard, for having bad milk" (blood), and so on. Almost forty years later this refrain had hardly changed: "Whoever stays sheepherding for long, soon he goes crazy, this is my opinion, fuckin' life, A. Alonso, 1965."

A lot of others agreed with him: "After coming here one must endure hell," an aspen announces. J. U. conveyed the same idea with an interesting metaphor: "If life is supposed to be what these fuckin' old-timers say it is, my balls are carnations," he carved. These are the words of a disappointed man. Pete Sarratea agreed: "Nobody told us that life was so tough here. I wanted to turn around and go back home."[146]

One individual in Plumas National Forest expressed a dim view of his fellow herders and the whole operation: "I want to know how many donkeys worked for this company, some with four feet, others with two. Adios." He did not say he was including himself in the list, but the carving illustrates the low opinion many herders had of their lifestyle.

Some carvers, borrowing analogies from the sermons they heard in church, saw their occupation as "a cross." Some carved crosses and prayed for God's help: "Lord, have mercy on me." Many inscriptions simply say "Gaiso arzaina" or "Pobre borreguero" (Poor sheepherder in Basque and Spanish, respectively). Others are more specific: "This is a very heavy cross." "This is one hell of a life." "Advice to whoever reads this . . . there is no more pitiful life in this world. It is the plain truth."[147] One carved: "It is better to be dead than to live this way." He had company, too: "Anyone who lives this way is stupid and deserves to have his head cut off."

One blamed his problems on the country: "Death to America." And in lonely Cherry Creek, eastern Nevada, someone in 1957 carved: "What a tough life!"[148] but immediately added: "Hurrah," to lift his own morale. Some saw light at the end of the tunnel. "My calvary is ending,"[149] one carved. He had made it.

Face-to-face, Basque sheepherders are neither as vocal nor as forthright about their years in the mountains. They readily acknowledge being lonely and crying at night, but few admit to harboring the extreme thoughts quoted above. It seems that many messages were carved during moments of particular depression or when the herders needed to let off steam. Since no humans were present to talk to, the carving became an outlet, a therapeutic exercise that helped them through the transition.

Basques were not the only ones to suffer on the range. Other sheepherders probably fared just as badly, if not worse, according to the following poem found inscribed on an aspen in Idaho by R. K. "Bill" Siddoway, himself a sheepherder:

> I have summered in the tropics,
> Had the yellow fever chill.
> I've wintered in the Arctic
> Know every ache and ill.
> Been shanghaied on a whaler
> And stranded in the deep,
> But I didn't know what misery was
> Till I started herding sheep.[150]

Although it seems a contradiction in terms, the solitary herders lived in a state of flux. During the summer months they had plenty of leisure time to think about the world and life in general. Many did a lot of soul searching and were profoundly affected by their years of solitude. Old taboos went "out the door." Religious beliefs also suffered. American culture trickled slowly, yet relentlessly, into the sheep camp. In the ensuing culture shock, many ideas were eagerly accepted or forcefully rejected.

Not only did the herders live in a transitional state, they also lived in a "liminal" one, as defined by Victor Turner, except that the herders were not "outsiders" but "marginals."[151] At this juncture the *baso mutilak*, or forest boys, of the Basque Country come to mind. Joseba Zulaika spoke of the special comradeship enveloping these men. Together they would spend weeks at a time in the forest making charcoal or felling trees. They enjoyed living away from normal society and its laws, playing cards together or improvising verses. Later the *baso mutilak* remembered these times as an exceptional period of their lives.[152] Not that the liminal quality of their lifestyle could be compared with that of the sheepherders. The *baso mutilak*, after all, were still in their own country and only an hour or two away from home, while the herders had traveled across an ocean and were surrounded by an alien culture. If that wasn't enough, miles of wilderness isolated them from any human contact.

Reincorporation into Society

A number of carvings address the paramount topic of leaving the sheep and returning to society. It was a crucial turning point in the life of the herder, a moment he had dreamt of every day for years. He had put in his time, paid his dues, made money, and saved some. It was time to rejoice, to tell the world about it and celebrate, wasn't it?

In fact, the great majority of herders were tight-lipped about their plans, other than to say good-bye. One exception was P. S., who was looking forward to a better life in town. On his last summer in the high country he carved: "Adios sierras, viva la pepa" (Good-bye mountains, hurrah, let us party). They had reason to be uneasy about the future. Many were not sure how things would work out back home. People change. His sweetheart might still be waiting, but she was no longer the young girl he remembered; nor was he the same young man who left eight or ten years earlier. The security that money provided came at a high price, for he had sacrificed in the mountains some of the best years of his life: "Youth, divine treasure that goes away and does not return," one tree in the Ruby Mountains of Elko reads. In another range, too, a herder who missed that old "kick" recited the same litany: "My youth is gone."

The herders who were not so young anymore were particularly haunted by the realization that they had spent their prime years in terrible isolation. Men who were young in years felt old in spirit. Take Fermin Ur-

57. A real philosopher, this one: "Juventud divino tesoro que te vas y no vuelves" (Youth, [you are the] divine treasure that goes away and does not return).

rezti, "Bazkito." He arrived in 1916 at age fifteen, and in 1922, when he was just twenty-three years old, he carved: "Good-bye youth." An unidentified herder in the Ruby Mountains was feeling the same pain: "How sad it is to grow old without having enjoyed life," he carved.

In the summer of 1954 A. Etchemendy was seriously considering retiring. He was no longer a young man, and he had seen plenty of wilderness, walked thousands of miles, and worn out many boots. He carved in Basque: "I have been sheepherder with donkey many years, I should quit." Some herders preferred burros to horses, but that meant they were on foot and had to walk most of the time. That took a toll. Their diet, too, heavy on meat and fat, was not the best, but the job was physically demanding and most herders stayed fit. However, smoking, substandard hygiene, lack of health care, and often excessive drinking wrecked the health of some herders. For them retirement was not an option, it was a necessity.

In the mountains the herder dreamt about the big day when he would reenter his hometown. He would start a business, buy a farm, or fix up the old one. In any case, he would not have to work so hard in the future. Most herders entertained such thoughts. Did reality match them?

A number of the herders in the 1940s and 1950s returned to the Basque Country to find that life was not at all what they had imagined it would be. Their old buddies were no longer around. Things were not the same as when they had left. They blamed society or others, but

58. Not many sheepherders are unable to remember every chronological detail of their careers in the mountains. In this carving, Fermin, from Nafarroa, as an afterthought, added "Vasco" (Basque).

actually the herders themselves had changed the most, although they refused to recognize it. Many ate and drank with little moderation. When someone chastised them for it, they would answer that for years they had denied their bodies the comforts enjoyed by even the cows in the stable. The propensity of Bizkaian repatriates to drink excessively was notorious. I used to hear that the bars and restaurants in the town of Gernika would go under if it were not for *Amerikanuak* (ex-sheepherders from the United States).

In another town of Bizkaia there was a famous ex-herder, a bachelor like many others, who rented an apartment in town. He and his fellow herders scandalized the citizenry by refusing to go to church on Sundays. Often in the mornings he would get up and head straight for the bar, where he would say to the owner: "I need some control this morning. Pour me a shot of *kontrola*" (referring to the French liqueur Cointreau).[153] In Bizkaia they have a saying about sheepherders who return from America with more money than humanity: "Asto jun ta mando etorri" (They left as donkeys and returned as mules), a clear indication that in Basque society donkeys enjoy a higher standing than mules. Some ex-sheepherders could not fit in and ended up returning to America to spend the rest of their years in some Basque hotel.

Basques who decided to stay in America took two different paths: some married, some didn't. Most of the former had steady jobs and raised families; many in the latter group lived in Basque boardinghouses until they died or relatives put them in a rest home. Basques who married usually fared well, but the divorce rate among those who took American wives was very high. Many of the bachelors in boardinghouses suffered from alcohol dependency. Most writers who have written about Basques avoid the more sordid aspects of Basque history in the West. The story of alcoholism, gambling, whoremongering, and the generally wasted lives many ex-sheepherders led in the hotels remains to be told. It was fortunate that hotel owners, especially the women, often went out of their way to nurse the old herders in their waning years.[154]

In this respect Jean Lekumberry of Gardnerville recounted an insightful story that occurred in a Basque hotel some decades ago. An old sheepherder was putting quarters in the jukebox and then pretending to play it like a piano. Just then a young sheepherder came through the door, and the hotel owner told him: "Young man, do not stay in the mountains too long or you will go crazy like this guy here." The old sheepherder turned around and said: "No, no, you've got that all wrong. I am perfectly all right in the mountains. It is here in town where you guys make me crazy with all the whiskey."[155]

The aspen carvings record the drinking, the whoring, and afterward the futile regrets over a wasted life. The fellow who carved the following in Plumas National Forest must have been one of them: "Good-bye my love, good-bye forever, now we will live here all of us."[156] The lyrics are reminiscent of a Basque folk song.

The Farewell Ritual

Most herders quit because the time was right for them, but quitting was not always the simple affair it was for Martin Larralde in 1925, who just carved "adios" and left. The herder had contributed his blood, sweat, and tears, and sentimental people like the Basques needed a more elaborate ritual than a simple good-bye. There are many carved pieces of what I call the "farewell ritual."

The following "adio" (Basque for *adios*) was carved by a herder who went crazy. In it he said good-bye to the range, to his boss, and to the whole country: "1949 Cherry Krik, B Paris, Adio Amerika" (1949 Cherry Creek, B[eltran] Paris [the sheep boss], good-bye America). On 31 September 1931, Michel Franchichena said *adio* to Nevada, hello to Colorado, and "Vive la France," all at the same time and on the same tree.[157]

The immediate final decision to leave the sheep often corresponded to a specific reason. A frequent one was a quarrel or disagreement with the boss or camptender. In 1970 a fellow named Juan apparently had a bone to pick as he carved his adios: "Good-bye [and] fuckin' shit for the Americans.[158] Commonly, however, the herder's accumulated savings met his expectations, and he knew it was time to return home or resume life in town. Each herder set his own savings goal, and once it was reached, his motivation to stay with the sheep waned rapidly.

When retirement was thus planned, the herder arriving in the high country for the last season did not forget to carve a good-bye somewhere. As early as 1902 in the Lake Tahoe Basin someone carved "Banoa" (I am leaving) in Basque. In 1936 Jean Azconaga carved one of the most inscrutable good-byes of all. After examining it for several minutes I decided that the correct reading of the inscription was "Bazfini hauduc miseria." I knew that he was probably talking about leaving, *fini*(sh), and the reason was clear: "hauduc miseria" (what a miserable life);

but what was "baz"? It finally dawned on me: Azconaga, who for a while worked for Laxalt, was talking to his boss, which to a Basque ear sounded something like "baz." He linked "baz" and "fini" as one word, which made it more difficult to understand. As it turned out, this was not really good-bye, for Azcona was still herding in 1943. Just a bad day, perhaps.

Saying good-bye to fellow herders and the animals was not enough; one had to bid farewell to the mountains, the canyons, and the sheep camp, too. The rugged Sierra that some had come to hate now became a subject of heartfelt good-byes. In fact, sometimes the mountains were the first to know. The herder who carved: "Good-bye Alpain for this year or (perhaps) forever"[159] was telling the mountain secrets that perhaps he had not confided to his own camptender or boss.

59. An example of the good-bye ritual: "MP June 28 1956 last year en Pinut goodby." Mateo Pedroarena had learned some English by the time he decided to quit his job. All the references to Pine Nut that I found were misspelled.

Ojer had no idea what he would do next, as he confided to the mountain: "Good-bye Bolcanon [Ball Canyon], I am going away but I do not know where." Another herder turned sentimental and addressed the creek like a person in his farewell. "Good-bye Juniper Creek, if God gives me health I will return to see you." In other words, he told the creek, "I will miss you." In 1933 Miguel Suasia of Bizkaia was even more emotional: "Good-bye beloved [forest] reserve."[160] Earlier in the same vicinity someone had left a much different message: "I shit on this place, good-bye forever, 1910."

Most herders simply carved "adios," of which there are a good number in just about every grove. The other classic message announcing the momentous decision consists of two short words: "No mas" (No more), normally without the appropriate accent, and it is repeated in many groves, but not as often as "adios"/"adio." Sometimes the herder carved his good-bye in English or its Basque equivalent, "agur."

The Pine Nut Range has more good-byes than other groves. Here, half a dozen herders took this ritual very seriously indeed, but I think it has to do more with the power of mimicking than anything else. As early as 1916 the herders in Pine Nut started carving "Adios Pinot," and surely they were imitating older herders whose good-byes have disappeared.

In a small area of Pine Nut I recorded the following eight messages, some carved side by side:

"Adios Nevada, Jose 8-5-53"

"M.P. 28 june 1956 last year en Pinut goodby"

"Miguel Recarte, 1959 Good-bye Pinut, so long"[161]

"Jesus Pedroarena, Navarra España Garralda last year with sheep good by adios, 1961, 60, 63"[162]

"1968, Adios Nevada, Alberto Alonso"

"Adios Nevada no more, Toribio 1969"

"Adios Pinot, no more"

"Jose Urquiaga, 1973 July adios Nevada"

These farewells were intended to be read by other herders and were carved in one spot for that special impact.

The Human Animal

Henri Breuil and Raymond Lantier claim that "the barrier separating man from animal was vague" in the Paleolithic period. They base their argument on the many

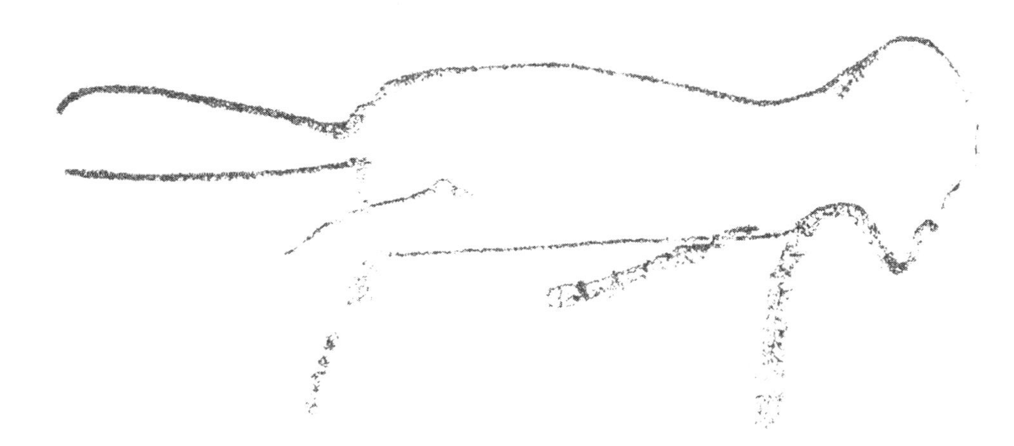

60. A bartender in Gardnerville, Nevada, taking a look at an old sheepherder in the bar, reportedly stated: "This is what happens when you stay too long in the mountains." I think that is exactly what D. Borel had in mind when he carved this four-legged, bushy-tailed animal with human face.

man-animal associations in the cave paintings of that time and on the figures of men with tails, women with horns, and so on.[163] This observation is relevant to an understanding of the sheepherder's mentality as well because he, perhaps even more so than the Magdalenian people, lived literally surrounded by animals who were his main concern and his only companions. Furthermore, Basque oral literature offers a credible basis for an animistic interpretation of the culture. In oral legends, animals and humans live side by side, interact normally, speak the same language, argue, and discuss important matters.[164] The total immersion of the herders in the animal world is revealed, for example, by their matter-of-fact description of zoophilic acts (although the topic is handled with so many jokes that one never knows if one's leg is being pulled).

In the extreme circumstances and the rarified environment of the mountains, the notions of human versus animal sometimes became blurred. Animals not only made up the material *communitas* of the sheepherder, they also played an important part in his work and survival. The herder knew he was lost without the dogs, and he depended equally on the donkey, horse, or mule to carry his pack. Thus we can understand the many hurrahs to donkeys and dogs. In 1945 Frank Erro carved: "Hurrah for me and the donkey of Petotegi." Somebody named Martin in Tahoe National Forest carved: "Los perros son amigos sinseros" (Dogs are sincere friends). Most pet dogs and cats would be lost without their masters—I am not sure about donkeys and horses—but on the range, it was the reverse.

A beautiful and detailed carving in Toiyabe National Forest shows a donkey with pack being trailed by the herder. The message in Basque explains the scene: "Here is Martin Larzabal about to pack the burro, 1929."[165] If Larzabal's scene was not expressing the connection and dependency between man and donkey clearly enough, another who came later filled that void. In the same language, below the word *donkey*, the later carver added: "Two of them." It is typical sheepherder humor that, consciously or not, blurs the boundaries between man and beast. The herder and the animals not only shared the range, they also enjoyed or suffered the same weather and drank from the same creek.

Do not try to convince the herder that dogs cannot think. Every herder had the smartest dog in the world. One dog, according to its owner, knew everything "except how to read and write." An intelligent, well-trained dog anticipates the thoughts of its master and goes to work even before the signal is given. In moments of action the eyes of the sheepdog are constantly locked on the herder's face and hands, his ears on the herder's whistle and commands. The concentration is total, but the dog can switch immediately if something else diverts his attention. They might build a human robot someday, but never a sheepdog robot. Mistreatment of dogs was not condoned, and sometimes it led to fights, not among the dogs but rather the herders.

What impressed the herder most was the dogs' fearlessness. When suspicious noises outside the tent aroused the sheepherder in the middle of the darkest night—a hungry bear, perhaps, or worse—the herder inside might be terrified, but not the dogs. Without blinking an eye they would exit, barking, to scout and confront the danger. We know how Jacque Battita felt about his dog when he carved "Viva le perro" with the French article. Someone who apparently did not think much of that carved below it: "Stupid bastard."

Luis Jayo of Goierri, Aulestia, Bizkaia was a famous dog trainer whose name was passed along the sheepherder grapevine. When it was time to start the fire, he would tell the dog: "Looks like we need some kindling," and the dog would take off and be back shortly with small, thin branches. Then, after the meal, he would say: "I am going to take a siesta so don't let anyone come into the tent." And sure enough, the dog would sit there motionless and guard the entrance until Jayo woke up.[166]

All would agree that the sheep was the most important animal in the herder's life, but one would not know it from the few figures in the carvings. From the herder's point of view, sheep and women were at opposite extremes: Sheep were everywhere and women nowhere; one could only think of them, imagine them, and carve them.

The Basques carved many weird figures on aspens that are difficult to interpret. I believe some are simply badly drawn, but others are deliberately distorted. One carving, probably dated in 1925, found in the Tahoe National Forest shows a horizontal figure with a human male's face and a hybrid body—half human, half (probably) coyote—with a bushy tail which looks like that of a fox. The figure is walking on four legs. So this is what happens when you stay too long with the sheep, I thought. As soon as I saw it, the song titled "Giza aberea" (The human animal), by Xabier Lete, came to mind.[167] Another man on the Cherry Creek Range carved an erect man holding a walking cane. The figure is quite gentleman-like except for the prominent tail curving upward from his derriere. Furthermore, the legs and feet appear to be those of a goat or some other animal. I found one other arborglyph of a human with a tail.[168]

The herder became used to talking to the animals. The dogs were never far away, even when resting, and he held regular conversations with them. According to the carved evidence, the herders talked to the donkeys more than the dogs. Perhaps the donkeys were better listeners. However, they would not listen to just anything. J. P. of Minden, Nevada, told me that his donkey, Felipe, not only was smart but also had a refined musical taste. Classical music was just fine, but whenever the radio played rock'n'roll Felipe would flip his ears and start braying.[169]

Donkeys are the subjects of many tales among herders, some tall, some not. Old donkeys were especially valuable because they could lead the new herder down the trail and keep him out of trouble. The sheep boss would entrust the veteran donkey to the newly arrived herder and say: "Just follow him." One ex-herder, recalling his first impression with American donkeys, remembered thinking to himself that they must be smarter than Basque donkeys after being told to follow one.[170]

The conversations between herders and their donkeys recorded in the arborglyphs commonly refer to sheepherding activities. Often they deal with moving or packing. The donkey is always represented with a pack and being led by the sheepherder. In the Mount Rose area near Reno there are many carved figures of donkeys with accompanying messages like "Arre astua" (Giddyap, donkey). On one tree a beautifully executed carving from 1940 by Dominique Harlouchet shows a typical herder with staff and hat pulling the donkey's lead rope. The caption reads: "Hello Jack."[171]

Two other herders addressed their donkeys in the same area. On 11 September 1956 Jose Almirante from Nafarroa carved: "Donkey, let us go to the desert to finish the month of September and then on to California." The other, carved nine years later, shows a herder pulling the lead rope of a loaded animal. The caption below it says: "Donkey, let us go to Reno to ship the little lambs, P. S., 9 July 1965." In a carving in Plumas National Forest a herder is riding his donkey, something I did not know many did in America. The caption says: "Giddyap, donkey, raise your foot when urinating."

One can make something of these conversations or nothing at all; it is up to the reader. I must add that *bertsolariak* often sing themes involving animals. For example, in the 1990 Kantari Eguna in Gardnerville, Nevada, two *bertsolariak* were given the following subject: one was to be a hardworking mule whose owner—played by the other *bertsolari*—wasted all his earnings at the tavern and did not have enough money to feed the mule properly.[172]

chapter four

Life Without Women

Ez dut bertze penik nesca batena.
(I have no other ache [but] that of [lacking] a girl.)

I have plenty of pasture for the sheep, I also have provisions, I lack a woman.

Here, a person needs a woman, that is the only thing that troubles the herders.

When the larger public "discovered" the tree carvings, they also discovered that the sheepherders had a sex life after all. Articles and books written about Basque herders emphasize the loneliness of their lives but say amazingly little about how the sheepherders coped with being deprived of contact with the female sex. Why such a void? The herders talk constantly about women, so those who interviewed them and wrote about them must have selectively excluded sexual matters from the discussion. Or, more likely, the sheepherders did not volunteer any information.

Today the situation is almost the opposite, to the point that arborglyphs are fundamentally equated with nude women and pornography. Certainly the tree carvings offer a glimpse of Basque sheepherders' views on women and sex and their sexual fantasies. By doing so, they shed much-needed light on the veiled topic of Basque sexuality.

Anyone wanting to know what the world would be like without women need only become a sheepherder. Herders lived that reality, and carving after carving loudly proclaims that it was not easy. But before we go any further, it is necessary to realize that, as far as the literature is concerned, Basque sexuality is an unknown quantity; it is the least-documented aspect of traditional—and nontraditional—Basque society. According to Rafael Castellano, sexuality is taboo for Basques. J. M. Satrustregui attributed the lack of information regarding the subject to Basque shyness.[1] It seems that yet another Basque mystery has been solved in the United States.

What the two or three extant works on Basque sexuality say is in stark contradiction to what the trees in the mountains of Nevada and California proclaim. The carvings make a statement about what Basques might say and do if social conventions and constraints were removed. It is important to remember, however, that the sheepherders' lifestyle was not a normal one.

When I came upon the three carvings cited at the beginning of this chapter I took them to be real SOS messages. But out there, in such a faraway location, who would ever read them? What good would they do? Did the herder think that his boss would happen into the grove, find the tree, read it, feel sorry for him, and bring him a woman? Then I realized—again—that the messages were not carved for public consumption. Outsiders were not supposed to see this raw erotic material, which for a century the mountain jealously guarded.

Although it was part of my research, I did not feel comfortable exposing the unabashed intimate cravings of the herders. Was I betraying them, meddling in their private lives? In the end I decided to bring the information back to town with me but to leave the carvers' names on the trees in the forest. My fervent hope is that the herders understand that knowledge is important and that it is good for people to know about the nature of their solitary lives, their yearnings, and their reactions to the lack of women.

The herders carved erotic material for personal use and to allay their mutual pain and their emotional and sexual hunger. All indications are that they enjoyed the exercise. The thousands of carvings on the subject of women—or, more exactly, their absence—strongly suggest the importance of the issue. After their own names and herding news—including loneliness—this was the sheepherders' favorite topic, one intricately linked to

61. Standard sheepherder musing: "Aki se necesita mujer eso solo preocupa a los pastores" (Here, a person needs a woman, that is the only thing that troubles the sheepherders).

isolation. And the herders did not carve women alone; to the contrary, men appear in many carvings as well, in a symbolism seeking to restore the balance of nature, according to which the sexes couple and merge.

I often admired—and was surprised by—the frankness of the herders. What had happened to Satustregui's shy Basque Country farmers? I was reminded of Rodney Gallop's incisive remark about the Basques: "It never occurs to them to mince their words even before their womenfolk, who nevertheless see therein no cause for blushing."[2] Basques do not seem to know what euphemism is, or at least they seldom use it.

Women in Sheep Camps

The longing for women is so strongly expressed in the aspen carvings that one wonders why women were excluded from the range in the first place. Wouldn't the herder do a better job if he had a woman with him? The answer according to the carvings would be a categorical yes. In fact, in Texas and the Southwest man and wife were sometimes hired together for sheepherding. He looked after the sheep and she drove the wagons and cooked.[3] I know of no rule whereby sheep bosses specifically forbade the herders to have women with them, but there was no need for one. For eight thousand years sheepherding has been overwhelmingly the province of lone men. In Europe, some Basque women take care of the sheep, but they do not stay alone in the high-country *txabola* (cabin); they come home for the night because the distances in the Basque Country are minimal compared with those in Nevada.

Although it is not common knowledge, there *were* women in Nevada sheep camps—not many, but some. They were the wives of small sheep operators who herded their own flocks. One such woman was Theresa Laberry of Eureka, Nevada. In the summers of 1928 and 1929 Catherine Coscarat Goyenetch of Bordaluzekoa farmstead in Urepel, Benafarroa, stayed in the sheep camp with her husband, Pete Etcheverry.[4] And these two women were by no means the only ones.

The whole Gallues family of Reno spent most of the summer months in Kyburz Flat at the Wheeler Sheep Camp. Martin was the manager for the Wheeler Sheep Company and later for Reginald Meaker, and his wife and children lived with him in a three-room cabin. I do not think this setup was very unusual, although family members of other sheep owners or managers might have spent less time at the camp. Of course, these women were staying at the camp, which was a halfway station between wilderness and civilization. The real sheepherder on the range was further removed and still without female companionship.

One woman who still has strong memories of sheep camps is Ana Hachquet of Elko, Nevada. She was in her mid-nineties when I interviewed her in 1992, and given her robust health, it appears that camp life did not harm her. When Ana married, she joined her husband in the mountains. They lived in a tent. "Life was hard then," she told me. Unlike the women mentioned above, who stayed with their husbands only through the summer,

Ana stayed summer and winter. She did most of the chores that camptenders do: cooking, baking, washing clothes, and so on. She dug trenches for the Dutch ovens. In the winter she broke the ice with an ax to get water. With typical Basque modesty she said, "I did the work of a man." Only when she was ready to give birth did she leave her husband to come to town.[5]

Such examples, however, must have been rare, for the tree carvings say nothing about them.

A Primitive View of Woman

It is utterly strange—amazing, in fact—to walk through an aspen grove and come face-to-face with figures of women—some fairly normal but proportionally exaggerated, some deformed or weird. Among the latter, some have no arms or the legs are short or unfinished; often the feet are missing. I have even seen headless women. If we want to find a similar art form we have to look to the Paleolithic female figurines that have been discovered in several parts of Europe and the Middle East. In northern Iraq, for example, the prehistoric Arpachiyah clay statuettes of women are headless and emphasize the middle part of the female body. Some are in a crouched position.[6] I recorded aspen carvings that appear to be copies of these ancient terracotta figures. But what was the herder's intent in carving a headless woman? I also found a pair of female legs with garters and high heels, and nothing else. It is much more difficult, however, to explain the headless figures. I suspect that some Arpachiyah statuettes are headless because of damage, but that cannot be the case for the tree art. Perhaps the head was too complex to carve; or perhaps it was simply a part of the woman's anatomy that held no interest for the herder. Or it may have been the case that the carver did not like his effort and left it unfinished.

Perhaps stranger are the several carvings of women with well-defined breasts *and* phalli. They are all nude, all standing and urinating male style, and all have arms—in this case holding the phallus. I do not remember seeing any other artwork quite like them. The arborglyphs consistently show big, plump women with large breasts, hips, and buttocks, like the famous Venus of Willendorf.[7] Some are squatting, and rarely urinating or defecating; but more often they are spread-legged.

The armless and footless figures on trees are relatively numerous, thus closely paralleling the prehistoric terracotta Venuses.[8] One of the constants of tree art is its simplicity and primitive look. Many of the figures on aspens in Nevada and California could be mistaken for Neolithic, if not Paleolithic, art. Marija Gimbutas's wonderful book on primitive European art shows women without heads and arms or with small appendages instead of arms, and women with exaggerated breasts, buttocks, and abdomens.[9] I have seen dozens of similar images carved on the aspens of the Sierra Nevada.

There are discernible parallels between aspen art and the cave art of the Pyrenees, Africa, and Australia as well. Scholars tend to contemplate prehistoric cave paintings and art with awe, accrediting the slightest mark or paint stroke with symbolism or magic rituals. Apparently no one has ever considered that some prehistoric art is simply doodlings or graffiti, although at least some of it must be.

The lifestyles of the artists offer some similarities as well. Herders and cavemen had a lot of idle time—the herder during the summer and the caveman during the winter. At least some cavemen used this time to carve and paint on rock walls. While most experts agree that prehistoric artists were obsessed with fertility (some say sex),[10] the Basques were simply lonely and horny.

The herders attempted to represent sex and females as forcefully as possible. One carving of sexual intercourse is so simplified that I call it "abstract intercourse." On one side is the trunk of the woman, with only the buttocks and the breast line to indicate her identity; she has no head, arms, or feet. Facing her, about an inch away, is a straight vertical line with a horizontal penis connecting the two figures at the midpoint. There is nothing else.

The Female: Virgin or Prostitute?

Basque society in Europe is based on a strong, unshakable family life centered in the farmstead kitchen and hearth. The mother is the center of this warm human environment. The church sanctified the mother's role in the figure of the Virgin Mary, the Mother in Heaven, and made her even better than ordinary mothers: Mary was mother and virgin at the same time. Earthly women can be only one of these things. In the eyes of the Church, virginity is the higher calling.

The carvings reflect a clear dichotomy with regard to how herders perceived women. The mother figure is present in a small number of arborglyphs, but the women featured in hundreds of other carvings belong to one of two major groups: Virgins and Prostitutes.

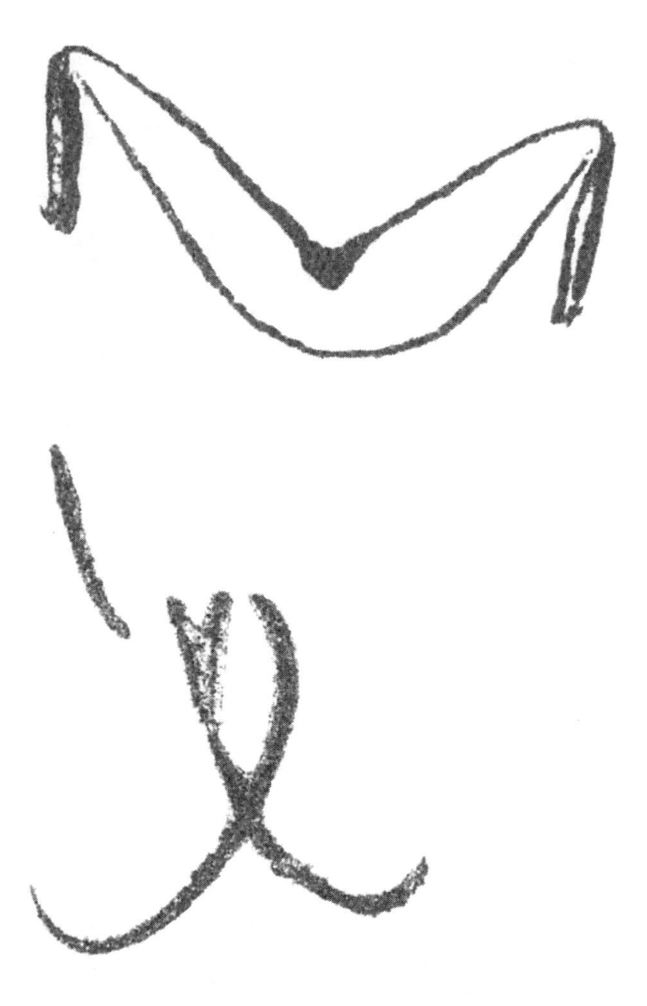

62 and 63. Two samples of schematic yet imaginative erotic art: *above,* female legs and nothing more; *below,* even less, representing the same part of the female anatomy.

The Virgin is personified in the fiancée left behind in the Old Country. In the isolation of their mountain retreats the herders idealized their girlfriends and made them into a prototype of all the good qualities that reside in the female before she engages in sexual activity. The herders, indoctrinated by the Church, sublimated the virginity of their fiancées in their carvings. At times the reader of the carvings is left with the impression that some of the Basques came to America to prove their worth to their fiancées, as the Quixotes of the twentieth century.

The expression "hurrah for my fiancée" is exactly repeated in every grove. It is impossible to doubt the sincerity and enthusiasm of these messages. Some are more elaborate and detailed: "Hurra for my fiancée M. R. A. from Errazkin, Larraun Valley, Nafarroa." A letter from one's fiancée was like cool rain in the hot summer, but some became all steamed up after reading it: "Today I received a letter from my fiancée and my balls are going crazy." Herders with fiancées were better equipped emotionally to deal with the isolation. As John Bowlby says, people with "strong bonds to a few other special and particular individuals . . . so long as the bonds remain intact . . . feel secure."[11]

The young men without a special girlfriend back home hailed all the girls in their hometowns: "Hurrah for the girls of Hendaya"; or: "Hurrah for the girls in France." And: "All the girls in Garazi are virgins." Virginity was a sign of youth and wholesomeness, one of the highest qualities a woman could possess.

There were also herders whose relationship with their fiancées had deteriorated because of jealousy or some other reason. In this case the Old Country girlfriend was no longer a symbol of goodness but a source of pain.

The Prostitute was the embodiment of an utterly "sexist" (not discriminatory) view of women, with heavy emphasis on the mechanics of the act of procreation. The Prostitute may not be as noble or as worthy as the Virgin, but sheepherders carved more "hurrahs" to them than to their fiancées back home. Conversely, I recorded at least one carving in which a prostitute from Elko thanked the sheepherders. And on another a woman (a whore?) shows her affection for a herder and defends him against unspecified charges.

Tree after tree shows sheepherders facing prostitutes—holding hands, simply saying hello, engaged in sex, or with the man in the act of displaying to her his prominent manhood. Woman had a double personality, a double face, in the herder's fantasy. The Old Country virgin was more often fully clothed; thus she was symbolic and nourishing to the spirit. The American prostitute was for the body—realistic and necessary for the flesh and the libido.

The Sweetheart in the Old Country

Many herders left girlfriends, fiancées, or even wives back home. A number of glyphs mention the names of girlfriends, but only a handful refer to wives by name. According to some oral sources, more married Basques came in the early decades of the twentieth century than later. In most cases the women left behind needed a great deal of patience because "a few years to make money" often turned into a few more years. While many women believed, and mourned, that their boyfriends and husbands had forgotten them, their men were carving hurrahs and tender inscriptions in the mountains of Nevada as proof of their eternal love for the women back home:

> "I ache for my sweetheart."
>
> "You are the ship and I am the sailor, you are the (stone cutter?) and I am the diamond."
>
> "My soul to adore you, my heart to love you, my life to live close to you."
>
> "My plump little girlfriend is called Margarita. She is cute but effeminate."
>
> "Hurrah for Miss Dominica, my fiancée, J. M., October 11, 1923."

One aspiring poet carved in the 1920s: "When I saw you coming, I told my heart: what a cute little stone for me to stumble on."[12]

Some herders with sweethearts waiting in the Basque Country forced themselves to formalize their commitment by carving reassuring statements once in a while: "My pretty loved one Lucita, I will pay you a visit soon, G. I."[13] Some rationalized their sacrifices, the daily deprivations: "I am doing this for you, long-necked one."[14]

One herder toyed with the idea of marriage. On one tree he carved that he was going to quit in two years. He had saved some money, but he apparently had doubts about his girlfriend—or himself—as seen in the following statement, which mixes several Basque dialects and borrowed English words: "Eusebia is his girlfriend, but looking at my dirt [literally, cow chips] they will say that I must be about 95 years old."[15] This herder's view of himself was confirmed to me one day by a woman who knew him. The point here is that some herders who stayed too long in the mountains often neglected personal hygiene, thus driving a wedge between themselves and the women who otherwise might be interested in them.

One herder from Benafarroa fantasized about being reunited with his lover, and his semiarchaic language leaves little room for the imagination. On another tree he noted that she was hairy. One, and perhaps two, Spanish herders from Cantabria carved sexually explicit figures and messages with family affairs mixed in: "The important thing is the chestnut that has hair." On another tree the story unfolds: "The important thing is the fiancée's chestnut, and let her parents die."[16] These herders fantasized about sex with a cousin or even with a mother-in-law.

Lost Girlfriends and Wives

We learn from the trees that sometimes the long-suffering sweethearts in the Old Country grew tired of waiting for their men. One herder's laconic confession begins sadly: "I used to have a beautiful girlfriend in Baztan" (Nafarroa). On several occasions the men blamed themselves for the breakup; for example: "I had a girlfriend but I have lost contact with her because I am a bastard, signed with rubric." But this herder was ambivalent about making such information public, so he added "the first who reads it commits a sin, P. G. 1935."[17]

64. "Mi linda pocholita se llama Margarita es linda pero es marikita." (My plump little girfriend is called Margarita. She is cute but effeminate.) The last word is normally used in reference to men as homosexual, but in this instance the herder is talking about a woman. Carved by E. I.

The percentage of lost fiancées seems to vary according to the decade. Herders who came in the 1940s and earlier often went back, and many returned to America with their wives. But in the 1950s and especially the 1960s and later, the number of Basques who returned to marry their Basque sweethearts declined steadily and drastically.

The news of a rift with the sweetheart usually arrived by mail, written by a family member if not by the woman herself. During the next several days, if the herder still cared about her, he lived in a daze, read the letter dozens of times, and did not care if the coyotes ate half of the lambs.[18] It was much worse when the bad news came through the grapevine—that is, when another herder received the news of the breakup before the involved party himself. The herder felt not only rejected, but humiliated as well. Someone who apparently had gone through that situation carved: "A lover's suffering is of a bad kind." Others eased the pain by singing a well-known Basque folk song that says the same thing.[19]

Breakups took many forms and had several causes, the major ones being that the herder seldom wrote or that the girlfriend found another man. Sometimes the young man had left town because the relationship went sour or because his girl became pregnant and he refused to marry her. I did not find in the carvings any indication of tears shed over such cases.

High-Country Soap Opera

In a desolate area of northwestern Nevada is carved the most detailed story of a love affair that I have found. The young herder from Nafarroa—let us call him Pete—was consumed for weeks by his emotions, and it is doubtful that he gave much attention to the sheep. A carving in White Pine County, Nevada, indicates that there were many herders like Pete: "The sheepherders are always thinking about girls and not about sheep." It may have been carved by a manager or the sheep boss, or perhaps by a herder who thought he was better than the others. Day after day Pete went up and down the canyons carving and dating the developments of a crippling passion for his fiancée. Her name was Mari, and she was his cousin (in the small villages many people were related, and it was not rare to date and marry a cousin). The dozens of messages spanned two years.

In the earliest carvings Pete enjoyed telling about their wonderful love. "Whenever she wanted, I would unload," he carved. Mari loved him because he had a large penis and she a tight vagina.[20] One year after he came to America she became upset and stopped writing to him. The carvings do not provide details about the problem, but he blamed himself: "I deserve it, because I am a jealous bastard," he carved. On several trees Pete carved his despair at the thought of losing Mari. Then, one day, she wrote again, and he regained his optimism; but it appears that her gesture was too little, too late. He finally revealed what was going on in Nafarroa: "My fiancée is in love with a Galician and she has stood me up."[21] After that, Pete carved fewer trees—he probably gave up on her. One aspen simply says: "How sad it is when the woman deceives the man in this manner." Surprisingly, one month later she wrote to him again, and he was so thrilled that his next comment sounded almost as if he had had a telephone conversation with her:[22] "Today Mari was very cheerful." That is the last dated inscription. The following year the Nafarroan herder did not return to that summer range, so I do not know how the love affair ended. Did he go back to her? Did they make up and get married? For the latest breaking news, tune in again next year to the arborglyph channel.

There were losers, like Pete, and then there were braggarts, like Antonio. Breaking all conventions, Antonio bragged about not having any current girlfriends, but read on: "I am twenty-two years old and I have had ten girlfriends and I married none of them because Antonio did not want to." Perhaps he lost count, because on another tree he stated: "Antonio has had eight fiancées, and he married none of them, because Antonio did not wish to."

Sometimes we cannot be sure if "girlfriend" refers to an Old Country girl or an American. A philosophical Bizkaian herder with imagination carved: "My sweetheart left. When will she return? When her belly is full of electricity."

I recorded half a dozen messages reflecting ambivalent feelings about fiancées. In the first part of the statement the girlfriend is usually described as pretty, but immediately afterward the inevitable contrast comes into play. The carver might add that he did not think she was faithful to him. J. P. I. carved: "JPI has a very beautiful fiancée but she is a whore and screws everybody."[23] And from another jealous herder: "My beautiful fiancée Mari, what a great whore she is."[24] In 1922 B. B. was over his jealousy; you might say that the relationship was beyond repair: "My darling Marie is a cunt, I am going to kill her."[25] It does not appear likely that any of these three fellows rushed home to marry.

Many if not most Mexican and South American herders were married, and their carvings reflect their anxiety regarding distant wives. "How sad life is, but the saddest thing is to sleep alone even though one has a wife, Luis." This Peruvian herder, named Luis Condor, elaborated the thought on the next tree: "It is very sad to sleep alone when one has a wife and one is ignorant of the fact that she is sleeping with someone else." This is a typical theme in the lyrics of *ranchera* music. An old Mexican who claimed broad experience in affairs of the heart and played the role of guru for the younger herders carved: "My friends, I am going to ask all of you who have wives for a favor. Don't leave them alone for long, because the devil never sleeps and they [the wives] are not stupid and make themselves available."[26] I recorded three more carvings by lamenting married herders.

Of a somewhat different type is the carving (dated in the 1920s or 1930s) on which two striking figures can be seen in a dancing stride, a man leading a woman by her hand. The caption says: "After one gets married, that is when horns come out." I do not quite understand the meaning.[27]

In a remote canyon of the Ruby Mountains there are about a dozen comments carved by a self-righteous herder criticizing an adulterous affair between a camptender and a married woman. He began by quoting the church commandment "Thou shalt not covet thy neighbor's wife," and went on to accuse the man of being "a destroyer of families," among other epithets. The unidentified critic appears to have been somewhat jealous, as indicated in the following carving: "How happy he must have been when he had her under his body." When the critic called the Don Juan "fucker," some other carver defended the adulterer by adding the aside: "He was a macho man."

65. "Que triste es la vida pero lo mas triste es dormir solo teniendo mujer Luis." (How sad life is, but the saddest thing is to sleep alone even though one has a wife, Luis.) On the adjacent tree this Peruvian recorded his full name and the date.

Masturbation

After finishing a sexually explicit and stimulating arborglyph, the herder might proceed to masturbate in front of the tree. One aspen in Humboldt County that has no sexual figures on it reads: "Go figure how many ejaculations this tree has received." In case someone was not up to speed, another carving matter-of-factly acknowledges the downside of sheepherding: the herder has to masturbate a lot. On another mountain range someone carved: "I am hot today, I am going to do it" (masturbate). Not far from this tree the herder left a follow-up story: the figure of a man arching wildly backward and ejaculating. In a remote grove another carved a rhyming statement: "A fucker, this is where I unloaded the donkey and my dick." And F. E. on 8 August 1967, after spelling his name fully, carved: "I jerked off here." Dozens of trees bear similar statements. An Iparralde Basque who insisted on using Spanish carved: "Aqui ice un jaco" (This is where I masturbated. Perhaps he was thinking of "jack off"?).

The herders also discussed their sexual prowess with one another by telling tales of sexual marathons on the range. One day, as I was talking with some ex-herders, one said that his record was eight ejaculations, one after another. The fellow beside him said that one day he did it ten times but developed a headache and had to quit.[28] Nobody raised an eyebrow at these stories.

Zoophilia

The range used to be full of stories and jokes about cowboys, horses, sheepherders, and so on. A number of the jokes told by cowboys allude to zoophilic acts by sheepherders. The Basques take those stories in stride by simply denying the charges or responding that cowboys do the same, but usually they just play along and laugh. In fact, a Basque restaurant in Nevada used to sell a bumper sticker that said: "Nevada: So many sheep, so little time."[29]

Because it was the herder's job to ensure that the rams impregnated the ewes, he necessarily had to witness copulation after copulation during the breeding season. The lives of the animals and their herders were so closely intertwined that herders did not hesitate to lament their own sexual drought and envy the satiated rams. One carving shows a scene in which a herder counsels a hesitant bull in verse: "Mugi fite zezena gozatu nahi baduk buztena" (Come on, bull, move quickly if you want to enjoy your penis).

To a point, the carvings reflect the impact of the everyday breeding of the ewes. For example, men portrayed themselves with phalli large enough to match those of the rams, although when a Basque brags about the size of his penis he refers to it as *asto pito* (donkey penis). In a good number of sex scenes humans are copulating not face-to-face but animal style. In one suggestive scene, a powerful black-headed ram faces a naked kneeling woman, who, however, is drinking and seemingly paying no attention to the animal.

If we take the arborglyphs as evidence, the issue should continue to be relegated to the realm of jokes. I have found only four carvings—all from different areas—that mention zoophilia at all. The fact that two of them are signed seems to indicate that they are not to be taken seriously. Three out of four of these informers claimed they had copulated with a jenny.[30] The fourth did it with a goat.

Sexual Revolution in the Sheep Camps

Uniquely representative of herders' view of sex is this carving in Elko County: "Bulsiet fok is good" (Bullshit, fuck[ing] is good), apparently carved by someone who had been enlightened in Nevada or was trying to convince other herders that there was nothing wrong about having sex. That short statement covers a lot of sheepherder and Basque history, and demonstrates the swift and profound changes that took place in the sheepherders' outlook. That this was wholesale revolution is illustrated by a message some forty miles from the previous one: "Up with the penis and down with the panties."

The sexual and erotic material recorded on aspen trees would seem out of place, if not offensive, to many Basques in Europe, who would not understand that the situation of the herders in the United States was very different from their own. In the twentieth century sexual commentary and activity were severely curtailed in Basque society, especially outside marriage. The Catholic Church sternly discouraged even talking about sex in public. Priests in the Old Country relentlessly preached against sex, labeling it "dirty sin." But one wonders if they ever really persuaded the people.

In earlier centuries the Church also attacked dancing, particularly the slow variety with couples embracing.[31] Modern Basque literature, often authored by clergymen, portrays ideal Basques as chaste farmers, sheepherders, and fishermen. It is sufficient to recall Joanes, the main character in Txomin Agirre's *Garoa*, or to browse through Orixe's *Euskaldunak* to be persuaded of such literary tendencies.[32]

Nevertheless, rural Basques are rather blunt and frank when talking of sexual matters. The output of many *bertsolariak* singing in taverns and popular gatherings exemplifies this tendency. The key is to discern when it is appropriate to talk about such matters and when not. If the priest or a stranger is present, for example, it is a bad time. This behavioral pattern, which Basque peasants craftily employ, may be responsible for the obvious contradictions in the understanding of this subject. In fact, there is a whole dictionary of Basque erotic vocabulary.[33] Ramon Etxezarreta's book lists 175 words to identify female sexual organs, 167 for the male parts, 90 words to express sexual activity, and 351 different ways to say *intercourse*.[34] How, then, could it be said that sex and sexual matters are taboo and never discussed? I think we have to dismiss this taboo.

Furthermore, if, as Satrustegui said, Basque farmers are too shy to talk about sex, we would be hard-put to explain the sudden metamorphosis in the behavior of the sheepherders in the United States. That there was a transformation is certain. No one would deny that in Nevada the sheepherders became more liberated in their ideas, including their ideas about sex.

Not one of the carvings so far found refers to sex as sinful. Moreover, no commentator castigated an excessively explicit figure or statement. It is possible, of course, that some of the religious statements are in reaction to objectionable material; for example: "Jesus came to this world to die on the cross in order to give us the grace to forgive our sins and open the gates of heaven, let us praise the lord." We can follow the carver's train of thought and believe that he may have directed this message to other less religiously inclined herders.

Some Basques were in their teens when they began their solitary lifestyle in an alien country,[35] and knew little about women and sex. We are lucky that a few arborglyphs inform us about these Basques whose eyes were opened in Nevada. One Bizkaian tried to express some of his newfangled ideas in English: "I now foky sgood evry pipol" (I know that fucking is good for every people). South of Reno another herder, also struggling with English, carved a similar message. Numerous unsigned inscriptions state the same idea.

Many carvers borrowed the Anglo-Saxon term, usually written "fok" (or "foky fok," "fok is good," "fokin," etc.), which the herders learned very quickly. Another word they learned was "pussy," which is always misspelled on aspens. On one trunk I read: "One pusse for every man." In a remote Nevada range a Benafarroan Basque carved in English in the mid-1930s: "We want a little buse pour shepee[r]s."

Antonio, the herder with many fianceés alluded to earlier, could not imagine himself living this way, but the unthinkable was sinking into his consciousness: "Antonio without woman." The same idea is expressed on a 1915 carving in the Sierra: "Aluha baterez" (No pussy at all).

Since the herders could not conceive a world without women, they proceeded to reinvent them or find a substitute for them. Thinking about the fiancée, the lost lover, or the prostitute in town, reminiscing about the past, and fantasizing about the future were not just techniques to deal with the absence of females and loneliness; they were a healthy remedy and pastime. Recently I asked an ex-herder if he ever carved naked women, and he said: "When you are twenty years old, what else do you think about?"[36] Some historians believe that Paleolithic artists painted animals on the walls of the caves as a means to control them or gain power over them, to be successful in the hunt, or for religious purposes. It is possible that the intent may have been simpler than that: the next best thing to having bison meat hanging in the cave was to have its figure painted in the back room. It is no different when we hang photographs of our absent relatives on the walls at home.[37]

Some sheepherders found solace with girly magazines, which can still be found abandoned in sheepherder cabins, but in the old days these were not so easily obtained. It was much cheaper and handier to carve figures of women on aspens. Carving gave the herder control over the missing females. He could carve them any way he wished. They could be thin or fat, short or tall, dressed in nice clothes or naked. Some figures that I recorded have feathers and flowers in their hair. The carvers could even try to make them look like their sweethearts back home.

Those who had not seen many naked women had to push their imaginations. This and the fact that they had no training in drawing account for some of the convoluted forms and body parts one finds carved in the mountains of Nevada and California. More likely the herders carved the likeness of some prostitute in town. Such women were identified on aspens by the term *puta* (whore).[38] The women the sheepherder carved obeyed his every whim. Most herders seem to have wanted their women stark naked and in compromising positions. One naked female figure carved by someone prompted the following ironic advice from another: "Cover yourself up, people are coming." Another carving says: "Wow! What a bang if I could catch her under these aspens."

One of the most striking examples of sexual fantasies is found in the southern Ruby Mountains, where a headless woman lies in a birthing position. This carving could be put to good use in an anatomy class, except that the accompanying labels are borrowed from the sheepherder's environment. The large close-up of the female pelvic area is explained in metaphors: The breasts are called "pretty mountains." The two thighs and knees are labeled "high mountain" and "the mountain of goodbye," respectively, while the genital area is described as "the canyon of memories and the tent from where live sweet water flows."

Much of the material seen on trees, such as busty women in high heels and fancy hairdos, is reminiscent of the Nevada whorehouse setting. The female figures I found most often are crudely drawn with open legs and exaggerated curves. Several of them are carrying a tray with glasses or a bottle; sometimes they have fancy hats or strange headgear. They are sometimes larger than men, with disproportionate body parts drawn according to the philosophical principle that when you want more of a good thing you display it large.

Jack Muldoon called some of this art "Mountain Picassos" because of its primitive cubist qualities, but arborglyph art was established even before Picasso created his characteristic style. The arms on the figures, male and female, for example, are almost always too small, like useless appendages. One day a herder eliminated them altogether, and after that many herders carved armless women, following the rationale that arms obstructed the view of the curves. The more refined artists drew the women more harmoniously, but for the majority erotic impact was the chief motive for carving curvy women.

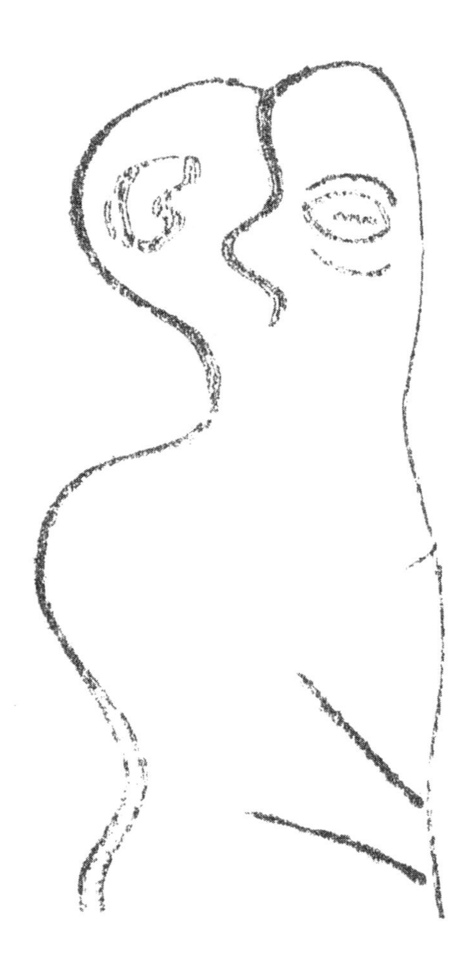

66. If you like an art that requires a quantity of imagination, try this very "Picassoesque" couple embracing.

Sex as a Game

In the Basque mind, doing something for its own sake is too simple; it is more fun if you turn it into a game. Many of the sexually explicit carvings must be understood according to this fundamental principle. The rules of the game stated that the greater the fantasy and its shock value, the better. Often the carvings have less to do with sex, or sexual drive, than with amusement. The herders competed to carve the most beautiful woman or the most striking and explicit sex scene, just as they competed to raise the fattest lambs. One Nafarroan completed a fair nude female figure and advised his fellow herders: "Stick it in as far as it will go. You will see how good it feels." In 1929 M. L. advised: "Chanza delaic silotic ongui sartu" (Whenever you have a chance stick it deep into the hole). In 1906 B. M. drew a female and next to it carved a message for a friend: "Javier, I am leaving you here my fiancée so you can enjoy her whenever you desire."

In this environment imagination was granted free rein, and the isolation provided the artist with complete freedom of expression. At times, in typical Basque fashion, and much like the "magic realism" of Latin American writers, the herder gave no hint that he was joking: "This is where I got her," it says on the side of a female nude figure, and then continues, "ask her if you do not believe me." Sometime later, I do not know how long, another herder came and carved: "I ask her but she does not answer."

Entertainment is what E. I. had in mind when one summer day in 1967, while sitting on a luxuriant grassy bank beside a crystalline Sierra brook, he drew *pocholitas*, as he called them, shapely girls with goatlike feet. All his figures (about five or six have been found so far) look nearly the same: a naked woman with a small bosom and with arms resting on either side of her waist, standing with a wide stance, looking straight out of the tree. In spite of the work it took to complete such a figure, the carver was not thinking of a long-term relationship with her: "She is a good one to spend a night with," one caption says.

A 1901 carving above Lake Tahoe is still clear enough to show a man fingering a woman with legs wide open. On the opposite side of the Silver State, exactly the same scene is carved with the English caption: "Nice looked."

The vocabulary associated with the sexual act itself is sometimes in Basque, even when the rest of the erotic statement is not. The male member is called *buztan* (or *buzten*), and there are quite a few short rhyming verses such as: "Lete, buztana tente" (Lete [has] an erect penis).[39] The old Basque word for sexual intercourse is *jo*, but in the carvings the herders used primarily *ziko*, followed by *sartu*.[40] The female genitals are called *motxa* and *alu* (the spelling varies). In one example, a depressed herder carved: "Ziko ziko alua ge" (Fuck, fuck, without a pussy). And on another aspen I read: "My penis is strong but I lack a pussy, if I had it I would screw it."[41]

As in other societies, part of this male game was to brag about one's virility and penis size. One message claims thirteen inches, but some of the carved figures are even bigger, as if anything to impress future herders was acceptable. "My dear, if you only knew about my penis!" is a classic statement.[42] Several herders claimed fantastic sexual prowess. One claimed to fear no man because he had "bigger testicles than the horse of Saint James." Of course, when such words came from the woman herself, it was more gratifying. In central Nevada I found the carving of a standing nude couple. The man, who is facing the woman, has an erect penis, which she holds and praises: "Esta [*sic*] bueno."

The herders candidly carved their sexual desires: "GI, 1928, feel like screwing." "I feel like screwing, but I have no one to do it with, P. E." In the following example the carver expressed the idea in Basque and Spanish: "Machain deseyua, eso es todo" (Pussy hunger, that is all). Standing some eighty miles from Elko is a tree with this message: "If I were in Elko where Mari is, what a fucking." The following is certainly different, and there may be more than one way to interpret it: "I am Mexican, I fuck cheaply."[43]

One herder in 1946 carved a scene that many herders must have fantasized: a naked couple touching each other. In a similar scenario by another carver the woman praises his manhood. In the vicinity of Gardnerville, Nevada, there is a grove heavy with graphic sexual content and scenes that look like replicas of erotic murals in Hindu temples: contorted figures of naked couples, threesomes, and even foursomes engaged in a variety of sexual acts. Most appear to be the work of one person who spent hours amusing himself in the shade under the aspens. He no doubt contributed to the enjoyment of a lot of herders who later came to this grove. It is places such as this that give the impression that the herders were sex crazed. But it was just a game designed to kill time and boredom. As one herder told me once: "The funniest thing happens to us. On the range the bulk of our conversation is about women, nothing else, but when we come to town we only talk about lambs, dogs, and pasture."[44]

I mentioned earlier the high regard that women enjoyed in the Basque Country and their powerful influence in the home, but you would never know it from the carvings left by the herders. The casual Anglo-American viewer might think that many of them demean women. Some carvings certainly appear to do so, but they are sexual fantasies rather than expressions of the men's actual views of women. I really wonder if they are intended to be much more than that. When a herder carved "hurrah for the whores of Reno," he was actually praising them, not demeaning them. Psychiatrists agree that such fantasies are normal for young healthy males. If no such material had been found among the carvings, then we would have to start asking questions.[45]

Prostitutes

Dozens upon dozens of carvings say "hurrah for the whores"; very few, on the other hand, express antagonistic sentiments. "I shit on the whores" was probably inscribed by someone who felt cheated or unfairly treated.[46] It is not easy to be objective about these intimate matters, but I think some of the hurrahs are meant to be thank-you notes. They were carved by ecstatic young men who had no reason to lie. Unfortunately, the women in town probably never knew about the devotion of many a herder. I do not suppose a sheepherder ever told a whore, "I carved your picture on a tree." It wasn't something he happened to be thinking about while he was in her company—his fantasies had been forgotten for the moment. In the whorehouse, he had the real thing.

Houses of prostitution were not lacking in the larger Basque cities, but the authorities never allowed them to operate too much in the open. It does not appear that prostitution existed in the smaller Basque towns, much less in the hamlets. The Nevada-style "red-light" district was certainly a shock to some sheepherders at first, but they recovered quite nicely.

Fifty to sixty miles from the nearest whorehouse a herder carved: "We cannot hold our dicks any longer. Tonight we are going to the whores. Hurrah." This message is in Elko County; according to the trees, Elko and Reno had the most popular prostitutes in Nevada. But the carvings talk about whores in California towns, too—Markleeville and Susanville among them—as well as in Lakeview, Oregon. According to one report, in the late 1930s there was a Basque lady (E. G.) who drove the herders of Castle Peak, California, into the town of Truckee for a visit with the ladies. She was asked if she felt comfortable with the job and if the herders did not harass her during the trip. Reportedly, she answered: "They were perfect gentlemen."[47] Prudery is not a Basque attribute anymore than euphemism is.

When a sheepherder was preparing to go to town, the news and the jokes spread fast among other herders, who carved their expectations for him: "J. G. is going to screw . . . he will not be sorry." The same herder carved in the same grove: "J. G. is getting ready to suck, 4 July 1930." When the herders finally made the long-awaited trip to town they were excited, but few raised hell like the Old West cowboys trailing longhorns from Texas to Montana did in Cheyenne, Wyoming, or Dodge City, Kansas.[48] Occasionally, a herder would get into a fistfight at the bar and maybe even be thrown in jail for a while if he was intoxicated, but anything more serious than that was rare.

The first thing the herder did in town was head for a Basque hotel or boardinghouse, take a bath, and put on clean clothes. In the old days most herders kept a suit at the local Basque hotel. Next, before he got a haircut, he had time for a couple of drinks and some animated talk with fellow Basques at the bar. After that, depending on the time of day, he would have dinner at the hotel or pay his first visit to the ladies, which he did preferably in the company of other herders or Basques staying at the hotel.[49]

One of the first things Basques learned about in Nevada was the existence of legal brothels in many towns. Newly arrived herders seldom had the chance or the money to visit the brothels. More likely they were sent straight to the sheep camp, where, after they recovered from the shock, they began dreaming about their first chance to see the lights of town.

By their own admission, a number of the herders were virgins when they arrived in the United States, so the first time they had sexual intercourse was a memorable event worth carving about. One herder described his experience in the third person: "He broke the *virgo*[50] in Carson City on the 20th day of September of 1967 and he is 24 years old, and he had never wetted it [?] before."[51] This man was proud of himself.

The herders had a system for rating whores and whorehouses, and most of those on the Nevada sheep ranges knew about it. One herder told me that during those rare visits to the brothel, they would stay a while, sometimes most of the night, "until they tried them all."[52]

There was one problem with easily accessible whorehouses: a herder could lose his hard-earned dollars in a hurry. The herders hated to admit it, so they did not carve much about it, but I recorded four or five carvings in which the fellows decried the dollars they had squandered in the brothels.

But that did not keep the majority from hailing "the whores of America." In 1968 I saw an aspen in Elko County proclaiming the whores in Elko as the best in the world. A lively argument was sometimes started when someone carved a statement about the whores in the Old Country. A herder from Iparralde who hardly spoke Spanish tried his hand at writing in it because he wanted to comment on an arborglyph. He carved: "They say that there are no whores in France but there are many."[53] Another tree reads: "The whores of Spain are no more . . . Hurrah for the American girls." I do not think he meant to say that they did not exist, but that they were no longer available.

At this point the reader may have difficulty believing the next two messages, but in July 1926 G. I. carved: "Without a woman and not missing it." And three years later E. G., talking about a fellow herder, inscribed: "Shit, Juan, you hate whores."[54]

Several people told me that it was not uncommon for prostitutes to go up to the sheep areas in the summertime. One enterprising Basque man recruited women and sent them to the high country of north-central Nevada, but all the herders who came in contact with them became infected with venereal disease.[55] Buster Dufurrena of Denio, Nevada, informed me that in the 1930s one sheep camp on Pine Forest (Humboldt County) was called Florence Place after a prostitute who visited the sheepherders during the summer.[56] My own findings in the Sierra Nevada aspen groves would be best explained by the presence of "traveling" women, especially from Reno.

Jess Goicoechea of Elko reported that "in the old days women rode the range. One time a woman came to Coon Creek. Didn't stay too long" (meaning that she did not have much business). Goicoechea's wife, Mariana, has been around sheepherders most of her life and had this comment on the woman's lack of success: "It took the fun away from going to town to celebrate, having some drinks, and getting drunk."[57]

Several carvings corroborate that view. When the workload was light or transportation was available, several herders might steal a fast visit to town. One did not need much of an excuse to go. The following message is located in the Marlette Lake area: "It is raining, so let us go to the whorehouse, 1942." If this herder made the trip on horseback, it was a good ninety-minute trip to Carson City—and then back again. Another herder was anticipating the good time ahead: "P. S., 20 August 1967 today is the day of the whores. Hurrah for the whores of America." Someone added: "And [hurrah] for the whores of Bizkaia as well."

"Jani"

Some of the prostitutes traveling the high country made a reputation for themselves, and their names and portraits are carved on aspens in some mountain ranges.[58] But "Jani" (not her real name; her last name was Hispanic) eclipsed them all.[59]

The story of Jani is connected to two, possibly three, Nevada sheepherders, but particularly to one whom I shall call Joe. Joe's stint as a sheepherder spanned some twenty years, beginning in June 1926 or earlier.[60] I do not know what Basque region Joe came from; his last name is not specific enough to determine origin. He carved a

67. Bust of "Jani."

68. Drawing of a female, "Jani," the most carved woman in Nevada and California.

lot of female figures, men and women coupling "sheep style," and stars, but he revealed very little about himself. Joe was a good carver, and the sex scenes are very realistic, which I believe is the reason some of them have been scratched over. As a rule, Joe did not sign the inscriptions that were sexually charged or those of nude women.

Joe was carving Jani-like figures as early as 1926, but it seems that he was new to the craft then, and his figures lack consistency. Apparently he met Jani in 1932, and in the following years he carved dozens of her portraits. I have recorded forty or fifty. I call them portraits instead of figures because the anatomical details are constant. I have found carvings of her from Alpine County, California, in the south to Modoc National Forest in the north.

70. "Jani" looking like a mermaid. She may appear to be totally nude, but a closer look always reveals that she is wearing high heels. She and many other female nudes lack arms for, as far as the sheepherders were concerned, they hindered the overall impact of the naked body.

69. "Jani" pregnant.

And until the aspen groves are thoroughly researched, no one knows where else she might appear.

Jani's physical attributes distinguish her from other nudes. The shape of her head; her prominent chin; a "flapper" hat or cloche, worn by women in the 1920s and 1930s with bobbed hair; the exaggerated breasts (over and under instead of side by side); her ample buttocks and prominent belly; and her high heels give her away immediately. Anyone who knows about aspen art and how easily the tree can alter the look of the carving must be amazed at the consistency of Joe's artistry.

We shall never know whether Joe's portraits captured Jani's physical characteristics faithfully. She appears to have been the motherly type, with ample body and strength in her facial features. The artist captured the sensuousness of her lips and aloofness reminiscent of Leonardo da Vinci's *Mona Lisa*. Whenever she is depicted without the cloche, her curly hair is treated in great detail and is never long.

The early figures of Jani are perhaps the most interesting. On a tree dated 25 July 1932, Joe carved Jani in full regalia (nude, that is) with a pearl necklace intricately sketched by pressing in a spent 30-30 bullet shell. Joe added: "I love Jani." On another nearby tree he carved: "1932 Long live Reno, Nevada. If she leaves I am going," but this message is not totally clear.

The 1932 and 1934 dates are certain beyond doubt, but almost none of the other portraits of Jani in the Sierra is dated. The carvings by the other two "pretenders" to Jani's affections are dated 1931 and 1934 in one area of the Sierra. In Humboldt County she is dated September 1933 and 1953 (unless the 5 is a mistake and should be a 3), but Joe's name is nowhere to be seen. This can mean that he lost her or that he preferred not to disclose his identity. We shall never know whether this was a one-sided love affair. It does not appear that it was, but it is possible that Jani was unaware of Joe's devotion for her. Did she know that the herder had immortalized her among his fellow workers? Or that tree after tree in Nevada and California bore mute witness to her charms? Jani, indeed, deserves some kind of title, perhaps "the madonna of the sheepherders."

The probability that she may have spoken Spanish adds another dimension to Jani's story, and perhaps another reason for her popularity. One can only imagine how uncomfortable it might have been for a herder to be unable to communicate with the woman he was having sex with.

The twenty-one-year relationship was typical enough: sometimes up, sometimes down. On several trees Joe scolded Jani, insulted her, called her "whore." One problematic inscription appears to say: "She never did it for free."[61] Other carvings tell her how much he loves her: "Te kiero mucho tanto como otros trampa [?] ai Jani" (I love you a lot as much as other tramps [?], alas Jani). If the size of her abdomen is any indication, Jani was often portrayed as pregnant. Once or twice the baby—a miniature portrait of Jani—is carved inside her.

Jani must have been a fairly liberated woman, for in a couple of portraits she is smoking a huge pipe. She has feathers on her hair and appears once wearing a hat. In some of the carvings dated 1932, a burly-looking man who is wearing a hat and smoking is copulating with her sheep style. In carvings of later years she is alone. The herder made sure that no arms obstructed the view of Jani's curves, which is why she is armless in most portraits, except when she is performing fellatio or smoking.

One aspen carving is a particularly powerful and original expression of the herder's devotion and love for Jani. Her portrayal is typical enough as far as the usual characteristics of the hairline, the high heels, no arms, etc., but she is not alone. On top of her head there appears a figure that resembles a baby and is slightly larger than her head. But this is no ordinary baby. He faces upward, resting his big head on hers, lacking arms and feet, and sports a conspicuously erect penis. The interpretation of this carving is open to debate, but my own opinion is that Joe was assuming the role of a baby, with Jani as the mother. Joe's own portrayal indicates that he was trying to get himself into her head in order to remind her not to forget his devotion. It is a thoroughly strange image, but intense and direct.

Most of the portraits of Jani are much larger than average, some as tall as five feet; and often her feet, clad in the distinctive high heels, almost reach the ground. One in the Sierra is exceptionally large, nine to ten feet high, with the added detail of Jani dripping milk from her breasts. These particulars were more meaningful to the herders than they would be to outsiders, because the herders knew how vital the ewe's milk supply was for the lamb. In fact, it was more than just the milk: without sustained contact with its mother the lamb's growth was stunted. One has to wonder if the herders without women sometimes felt as deprived as the motherless lambs.

As I mentioned above, Joe's was not the only name associated with Jani, and it was suggested to me by a sheepherder that she was probably a "traveling" lady. That would explain her near monopoly on the attention—not to mention the money—of the men of certain sheep ranges. The other herders' portraits of Jani are dated in the Sierra from 1929 to 1932, and perhaps even earlier. One carving clearly says: "She is Tom's whore friend."[62] If this scenario is correct, we can understand Joe's jealousies and the fluctuations in his feelings toward Jani.

The American Woman

In this section I will discuss carvings that refer specifically to "American" women, as opposed to messages mentioning "whores." It is not easy to ascertain how the herders regarded American women. Certainly they did not regard them as whores, but because the men's sexual and emotional deprivation weighed so heavily on them, the distinction may be academic. One arborglyph in the Ruby Mountains reads: "The American woman works well in bed." And another: "I like these Americans, they screw well."

One time, talking with a Nevada Basque, I mentioned that I had seen a few carvings that indicated the presence of women on the range. "Did you ever see a woman in the mountains?" I asked. "Yeah," he said. "One time this woman came up speaking American, I could understand some of it, but I could not respond. She said that she wanted to pet the cute little lambs, but hell, what she really wanted to pet was my dick. We screwed several times. She came back one more time. She was a teacher. I still know who she is."[63]

Beginning in the 1950s, but especially after the 1960s, urban Americans jumped into their four-wheel-drive trucks and the invasion of wide expanses of the Silver State began. Nothing was "remote" for these people, not even the sheep camps. Today only the roadless aspen groves and the inaccessible *harri mutilak* remain safe from the "off-roaders." We can no longer say that the herders are isolated. The few people who still herd sheep encounter more and more people—weekend hikers, bikers, campers, and civic groups. This is reflected in the arborglyphs.

By a popular trail that goes from the eastern Sierra over to Lake Tahoe I found two trees close to each other that echo encounters with hikers: "Hurrah for the blonds that pass through here, J. A., June 25, 1966." The next year another herder added a message of his own: "Here I am cleaning the trail for the blonds who go through here. I hope they are hot, P. S., 20 August 1967."

Far from the Sierras, in the Toe Jam Mountains, a Peruvian herder chanced to see a girl go down a dirt road. He must have been deeply affected because he had to wait until the next day to recover sufficiently to carve his impression: "Yesterday I saw a broad go by. She looked like a tramp and I wanted to screw her."

A Nafarroan in the Sierra carved a hefty-looking woman—his girlfriend—and then described her in two languages: "Una americana es fine for me culito tumach es vaquera me ama bastante" (An American woman, she is fine for me, her derriere is too big, she is a cowgirl, she loves me enough). In the mountains to the south of Reno happy herders recorded several visits by women. One in particular left an account on an aspen located in a shady spot where grass grows particularly tall and luxuriant: "Jani [for "honey"] today we are going to screw here, tomorrow it will be another day."

Since there were not enough Basque women to go around, many herders in the 1960s and 1970s took American women as wives, but most such marriages ended in divorce. One rancher in Winnemucca told me that he always wanted to marry a Basque woman, but they were always too aloof so he married a non-Basque. Several ex-sheepherders have commented that American women of Basque ancestry are tramps and can be trusted less than other American women. I am not suggesting that this is a characterization of all Basque American women; the statement should be taken as the opinion of several men who may have had unpleasant personal experiences with them.

Basque "Piece" Sign

In every aspen grove the visitor comes face-to-face with an intriguing and constant motif or design that has an uncanny resemblance to female genitals. They can be large or small, and more or less detailed, but the likeness is difficult to overlook. One cannot ignore them because they are everywhere, sometimes more than one on each tree trunk. After wavering back and forth several times, I finally became convinced that most of them are nothing more than a quirky natural growth, usually the scar left after a small limb was broken off from the main tree trunk.

As it happens, though, similar-looking female genitals—which sometimes include phalli as well—carved by the herders also abound, and it is not always easy to tell them apart. During one joint research field trip we were discussing the puzzling and ubiquitous motifs and a BLM archaeologist, a woman, came up with a name for them that everyone agreed was appropriate: the Basque "piece" sign. They are far more numerous than the "peace" signs that were also carved on the aspens in the 1960s and 1970s.

Porno or Not Porno

The Basque artist Jorge Oteiza would be happy to know that the sheepherders carved many erotic figures because he thinks that chastity is the greatest sin of the Basque people (in Euskal Herria). The question here is: Do the arborglyphs qualify as pornography? It is virtually certain that many figures were carved to cause sexual arousal, or at least to entertain, and that, according to *Webster's New Collegiate Dictionary*, renders them pornographic. On the other hand, pornography is peddled—sold and bought—and that is not the case here. Pornography happens in town and in the community, not in isolation in the wilderness. It can also be argued, however, that pornography is in the eye of the beholder. The Basque artist Ariane Lopez-Huici has said that the beautiful is erotic, the ugly, pornographic. Erotic is imaginative, pornography is mechanical. Anglo-Saxon culture, she believes, suppresses the fresh and candid view of the body.[64]

The mechanical aspect of the sexual carvings is undeniable, but how many people today are sexually aroused by them? Viewers' chief response, I would say, is more amazement, surprise, and puzzlement than arousal. That was the intention of the herders who created them.

I have been told that in the Lake Tahoe area children are sometimes barred or discouraged from walking the Rim Trail lest they be shocked by the arborglyphs. The figures are now much less offensive than when they were first carved, as most of the original details have been scarred over and blurred by the growth of the tree, to the degree that the viewer today must imagine the "dirt" in the details. At first sight it seems that some of them leave little for the imagination, but the fact is that they are very rough or sketchy. The herder should not be held accountable for other people's dirty minds.

Few of those who carved these images ever thought that someday people would find their way to these remote "galleries." They carved in isolation, when only a few hunters visited the high country, before four-wheel-drive trucks came along. This is important to remember if we do not want to misinterpret Basque erotica and the sheepherders' intent.

A herder in Humboldt County who may have thought that some of the statements his fellow herders had carved demeaned women inscribed: "Oi a(n)dreak norc baditu laudatu" (Ah, someone must praise the ladies [or speak up for them]). The meaning is not totally clear, but if he meant to rebuke his comrades for their sexually explicit arborglyphs, he was alone.

Two Porno Groves (at Least)

Table Mountain Wilderness Area, in the middle of Nevada, boasts the state's largest aspen forest. One particular grove there is known as "Porno Grove" for the abundant sexually explicit material it contains. Most of the figures date from the 1940s and are still relatively sharp. That is one reason they make such an impression on the viewer.

For years it was believed that Basque sheepherders carved the erotic material here, but it turns out that the author of most of the sexual scenes was J. A., a Shoshone Indian. He was a cowboy and a sheepherder who for years worked for Bert Paris and Jean Uhalde of White Pine County. According to one source, he even spoke some Euskara.[65] J. A. is the star of Porno Grove.

Al Ashton, a retired supervisor of Toiyabe National Forest; BLM archaeologist Roberta McGonagle of Battle Mountain, Nevada; my son, Erik; and I recorded the grove in 1991. At that time we still did not know that Basques were not responsible for many of the explicit carvings. None of the sexual scenes is signed by J. A. or by anyone else. However, J. A. carved his name on adjacent trees, which at that location are dated 1940.

What sets Porno Grove apart from other sheepherder erotica is the superior drawing technique combined with a realistic form of the human body similar to, but not quite as exquisite as, D. Borel's. The figures are not "Picassoesque" at all. I suspect that at least some of the sexual scenes were copied from pornographic magazines. We recorded a few that appear to be the work of other, less skillful, carvers, but the best and the majority are by J. A. I began suspecting his paternity when I found similar sexual scenes on other parts of Table Mountain. His name, and his only, always appeared on nearby trees.

A month after my visit to Porno Grove, much to my surprise, I found almost identical carvings in White Pine County, and J. A.'s name was there as well, dated 1942. In two years he had perfected his technique considerably. There is no doubt in my mind that he was the carver. He carved other motifs as well—such as bucking broncos—but always with the same unmistakable flair and style.[66]

71–75. Five different representations of women:

71. Dated 1920s to 1930s.

72. Female with fancy clothes carved by Anzas in 1935.

73. *(Left)* Originally dated in the 1950s or 1960s, someone added eyelashes, earrings, and the "baby." Observe the tiny arms. The original caption says: "Corazon mamacita por ti sufro" (Dear Mommy, I suffer because of you).

74. *(Facing page, top)* Probably dated in the 1960s or later.

75. *(Facing page, bottom)* Leggy female with a catlike face, dated in the 1970s or later.

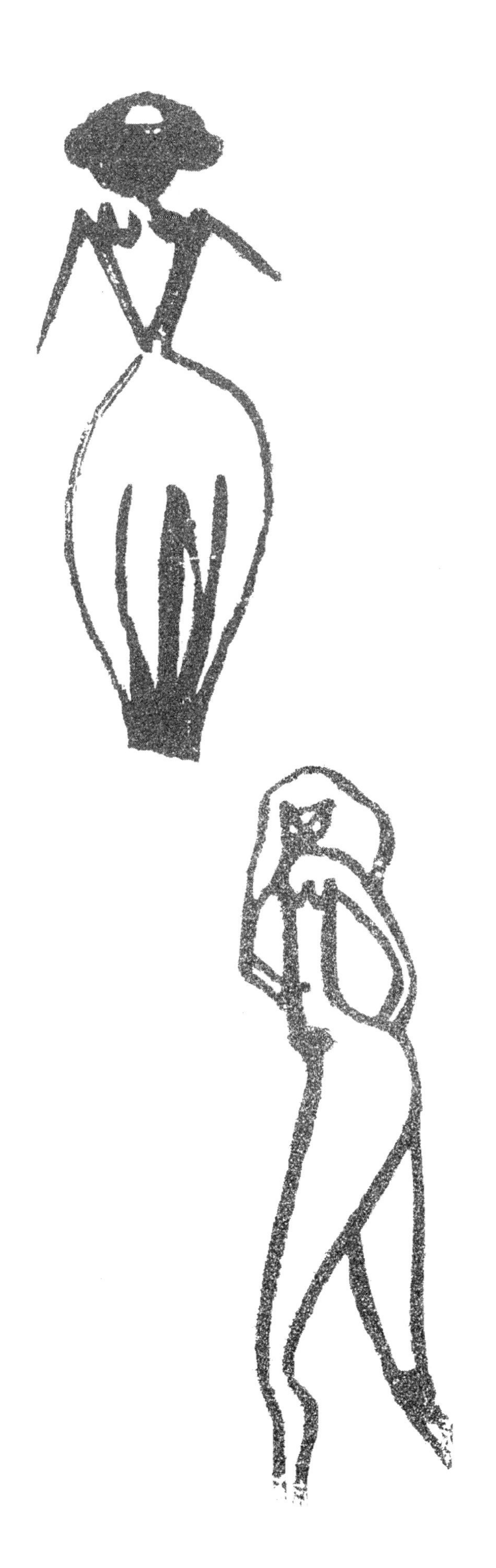

Reactions from Women

When I showed this erotic material to several Basque women, they responded in one of two ways: some dismissed it with: "Horrek edozer gauza" (Those guys are capable of doing anything), but without really condemning the material; others excused the behavior of the herders on account of their isolation. As women, none of them felt humiliated or demeaned by anything that the herders carved; however, they did not want the public to know what their relatives or husbands had carved. It was clear to me that among the Basques, sexual matters are taboo only when the place and time are inappropriate.

I showed the same material to an American woman married to a sheepherder, and she said: "That is their life . . . I don't find it offensive . . . it is only natural . . . I don't have any negative thoughts about it . . . people who don't understand [sheepherding] find it offensive."[67]

The arborglyphs are unique. There is no other literature quite like them. Therefore, logically, there is nothing to which they can be compared. They stand alone, like the trees on which they are carved.

chapter five

Artists, Sheep Camps, and Bread Ovens

Some carvings do not fit into the major categories of arborglyphs described in the previous chapters—weird figures, for example, or totally unintelligible sentences, of which there are more than the researcher would prefer. In this chapter I deal with these additional aspects and manifestations of the aspen carvings.

Cryptic Arborglyphs

Many of the carvings labeled "fantastic" by previous writers may simply be mistakes that the carver, obviously, could not correct; others may be doodlings. When I asked the herders what they did when they made a carving mistake, the answer was: "Go to the next tree."

On the other hand, some of the carvings are nothing short of enigmatic or puzzling. Are they the work of deranged minds, of "crazy Bascos" and "sagebrushed" men who stayed too long in the mountains? We will never know for sure. On the eastern side of Lake Tahoe, for example, are a number of figures of people with weird hats or headgear and human images in costumes that may be those of Old Country folk dancers. Some sport military uniforms. In fact, the caption next to one of these figures says "Cadet." The carver of such figures was almost certainly an Iparralde Basque, because many northern dancers wear costumes that vaguely resemble Napoleonic military attire. Remember, too, that Iparralde Basques were imbued with a special brand of nationalism, much of it instilled while they served in the military. It is a mixture of Basque and French pride based on a combination of former military exploits and school indoctrination.

Arborglyphs that are simply difficult to interpret are much more numerous than the so-called fantastic ones. They are often unsigned and undated. One of the most interesting and unusual of these shows a couple. Standing on the left is a shapely nude woman facing a much smaller, completely clothed man, almost positively carved by J. M. So far nothing unusual, except that the man is a priest in church vestments. His head is covered by a three-pointed biretta like those that Catholic priests wear in Europe. He wears heeled shoes. A cingulum girds his round middle, and the maniple hanging from his left arm is also clearly visible. The clergyman is facing the woman and reading from a book.[1] The book appears to be touching her breasts, but that may be due to scarring rather than being the artist's intention.

The disparity in the size of the male and female bodies is noteworthy, but not uncommon in the carvings. The question is, what did the herder intend to represent in the scene? Priests in the Basque Country (and elsewhere, of course) led celibate lives; only rarely did one hear stories of womanizing priests. Some parishioners made fun of it, while others criticized clerical hypocrisy. The likelihood that the carving represents a satirical comment on the sex lives of the clergy is fairly high. An alternate explanation is that it may represent a purification ceremony. In Euskal Herria until the 1950s and 1960s, women went to church to be purified after childbirth. It was a private ceremony, usually with no one present but the priest and the woman and perhaps the child or another close relative.[2]

Another relevant question concerns the personality of the carver. He was familiar with church vestments, so he was no doubt an ex–altar boy or someone who had studied in a seminary. There is no date on this arborglyph, but from the character of the scar, and keeping in mind the candidate who probably carved it, my guess is that it dates between 1925 and 1935.

In the Sierra, a few feet apart, are two aspens carved with figures of trucks. One of these is pulling a small trailer. Next to it is the inspired sketch of a winsome woman with long, straight hair. She has a beak for a nose and a long neck. On the other aspen there is a similar representation of a truck, and below it male genitals.

76. Intriguing carving of a fiendish-looking figure with two horns and pitchfork guarding, apparently, a little person. Among the details are a cross between the horns on his head and an arrow mark on his chest, penis, and webbed feet. The smaller figure is a woman.

The carvings, which probably date from the 1950s, are strikingly black against very white bark. Was the herder dreaming of the day when he would buy a truck and take off for a wild weekend with a woman?

Then there is the scene with two anthropomorphs facing each other. They have feminine bosoms, but one appears to have a penis as well. Both have strange-looking headgear with antennae-like protrusions. If the figure on the right were pouring liquid from a bottle the scene might be explained, because the other person is holding a tray with a glass; in that case they might be a couple of waitresses. However, the glass and the "bottle" are suspiciously low, at the genital level. Consequently, the artist may have been portraying ejaculation or coitus euphemistically. This arborglyph probably dates from the 1930s.

Snakes and Basque Culture

I have already mentioned recording figures of snakes, but at this point I want to consider other types of snake carvings. To an American reader of the third millennium nothing could be more bizarre than several snake scenes found in the Sierra. Basques generally dislike snakes—so much so that there are certain taboos about them. For example, as a child I was told not to use the normal word *hil* (to kill)[3] in reference to snakes, but *akabatu* (to finish off). And anyone who finds a dead snake is supposed to make the sign of the cross and spit on it.

Most Basques are Catholics, and their abhorrence of snakes is usually explained by the biblical tale of Eve's temptation in the Garden of Eden by the devil in the form of a snake. But the oral literature has preserved ancient beliefs that the Christian Church could not totally eradicate from the Basque cultural memory. The sheepherders originating from rural areas were likely to cling the longest to autochthonous beliefs, and in the Sierra high country I found traces of the old stories expressed graphically.

On several aspens overlooking Lake Tahoe, M. L. of Benafarroa carved entwined snakes fighting and hissing at each other with forked tongues. On a couple of other aspens M. L., drawing inspiration from Basque folklore, carved the bizarre spectacle of a female donkey nursing a huge snake while the helpless foal looks on. On another tree with the bark in bad shape it is possible to discern the same scene, but in this one the foal is nursing alongside the snake. I was quite puzzled by these carvings and wondered what was going on in M. L.'s mind.

A conversation with P. L. of Bakersfield, California, shed some light on the matter. When I alluded to M. L.'s carving, he excitedly volunteered some interesting information. P. L. said he knew of one beautiful cow in his hometown in Benafarroa that began to lose weight. No one could figure out why until one day she was observed nursing a snake. P. L. told a similar story about a woman whose child was losing weight because at night, while the mother slept, a snake would drink her milk. P. L. said that snakes have very smooth mouths and cows like to nurse them.[4] Another Basque, J. P., confirmed the accuracy of these stories and said that he had heard similar stories in his hometown of Garralda, Nafarroa.

According to Basque mythology, the Great Witch mated with Suar or Sugoi (a male serpent) and sired the first *jaun* (lord) of Bizkaia, Jaun Zuria.[5] Many readers will find these snake stories culturally as remote as the Minoan civilization of 1500 B.C., but it seems likely that the sheepherder who carved the snake images was still attached to these pre-Christian and perhaps prehistoric beliefs.

Outstanding Carvers and Carvings

In most groves there were two or three herders—sometimes more in the larger groves—who carved more and better than the others; these were the masters. When several top carvers were contemporaries, they sometimes competed, trying to outdo each other. The goal was always the same: to inscribe better depictions or funnier messages. The younger herders copied the work of older ones shamelessly. Carvers rarely criticized each other's techniques. I did find: "Melchor you do not know how to draw" on one tree, although I did not find anything below par in Melchor's carvings. Under normal circumstances Melchor would have answered this message, but I found no such response. Perhaps he had gone to another range or another outfit and had no chance to defend himself.

Some carvers excelled at carving certain types of figures, while others left a lot of messages with a common theme. Many herders had a definite preference for certain areas of the forest, as if wanting to "stake out" a territory for their carvings, much as an artist would choose a canvas or particular gallery to exhibit his paintings. Clearly, this was Maizcorena's case (see below). These personal territories may have corresponded with the sheepherders' campsites. Some offensive messages appear to have been carved deliberately in out-of-the-way places—at the edge of the grove or in a small clump of aspen separated from the main body of the forest. Such messages tend to be located away from the creek as well, because that was the herders' high-country "highway."

The theme a carver favored might be a gauge indicating his nature. Introverts were likely to dwell on themes dealing with loneliness, lost lovers, nostalgia, the harshness of life, and so on, while extroverts were more apt to address sheepherding issues, humor, their future in America, and stories of prostitutes. Furthermore, each individual had his own particular technique by which he can be recognized, even if the carving is not signed. The depth of the incision, the style of the letters, and the flair of the capital letters all give him away.

The carvings also offer clues regarding the carver's background and education. If a herder used erudite terms or Latin words (yes, there are a few), if he spelled correctly, if he included accents, and so forth, he clearly had more education than the usual herder. The average herder might have attended elementary school for four or five years, but only a minority had gone to intermediate school or further (usually to a seminary).

One of the most intriguing aspects of the carvings are what I call "dry spells." Herders, even prolific ones, seldom carved regularly. The usual pattern was one or two years of high production followed by a silence of several years, sometimes five or ten. Candido Olano came to northwestern Nevada and eastern California in 1948 and remained a herder until his death in the mid-1990s, but he carved—a lot—only in 1960–64. Frank Erro, too, worked in northern Nevada for several decades until his retirement in the 1980s, but the carvings I found by him are all dated in 1945. Why? Obvious explanations are that my recordings are incomplete or that the other trees, and their carvings, did not survive. But these dry spells are so common that another explanation is needed. Perhaps their interest in the art waxed and waned.

After reviewing thousands of carvings, I selected a short list of outstanding carvers. In most cases they receive the distinction chiefly on grounds of their prolificity. These carvers excelled for their artistry, their messages, or both. Additional herders who merit distinction are mentioned elsewhere in the text. I do not pretend to be absolutely fair or exhaustive. Some arborglyphs are in a class by themselves. One of the most unusual is a sunrise carved in the 1940s by Pierre Laxalt.[6] Although the carving is located above Lake Tahoe, the carver seems to have tried to capture a scene from his native Zuberoa in the Pyrenees: the sun, rising amid steep mountains, sheds light on a small alpine valley. A sheepherder's hut can be seen on the right corner of the vista. As an added detail, the whole carving is framed, as if the artist were watching it from the kitchen window. In keeping with the sheepherder tradition, its execution is fairly crude, but what it lacks in finesse it makes up in quaintness, intimacy, and imagination.

D. Borel

His carvings are all signed that way. I have no idea what the D stands for. There is a definite similarity between Borel's carvings and those of J. A. in that they look more "realistic" and photograph-like than 99 percent of all the other arborglyphs. Both men carved nude women, but only one sexual scene by Borel has been found so far. Borel began herding in 1916, probably at the same time as another Borel (his brother?), but apparently he carved

77. An exquisite rendition of sunrise (or sunset) as seen through a window, carved in 1946 by Pierre Laxalt. On the right side is an *etxola* (sheepherder's hut) or a *borda* (farmstead), and on the left is what appears to an animal.

very little until 1925, when he began experimenting with female busts. Several have an unmistakable "Marilyn Monroe" look—ahead of her time, of course—while others have shorter hair.

At least seven of D. Borel's carvings are located in one small grove in Tahoe National Forest in California, and there are more elsewhere. One of Borel's masterpieces is a three-inch-tall full-body nude, barely delineated. There is nothing quite like it carved on other aspens. Its quality makes it almost certain that Borel was educated or had studied drawing and art.

Adding to the intriguing mystery of Borel's personality is the fact that he also left one "Gora Euzkadi" (Long Live Independent Basque Country), carved in small, exquisite Gothic letters. It is the oldest "Gora Euzkadi" (1925) and the second-oldest Basque patriotic statement found so far. If Borel was from France, as Albert Gallues remembers it, then it is difficult to explain his sympathy for Basque nationalism.[7] Certainly that sentiment would be more understandable in an educated person. Borel was sensitive to the situation of the sheepherder as well, for nearby he carved "Pobre Borreguero" (Poor Sheepherder), spelled correctly, in Spanish.

Paulino Uzkudun

Paulino Uzkudun was not a sheepherder, of course; he was a famous Basque heavyweight boxer. Born in 1899 in Gipuzkoa, he first gained notoriety as an *aizkolari*, or woodchopper. In 1924 he became the boxing champion of Spain without a fight. After taking the European heavyweight title, he came to America and fought against the great ones, including Max Schmeling, Primo Carnera, Johnny Risko, Tom Heeney, and Griffith, among others. He lost to Carnera and Joe Louis, but the

78–81. Four samples of D. Borel's art carved in 1925:

78. As seen in this female bust, Borel had a professional touch. No other herder came close to matching the realism and finesse of his portraits of female nudes, which may have been copied directly from a magazine.

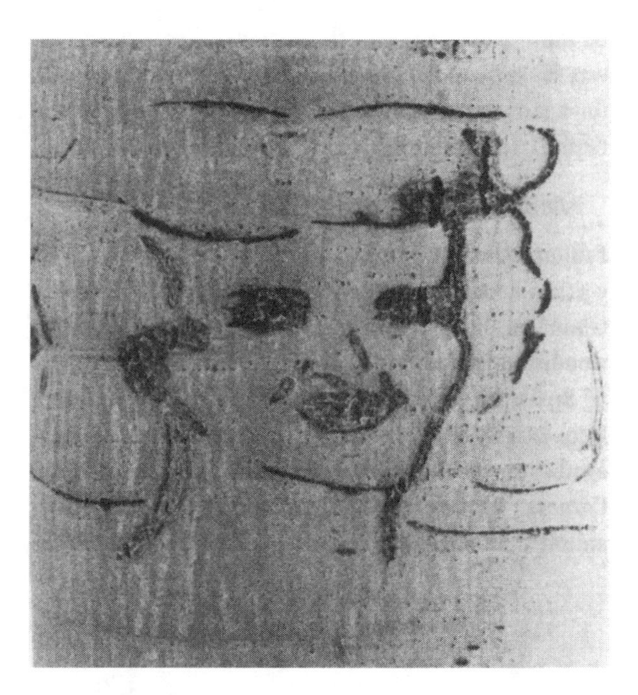

79. Female face.

81. This photograph has been slightly magnified to show Borel's unmatched craft with the knife. The actual carving is about three inches tall. The tree has been growing for seventy-five years, yet the distortion is minimal.

80. Female bust.

latter was the only boxer who ever knocked Uzkudun down.[8]

I was not aware of Uzkudun's impact on the Basque colony in Nevada until I read of it on the aspens. Apparently he was a hero of the herders even before he came to the United States and Reno, because the earliest Uzkudun arborglyphs I recorded were in the year 1929. On 4 July 1931, presented by Jack Dempsey, Uzkudun fought a well-publicized match to twenty rounds with Max Baer in Reno, Nevada, watched by some three thousand Basque fans.[9] According to one local newspaper, the total number of spectators numbered between eighteen and twenty thousand. "The Bounding Basque Awarded Decision," read the headline on the following day.[10]

Ethnic awareness fermented to an all-time high as fellow countrymen streamed to Uzkudun's training camp in Steamboat Springs. Nothing like it had ever happened in the Basque colony. The many carvings that refer to Uzkudun indicate the magnitude of his impact on the herders. In various groves of the eastern Sierra Nevada and in remote groves of Humboldt County, the great boxer still stands tall with fists ready. It is likely that more Uzkudun carvings will be found in the future.

The unskilled carvers were unable to portray Uzkudun accurately. His arms and biceps appear puny, and his head is huge, as are his waist and feet. On several aspens he appears boxing with an unidentified opponent. In a carving located in Humboldt County, Uzkudun is holding his huge erect penis with two hands as if it were a baseball bat. On the Nevada-California border south of Gardnerville, he and Joe Louis appear nude. An inscription dated 1915 identifies one of the boxers as Paulino, but it must be a case of overcarving. Another tree says in Spanish: "Paulino Uzkudun defeated by Schaaf, they fought on 24 July 1932."

What Uzkudun did to awaken Basque consciousness among the herders is the more remarkable given the fact that he was from Gipuzkoa, a region that had sent very few emigrants to the United States. But he was Basque, and that was enough for the sheepherders to support him, just as in the 1980s and 1990s the cyclist Miguel Indurain was cheered by all the Basques in the American West. One of Uzkudun's greatest fans was Martin Larzabal of Lapurdi, who carved several portraits of him.

Erramun (Ramon) Borda

Most of the time he signed as Ramon Borda, but in several carvings he used the Basque spelling, Erramun. My reason for including him among the distinguished carvers is twofold: first, the Bordas became numerous and well known in the Carson-Gardnerville area, and,

82 and 83. This is a problematic carving. The style is thoroughly reminiscent of the nearby arborglyphs carved by D. Borel in 1925, but the message, "Gora Euzkadi" (Long live Independent Basque Country), seems improbable for this date, especially if Borel was a Frenchman.

84. Portrait of Paulino Uzkudun, the Basque heavyweight boxer who fought against the greats of his time, including Max Baer in Reno on 4 July 1931. He was already famous among the sheepherders at that time, for his carvings are dated from the late 1920s and lasted into the 1940s and 1950s.

second, a handful of his arborglyphs show great originality. Borda was born in Bidarray, Benafarroa, into a family of about a dozen brothers and sisters; four of the brothers came to the United States around 1910. Gilen Borda's earliest carving is dated 1909, as is Batita's. Most of Erramun's arborglyphs are from 1915.

In self-portraits Erramun portrayed himself as a big, burly man with a prominent nose and girth. In one scene he carved himself drinking from a bottle and joking about it. In two others he is shooting a rifle or holding a gun. A couple of his arborglyphs are outstanding. What herder would see himself relaxing on a chair and reading a book? Erramun did. In a similar scenario, he is sitting and holding up a baby. As an example of his attention to detail, his hat is hanging from the back of the chair. These are unusual carvings.

If we were to analyze the man behind the carvings, we would say that Erramun was a fairly typical Basque immigrant who took the sheepherding job on a temporary basis. He did not intend to stay with the sheep for life, so he looked ahead to a time when he would be through with living in the mountains. He saw himself settled down with a wife, raising a family, playing with the children, and even reading books, all of which, in fact, he did.[11]

Etienne Maizcorena

Maizcorena was born in 1890 in Iparralde near Baigorri on a farm called Xorrotz;[12] he died in 1963 in Elko, where he is buried. Not surprisingly, his tombstone, which reads "Steve Maizcorena," includes a cross carved on

85. Thinking ahead of activities suitable for retirement. A 1915 carving by Ramon Borda portraying himself sitting on a chair and reading.

86. The great carver Etienne Maizcorena was born in Baigorri in 1890 and died in Elko, Nevada, 1963. Photograph courtesy of Dolores Taylor, Reno, Nevada.

one side.[13] He must have been a religious man, for he carved a number of large and visible crosses on trees.

His carving activities are divided in two mountain ranges and two periods with extended lapses in between. In 1931–32 he worked in the Lake Tahoe area, and from 1940 to 1942 he herded east of the Summit Lake area in Humboldt County. He seldom inscribed his name in full, very probably because it was so long; instead, most of his carvings are initialed.

He carved women, but never nude ones, though one of his glyphs depicts a pair of large legs with garters. His favorite subjects were Old Country couples dressed in detailed costumes, facing each other or shaking hands. The men wear a *txapela* (beret). He also carved a herder leading a donkey—a classic theme—a big-hat cowboy riding a horse, and a unique portrait of a *harrijasotzaile*, or stone lifter, holding a large stone on his knee and drinking while his dog watches him intently.

Maizcorena was not the most prolific carver, and he left no carved statements, but the variety of his figures makes up for it. His art is distinguished by the details in the clothing, like pockets, and the relative harmony of body forms. There is, however, one anomalous figure of a man with a huge head and hat with tiny stubs (arms) that protrude not from his shoulders, but from his neck!

Maizcorena showed his love of animals in several glyphs. Two carvings show big bucks with huge antlers (the bark of one tree is too deteriorated for the carving to be fully appreciated). A third shows a buck with its front hooves leaning on the trunk of a large tree, stretching up and nibbling on the branches; a lamb or a fawn lying on its side and watching completes the scene. This arborglyph is now greatly deteriorated and barely distinguishable. Maizcorena also carved several horses.

Aware of the merit of his art, Maizcorena chose a site for his "gallery" near a *kanpo handia* in Humboldt County, where all his arborglyphs "hang" some eight to nine feet above the ground. Obviously, he did not want anybody touching, overcarving, or disturbing them. People often ask how he did it, or whether the trees have grown that much. It cannot be the latter because trees grow at the top. Thus Maizcorena either stood on his horse, or he used a ladder.

Non-Basques

It would be easy to overlook non-Basques' output because they were a minority. But one cannot avoid being affected by the strong personality of the Spaniard Frank Rodriguez or the more recent carvings of the Mexican Jesus Hernandez Lopez [*sic*]. The latter is noted for his long messages, although he had little skill at carving figures. He was pessimistic and philosophical, as the following reveals: "The north is pretty but how many of us have wasted here while others are eating the money we sent. . . . My friend the dollars are very green, but we have no way [to get ahead]." I have considered only a few carvings by Peruvians and Chileans in this research because not enough of them have been cataloged and evaluated.

In the 1960s there were a few Spanish herders from Santander in the Sierra Nevada, and they carved quite a lot. I was told that Huesca (Aragón, Spain) was also represented among the sheepherders, but I recall just one carving by an Aragonese.[14] There were a few Galicians and Asturians, and even one or two herders from Catalonia. I also recognized a few names from Old Castile, León, and Andalusia. In the 1920s and 1930s, several herders working in eastern California were natives of New Mexico and Colorado.

Overall, the messages by non-Basques are distinguished by their outspokenness, loquaciousness, and complaining. For example, although it is virtually certain that more than one Basque objected to the arrangement or contract under which herders were imported in the 1950s and 1960s, only a man from Santander carved: "Down with the contract."

The anti-American slogans are another issue in which non-Basques parted with Basques. I do not recall recording a single "Long live America" carved by a non-Basque. The Mexicans were by far the most vocal and were more likely to carve "Gringo" or derogatory and anti–United States messages. Non-Basques were consistently more imaginative in what they had to say as well. The "Navarros," too, were wordier than their bilingual countrymen.

Frank Rodriguez

Most herders had grievances against their bosses, but the undisputed leader of the disgruntled was Frank Rodriguez. He had the personality of a union man, the social conscience of a Communist, and the restlessness of a perpetual wanderer. No one escaped his criticism, not even the bishop of his hometown in northwestern Spain six thousand miles away. He was a cynic, always unhappy with the world, and probably with himself.

Rodriguez signed his carvings in different ways—as Francisco or Paquito, for example—but normally as Frank. Each name showed one of the many faces of Frank Rodriguez, but all of his messages and comments have a common, unhappy denominator, and his carving technique is readily identifiable.

We know much about Rodriguez because he was one of the most prolific carvers in northern Nevada and the eastern Sierra—he carved both figures and messages—yet, we do not know enough. In fact, his very identity is in contention, and we cannot be certain whether there was one Frank Rodriguez or two. The question is a legitimate one, and it is based on his own contradictory carvings.

BLM archaeologist Roberta McGonagle of Battle Mountain, Nevada, and photographer Deborah Hoback studied 138 trees carved by Rodriguez at Steiner Creek (Bates Mountain, Simpson Parks Range, Nevada) roughly between 1922 and 1926. The messages were translated from Spanish by archaeologist Patricia DeBunch. One of these carvings states: "Some say my father was French, others say my father was a Spaniard. Usually I say that I feel ashamed to see such a lie. I am a Mexican Indian."[15] On several other trees he carved statements such as: "August 1926 Frank R. Rodriguez, Mexican from Sonora Cananea."[16] I have not visited Bates Mountain, but I have found the word "Mexico" carved under Rodriguez's name in the Toiyabe Range of central Nevada and in northern Elko County.

87 and 88. Two extraordinary samples of Maizcorena's art:

87. A couple shaking hands and getting acquainted, dated 1940, by Maizcorena. Observe the detailed clothing.

88. Man on horse by Maizcorena.

Rodriguez obviously possessed a complex personality. According to the information I have gathered in the Sierra he was actually born in Orense, Galicia, in the northwestern corner of Spain, in 1894. He apparently began herding in Elko County, where his earliest carving in the Independence Mountains is dated 1918. In the 1950s he was still herding sheep along the Nevada–California border south of Gardnerville, where his last recorded carving is located. In the mid-1950s he also worked for Miller and Kuhn of Rocklin, California. According to Severino Ibarra, his camptender, Rodriguez, was a little old man (a Galician from Spain) with a bad temper. His donkey always ran away from the camp, and Ibarra wondered if he did not mistreat the poor beast. Rodriguez told Ibarra that he had a wife in San Jose, California.[17]

After comparing data from my study areas with carvings from Bates Mountain, I have been able to find similarities. For example, on Bates Mountain he carved a headless nude woman with arms crossed on her back, and a few years later he inscribed almost an identical figure in the Hirschdale area (California), except that he labeled the latter: "Your mother."

Rodriguez's diatribes against the various sheep owners he worked for during his long career as a herder form a prominent part of his legacy. On Bates Mountain he left slurs against his bosses Jose Zabala (Americanized as Joe Saval) and Pete Arena. I suspect that Rodriguez was forced to switch outfits more than once because he ran into problems with his bosses. He may have been fired, or perhaps his restless nature did not allow him to stay long in one place. For whatever reason he moved a great deal and he came to know most of the sheep ranges in Nevada and Northern California.

The fact that Rodriguez was antiestablishment does not mean that he was prolabor. On the contrary, he attacked fellow herders as much as the bosses. Some of his carvings near Reno accuse Pradera's herders of being sons of bitches. He disliked Basques, and especially Bizkaians. He also attacked religious leaders. Rodriguez had a sharp mind, and already by 1926 he apparently had foreseen the demise of the sheep industry. "No more sheepherders after a hundred years," he carved on Bates Mountain.[18]

Strangely, Rodriguez did not favor political slogans as much as other carvers did, although he carved a couple of flags. I was told that Frank Rodriguez died near Winnemucca, Nevada, squashed between two train cars.[19]

Sheep Camps

In the old days, camps were the sheepherder's motels along the transhumance highway. There were many campsites along the hundreds of miles of stock trails that connected the winter range in southwest-central Nevada with the summer range in Jarbidge—that is, along the five main stock trails (and other minor ones) that stretched from Gabbs Valley, bordering Nye and Mineral Counties, to Charleston in northern Elko County.[20] These sites were usually choice spots near a road, a creek, a spring, or a grassy area—or, when all else failed, just any old place. Granted, the herders traveled light and an archaeologist's curiosity would starve in most of them, but the major camps—the lambing corrals, the shipping sites, and the *kanpo handiak*—contain some cultural debris.

In the old days, federal law required the sheepherders simply to bury their garbage, although sometimes they just dumped it in a pile and left it in the open. Today more stringent environmental rules are enforced: they can bury their garbage while on the trail, but once they arrive in the summer reserve, everything—bottles, cans, plastic, etc.—has to be carried out.[21]

The sheep camps were the primary sites where the sheep industry and its players operated. Until sheep numbers dwindled drastically in the 1960s, Nevada had many working sheep camps. Even today one can still find them, usually identified by the cemetery of cans and broken glass half buried, half decomposed, half hidden in the brush. Occasionally an old boot is present or a piece of rusty wire grown into a tree trunk. The old fireplace is likely to be visible. The abundance of aspen carvings often gives these sites away as well. Today most of the cabins are in shambles, if they have not disappeared altogether. In the 1980s Forest Service employees began burning the standing cabins in old sheep camps as a winter project. The old Ellison camp on 76 Creek in Copper Basin, Elko County, was destroyed in the early 1990s by Humboldt National Forest personnel.[22]

Preserving Whiskey Creek Sheep Camp

As more Americans become committed to preserving our environment, the federal government is following suit by protecting the land and historical resources that it manages. In the recent years of budget cutbacks, forest historians and archaeologists have been pressed to show what exactly is worth protecting in the mountains. One of the most constructive approaches that the Forest Service has discovered involves signing up community volunteers to help with cultural projects, thus fostering community involvement and interest, and saving taxpayer money as well. The preservation of original sheep camps and the videotaping and cataloging of aspen carvings fall squarely within the range of these efforts. After all, the federal agencies—the Forest Service and the Bureau of Land Management—control the land where most of the arborglyphs and sheep camps are found. Whiskey Creek is a one-of-a-kind sheep camp located in Granite Chief Wilderness Area, Tahoe National Forest, California.[23]

Whiskey Creek is a unique showcase of Basque-American culture—or rather the adaptation to American culture by immigrant sheepherders. The camp was built in 1954 by two Basque sheepherders, Severino Ibarra and Pedro Bengoetxea, who worked as camptenders for the Miller and Kuhn outfit. It comprises two log cabins and a bread oven.[24] Ibarra and Bengoetxea had never seen a log cabin, so they went to nearby Squaw Valley to take a look at one. "We can do that," they said to each other. With that knowledge, they returned to Whiskey Creek and, using only axes, in the summers of 1954 and 1955 built two cabins, a larger one for cooking and sleeping and a smaller one for a food cellar.[25]

Fausto Lavari of Aoiz, Nafarroa, an experienced oven builder, built the oven in the same year and roofed it with corrugated metal nailed to a hand-hewn cedar log. The oven has granite walls with a brick interior, and a four-inch clay sewer pipe serves as the chimney.[26] At first sight, Whiskey Creek Camp looks like a little mountain village. The oven's setting in front of the cabins gives the appearance of a tiny village square or plaza. In 1991 the State of California was planning to raze Whiskey Creek Sheep Camp because its location in the Granite Chief Wilderness Area violated the zoning law, which states that such areas are to be allowed to revert to nature. That is, no man-made structures should stand within their boundaries.[27] At this juncture Forest Service archaeologists were made aware of the significance of Whiskey Creek in the context of the Basque presence in the Sierra Nevada. They in turn persuaded their superiors that the sheep camp was unique and eligible for the National

Register of Historic Places. Sheep camps, a major component of 120 years of the sheep industry, are rapidly disappearing, and according to Tahoe National Forest Supervisor John H. Skinner, Whiskey Creek contains three classic sheep camp structures of "excellent integrity."[28]

The National Park Service agreed, and in a letter dated 25 October 1991 determined the eligibility of Whiskey Creek Camp for federal protection. The camp has already enjoyed more publicity than its builders ever dreamed of. On 1 June 1992, a crew from Euskal Telebista (Basque Television) videotaped Whiskey Creek. The report was shown on Basque television in Europe and is rerun periodically. But that was not the end of it. The members of the North American Basque Organization (NABO) in their August 1992 convention in Reno voted to enter into a partnership with Tahoe National Forest by agreeing to assist in the restoration and maintenance of the camp. This was the first time that an organization representing most of the Basque community in the United States had signed a voluntary cooperative agreement with a major federal agency.

On 15–17 August 1997, Tahoe National Forest arranged a working weekend at Whiskey Creek with volunteers from NABO, Winnemucca, Carson City, Reno, Stead, and San Francisco, assisted by the Backcountry Horsemen, who packed the supplies and the tools up and down the mountain.[29] The roof of the smaller cabin was repaired, and doors were hung, windows fixed, holes patched, and dirt removed from around the sills.

The second and last part of the restoration project took place in the week of 25–31 July 1999 and entailed the crucial operation of jacking up the cabins in order to replace the rotten sills at the bottom. Similarly, the four posts that held the roof cover of the oven were replaced and set on top of large rocks above the ground.[30]

Retired Tahoe National Forest supervisor John Skin-

89. Whiskey Creek Sheep Camp, Granite Chief Wilderness, Tahoe National Forest, California, built in 1954–55 by Severino Ibarra and Pete Bengoetxea. In 1991 the camp was nearly torn down in compliance with the rules governing wilderness areas. Whiskey Creek is now the only federally protected sheep camp in the United States. Photograph courtesy of the National Forest Service.

ner, who was among the volunteers, told me that the 1991 decision to preserve Whiskey Creek had been opposed by people both outside and inside the agency. The Sierra Club, for example, wondered why historical cabins in the Granite Chief Wilderness had been destroyed while Whiskey Creek was allowed to remain.[31]

In the fall of 1998 the archaeology department of Tahoe National Forest, Truckee Ranger District, was looking at the possibility of protecting the sheep camp in Russell Valley, which is larger and more typical than Whiskey Creek (the cabins are built with lumber).[32] The site has a long history as a sheep camp, and it was still in use in 1999. The cookhouse dates from 1909, when apparently the site was a cow camp. The Russell Valley Sheep Camp has a unique feature that proves the builders' ingenuity. In the hot summer, keeping food from spoiling can be a problem, but the food-storage cabin in Russell Valley was built right on top of the creek. The cold running water keeps the inside of the cabin refrigerated. It and the present oven were built in 1944–45. The cabin where the herders slept is on skids and was brought over from the Hobbart Mills lumberyard.[33]

The archaeology department of Tahoe National Forest, Sierraville Ranger District, has begun studying the possibility of protecting other sheep camps under the guidelines of the National Register, including the Wheeler camp and the McPherrin camp in Carmen Valley, which was still in use in 1999.[34]

Bread Ovens

Do not look for original designs in the Basque bread ovens of the West. They were built quickly to solve the everyday problem of feeding the herders fresh, tasty, and wholesome bread. Many Basque farmsteads in Europe—my own included—have the same type of ovens, which are also similar to those found in the California missions except that the latter were larger and more elaborately built. The ovens in the *kanpo handiak* were larger and more elaborate than those in the more temporary camps.

Most Basque bread ovens are as round as they are tall (usually four to five feet, but I have seen some three feet or smaller) and are constructed with brick. Some are built on a base of rough boulders and for added durability also have an outer lining of stones. The inside is always brick. They have a chimney at an angle toward the back, and an opening in front.

Just about every major sheep camp had an oven. You can still find them in northern Nevada and eastern California—and, I suppose, in other states as well. There is one in Secret Pass in Elko's Ruby Mountains. Sometimes you find them in unsuspected places, like the one in Palomino Valley, east of Sparks, Nevada, near the highway. Others are located in the Landa Sheep Camp of Blackwood Canyon, near Tahoe City, California, and at Page Meadows, both built by Steve Landa.

A few of the ovens I have seen are unusual. One in Copper Basin, Elko County, used by Ellison Sheep Company herders is cemented smooth on the outside and built on top of a large pile of rocks. It looks as if it had been dumped there. It leans to one side and is usually stuffed with beer cans and garbage left by deer hunters. The oven in the Mahogany Creek area of Humboldt County is more elaborate. It has an inner space for kindling under the floor, which is now collapsed. The date and the names of the masons are incised into the mortar above the doorway: Jean Battita Barrenetch and Michel Urrels, 1954. The oven is a blend of brick and cemented stone, as most of them are. In March 1994, during a presentation on tree art that I gave to several dozen Nevadans and Californians involved in forestry work and management, I found out that this beautiful oven had almost been torn down by a BLM official.

The oven at the Allied Sheep Camp in the Columbia Basin is completely round (most are beehive shaped) and cemented smooth. It exceeds normal size, sits on a large cement platform, and has a chimney with a hood and a fancy iron door that reads: "Hanscom & Co., Aetna Iron Works, 1874." This oven dates from 1960–61, when the camp's old oven was rebuilt by Luis Monaut and his cousin both from Nafarroa, and by Jose Manuel Jayo, from Munitibar (Bizkaia). When the old oven was torn down, it had a huge snake den inside.[35] In 1998 the Yellow Jacket Gold Mine was about to bulldoze the sheep camp and the oven, but the Elko Basque Club was alerted and saved it. Today the oven sits on the grounds of the Basque clubhouse in Elko, Nevada.[36]

Usually the camptender baked the bread for his herders, but sometimes the individual herders baked their own. To prepare the oven it was necessary to build a hot fire and keep it going for several hours or until the oven achieved a temperature of at least 350 degrees. The herder's thermometer was his arm. He stuck it inside the oven and determined whether the oven was hot enough by seeing how long he could hold it there. During bak-

90 and 91. Two samples of *labe*, or bread ovens:

90. Oven in disrepair at Jenkins Point, Plumas National Forest, California, built by a sheepherder named Urrutia.

ing, the opening in the front was covered with a removable door, which was secured with a stick and sealed with wet burlap. The chimney, too, was plugged with wet burlap in order to prevent the heat from escaping.

The famous sheepherder's bread is a simple affair. It normally contains just four ingredients: regular white flour, water, salt, and a yeast starter; in the old days herders used a sourdough starter. The Steens Mountain Café in Burns, Oregon, owned and operated in the 1990s by Tony Diaz and family, bakes the best bread in the West using starter from Tomas Zabala, a sheepherder who worked on Steens Mountain. Zabala, in turn, got the starter from earlier herders. Gradually, at least since the 1950s, the herders have switched from sourdough to dry yeast. The dough is placed in Dutch ovens and allowed to rise twice, then it is ready to go into the oven.

In the spring of 1993 I videotaped one of the last Basque sheepherders in Nevada baking bread the old-fashioned way: Candido Olano, manager and partner of Cook Sheep Company of Cedarville, California. He said his company had three outdoor ovens, most—if not

91. Unusual oven at Summer Camp, with a space underneath to keep kindling dry. The metal door is engraved, and the names of the builders are incised into the cement: Jean Battita Barrenetch and Michel Urrels, 1954.

all—built by Frank Erro of Erro Valley, Nafarroa. When my friends and I arrived at the sheep camp at about ten o'clock in the morning, we found six Dutch ovens of various sizes sitting in the sun on the bed of a beat-up Ford truck. Olano said that it took thirty-six pounds of flour to fill the Dutch ovens with dough.

As the yeast worked the dough, Olano worked the oven. The oven looked smaller than others I had seen, probably because it lacked a platform and was built into the hillside. The floor of the oven was cemented brick. Olano said that the special stones able to withstand the heat had been brought from a place near Via, Nevada.

After the dough had risen twice, the old herder pushed the Dutch ovens, one by one, inside the hot oven with a steel fence post. But first he had to make sure that the temperature was right. Olano had devised a unique "thermometer" to determine the right moment to close the oven: he placed a match on the chimney opening and counted the time it took to catch fire. If he counted to twenty or twenty-one, the temperature was perfect and he would proceed to seal the oven. On that particular day he counted only to seven, which meant it was blazing hot, so the door was left open for a few more minutes.

After that there nothing else to do but sit, talk, eat and drink, and wait for the bread.

"What do you use for fuel?" I asked.

"*Txamizu* [sagebrush], there is nothing better, but in the high country we use pine, or mahogany, if we can get it," he answered. Olano spoke in a curious mixture of Bizkaian and northern dialects. Most of the herders he associated with were from Iparralde, he told me, and, as he put it, "the French Basque language has stuck to me."[37]

We got a bit too involved in the conversation, and after some thirty to forty minutes the young Peruvian herder reminded Olano to check the bread. This he did, and one after another six big, round, golden loaves came out of the oven. Olano smiled with satisfaction as he held them up for the camera. This was the authentic sheepherder's bread; this was Old Nevada still alive in 1993. Sheepherder's bread is sometimes auctioned during Basque picnics, and it can fetch fifty dollars a loaf or more.

We did not taste any of the fresh loaves, which were left to cool on wooden shelves inside Olano's trailer. We ate old bread.

"Do Peruvian herders like your bread?" I asked.

"Like it? Some of them could eat one whole fresh loaf a day," he said.

The Kyburz Flat Interpretative Area (Wheeler Sheep Camp)

Archaeologists and officials at Tahoe National Forest have pioneered efforts to preserve cultural artifacts related to Basque sheepherders. The rebuilding of the oven at Kyburz Flat is the first chapter of that story.[38] In 1991 archaeologist Michael Baldrica of the Sierraville National Forest District initiated plans to restore the Wheeler Sheep Camp located at Kyburz Flat. The camp is situated in a designated "interpretative area," or historic zone, in the forest. Within a half-mile radius one can see prehistoric petroglyphs, the sheepherders' oven, and an 1860s stage stop excavated by Donald Hardesty of the University of Nevada at Reno in the summer of 1992.

The sheep camp was built and occupied by Martin Gallues and his brother Felix of Aibar, Nafarroa, before 1921. Martin came in 1910 or 1912 and at first worked for Miller and Lux in California. He came to Nevada in 1915 and was hired by Wheeler. In 1920 he married Margarita Amostegui, and today several of their descendants live in Reno.

The Wheeler camp originally comprised various outbuildings, an oven, and a spring. The buildings posed a fire hazard and were burned down. Of the oven, built in 1929,[39] only the rock base remained. With the assistance of interested citizens—staff and students from the Basque Studies Program of the University of Nevada at Reno, members of the AM-ARC group of northern Nevada, and Tahoe National Forest employees—the oven was rebuilt; the work was completed in October 1992.[40]

On 10 October 1992, under a friendly Sierra Nevada sun, an enthusiastic crowd watched as the oven was fired and, once again, bread was baked there the old-fashioned way. Adding a contemporary twist to the event, some people baked pizzas, hot dogs, and Basque sausages. U.S. Forest Service people from Sierraville were there in full force, and some of them—unable to resist the lure of the Zenbat Gara Basque Dancers—found themselves dancing Pyrenean folk dances, to the cheers of the crowd.[41]

Retired Reno sheep owner Abel Mendeguia and his wife, Judy, who had helped rebuild the oven, attended the celebration. For a couple of decades they had run sheep in that same forest, and they seemed genuinely amused by the significance of the event. Mendeguia remarked that it was ironic that old foes, as Basques and the Forest Service people used to be (for such appears to be the perception of many sheepmen), should now come together and work in this type of project.[42]

Mendeguia's assessment was right. Perhaps no other immigrant group in the American West has had more run-ins with the administrators of public lands than the Basques. For decades the officials of the federal government regarded the Basques with suspicion at best, and as violators of government land policies at worst. Whenever there was a forest fire, the Basques were the prime suspects.[43]

Sheepherders, on the other hand, commonly felt that they were harassed by rangers, foresters, and federal officials, who—the herders felt—did not know what they were doing because the turnover among them was so high. As one sheep owner put it: "Beti jela ematen ziguten" (They were always giving us hell).[44] But times have changed, and it is only fitting that old rivals should come together to record their common history.

Harri Mutilak

Northern Nevada ranchers who know the range well talk about the "sheepherder's monuments," which Basques call *harri mutilak* (singular, *harri mutil*). These crude stone structures marked the borders of the range allotments. They helped to prevent the mixing of sheep bands as well, and they were particularly useful to newly arrived herders who did not know the high country. In treeless areas, according to veteran herder Tomas Zabala, *harri mutilak* were especially helpful as guideposts on foggy days. He built a few himself during his lifelong sheepherding career on Steens, Trout, and Pueblo Mountains in the Nevada-Oregon border environs. A cluster of these stone piles can be seen from the Burns-Winnemucca highway in southeastern Oregon.[45]

Most *harri mutilak* that I have seen are less than six feet tall and nearly three feet across at the base. They are located in stony areas and built normally with large, flat rock slabs. The splattered bird droppings usually visible around the base and the top indicate that they are good observation points for hawks and other larger birds as well.

Some of the *harri mutilak* appear to have surpassed their primary purpose of range markers and taken on some other, unknown, symbolism. I have found several on lofty crags or overlooking breathtaking panoramic views at elevations exceeding eight, nine, and ten thousand feet. I have recorded these "special" *harri mutilak* in the eastern Sierra Nevada, in Elko, and in the Toiyabe Range in central Nevada, and all exceed seven feet in height. The literal translation of the term *harri mutil*, "stone boy,"[46] might also explain the meaning of some of these oversized piles that, from a distance, to a lonely herder may appear to be a companion.

Just as some arborglyphs are doodlings, a few of the *harri mutilak* may have been erected for no special reason. Perhaps as a way for a young, well-fed, and rested herder to exercise his muscles, or maybe as part of the training for a *harri jasotze* (stone-lifting competition), a popular Basque sport. But I do not believe that any sheepherder carried the rocks very far—every *harri mutil* that I have seen stands on a rocky area.

In Jarbidge, Elko County, I found a strange pair of *harri mutilak* that vary considerably in size. The smaller one is topped by a long, narrow stone strategically flanked by two smaller roundish rocks that hold it up. I suspected it to be a phallic symbol, in which case the larger structure would be the female, for in many tree carvings the females are larger than the males, although I realize that there is room here for other interpretations. I have seen two impressive *harri mutilak* in the Arc Dome area of central Nevada, but I have been told that there are at least three.

The *harri mutilak* built by Basque herders in the United States are in some ways reminiscent of the structures built by herders in Europe. In the Pyrenees one finds many megalithic monuments, or dolmens, which the Basque call *jentilarri* or *sorginetxe*, dating from the Neolithic Age that are believed to be the work of sheepherders. Joxemiel Barandiaran counted several hundred of them and believed that each belonged to a different family or clan.[47] Archaeologists ascribe diverse significance to these monuments, but most agree that they had religious meaning and marked burial grounds.

In 1968 I had a conversation with a Native American tribal leader of Lee, Nevada, who told me that he had destroyed a number of *harri mutilak* in the area surrounding the reservation, which is in the foothills of the Ruby Mountains.

"Why?" I asked.

"Because their gods are trying to take over our tribal spirits," he answered.

This adversarial view of the stone monuments may have arisen from the sheepherders' competition with the Indians for the range. I explained that the stone piles

92. A pair of *harri mutilak* in the Jarbidge area, Elko County, Nevada. They could be a "couple."

93. A most impressive *harri mutil*, nearly ten feet tall, located on an impressive summit.

had nothing to do with religion, and the tribal leader became apologetic.

In the Basque Country one also finds markers known as *autsarri* (literally "dust stone") in Bizkaia (called *austarri* in Nafarroa). The *autsarri* is a stone placed by a herder at the center of a circular allotment or property.[48] Although at first sight it does not appear that the *harri mutilak* in Nevada had anything to do with the *autsarri* concept, it is well to remember that both are linked to range distribution.[49]

Finally, unlike tree carvings, which are signed, we cannot know who built the *harri mutilak*. They may, in fact, have several builders as successive herders may have piled additional slabs on top of an existing pile.

chapter six

Conclusions

This book constitutes the sheepherders' own history. Further, it is a true history (any errors in reading or interpreting the arborglyphs are mine) because the carved statements constitute an uncensored literature.[1] The carver had no one looking over his shoulder, and the coyote or donkey watching nearby certainly did not care what he wrote.

However, this book is about only a few sheepherders—those in Nevada and northeastern California —not all of them. The twenty thousand or so arborglyphs on which it is based may seem like a lot, but they are only a very tiny sample of those that remain to be recorded across the western landscape. We still lack a general picture, an atlas of the arborglyphs in the western United States.

This book contains a large, previously unpublished body of information on a unique immigrant group, the Basques. Although the Basques are among the oldest ethnic groups in Western Europe, their written history is anything but ancient because culture in the Euskaldun rural society was transmitted orally. Unexpectedly, however, many of the peasants who immigrated to the United States as sheepherders began to write in order to document their lives.

Their choice of recording medium—tree carving—was rather unusual, but perhaps we should not make much of that. The aspens were available, and they provided a good surface for writing. Any other group of people would have done the same thing—or would they? In fact, other groups of sheepherders appear not to have carved nearly as much as the Basques. There were Irish herders in the Lakeview, Oregon, area, for example, but they made few carvings.[2]

There is no denying that the laconic arborglyphs offer only a sketchy and partial record, but the great majority of the messages are so personal that they provide an unparalleled picture of sheepherders' lives. The fundamental issues they treat are saturated with humanity. One who reads them sees

> assertions of individuality—the ultimate answer to the question "to be or not to be"—reflected in the thousands of names, surnames, and dates;
>
> tidbits about the simple, primitive lives of sheepherders, who sometimes felt like "lords of the range," but more often like displaced souls;
>
> the insufferable burden of loneliness that anyone physically removed from society feels;
>
> the insatiable hunger for female companionship—the missing link—which the fleeting visits to the brothels only aggravated;
>
> the pride of country and Basqueness, with all of its virtues and flaws; and
>
> the humorous testimonies to optimism and the will to succeed in an adopted country and the shock of the American culture.

The surprising absence of certain topics must be noted as well. In general, the carvings represent absent commodities and avoid depicting everyday, pedestrian objects.

The arborglyphs are a faithful reflection of Basque idiosyncrasies such as love of homeland, pride, a sense of conformity, and guarded optimism. Sometimes the nature of a culture is best observed when challenged by a new situation. It is then that it must fall back on its most basic foundations as it tries to salvage some of its most cherished elements. The arborglyphs are a product of just such a situation endured by the herders, who spent more of their time interacting with animals than with humans.

The arborglyphs tell about similarities and differences among the Basque groups, although the former are much more prevalent than the latter. It is significant that Iparralde Basques carved more in Euskara than all the other groups combined, but here an observation is also mandatory. Traditionally, the language has been the fundamental mark of Basque identity, and today it is the principal symbol of Basque nationalism. However, the Iparralde herders did not equate the use of Euskara in speech and carvings with Basque nationalism. To the contrary, they hailed France almost as often as they hailed their own hometowns. The reason? One Winnemucca Basque told me: "France treated us right. We don't have any complaints." The Bizkaians behaved conversely. They carved infrequently in Euskara but left numerous pro-Basque and anti-Spanish political slogans.

The herders from Nafarroa carved the most humorous messages, followed by Bizkaians. Those from Iparralde seem to have been the most conservative. The house motif was predominantly the province of herders from Iparralde and Nafarroa. Men from Benafarroa preferred elaborate calligraphic initials. Their As, Bs, and Ps are especially ornate with elongated tails and curves.

Whenever I gather with my countrymen I am reminded of the profound transformation that has affected rural Basques in the last fifty years. European Basques who visit the American West often say that the Basque culture found in the United States is unique. We even speak a different type of Basque, they say. In Europe the subsistence farming culture waned fifty years ago, but the herders who emigrated before that time brought its mentality to the United States, although this worldview is gradually disappearing as the older immigrants die off.[3]

The ex-herders themselves, especially those from very rural areas, are keenly aware of the situation. Fifty years ago, in the small villages of Benafarroa, 99 percent of the people spoke Euskara as the vernacular. Now, older Basques from Nevada or California who return to the Old Country for a visit are appalled by the fact that everyone speaks French. The same situation occurred in many small villages in Bizkaia, Gipuzkoa, and northern Nafarroa twenty years ago, but the *ikastolak* have been instrumental in reversing that trend to the point that today more children attend all-Euskara schools than any other type.[4]

If modern Basques came today to herd sheep in Nevada, their carvings would reflect the recent changes in Basque society, particularly the onslaught of the urban lifestyle and the loss of rural culture. On the other hand, probably the majority of the arborglyphs would be carved in Euskara rather than French or Spanish.

Above all else, the arborglyphs are a fair and democratic history. Once, people left evidence of their passage in the form of detritus, artifacts, and other refuse that gave little indication of their class. After writing was invented and history was born, the human record changed significantly: the populace no longer made or "left" for posterity their own history. Instead, a small elite of official chroniclers emerged, and the focus of their attention narrowed so drastically that all but the gods, kings, and highest leaders were excluded from the pages of their texts. For the last four or five thousand years, that is the type of record to which we have become accustomed, and, with some exceptions, this is the history that is taught from kindergartens to universities.

History has been elitist since its inception. Only since the 1920s and 1930s, with the appearance of the *Annales d'histoire sociale* school, have historians approached the study of the common people systematically.[5] It was a step in the right direction, but these French scholars depended on highly selective data. This book, in contrast, includes all the patterns discernible in the carvings so far discovered.

The carved evidence the Basque sheepherders bequeathed us deviates considerably from standard history. The herders appointed no historian and no scribe to chronicle their lives. Instead, each recorded his own activities (some, it seems, took the task seriously, while the majority were rather tight-lipped). Consequently, the sheepherding history told by the tree carvings is much more democratic than it would be if a professional had written it. Technically, the arborglyphs comprise the most democratic history ever written in the Western world because it is a history drafted by the actors themselves with virtually total freedom of expression. In doing so, the herders bypassed five thousand years of historiography to regale us with a "prehistoric" type of history.

The roots of the carvings can be found in the herders' rural culture, which was predominantly an oral one. The mentality of oral cultures—at least in the Basque Country—appears to be considerably more democratic and egalitarian than that of literary ones. There are no kings, queens, or princes in Basque folktales, only ordinary men and women. Some might be witches possessing greater power, yes, but even they are never absolute in the manner of other deities of the Western world.[6] The status of the individual in this quasi-egalitarian world was higher than that of the commoner in a traditional stratified society. No wonder that "counting heads" is a fundamental aspect of rural Basque culture.

When two Bizkaians from the same village meet at a Nevada picnic, they are expected not only to recognize each other by name or nickname, but also to remember the names of their farmsteads, the names of parents and their relations, and a host of other things. If one of the two fails the memory test, it is a bad day for both parties involved. While one feels bad for not remembering, the other feels humiliated. Is it any wonder that in a *baserri* setting the most frequent topic of conversation is people? Villagers, neighbors, names, and ages—but especially kinship—constitute the favorite fodder. A book can be closed and left forgotten on the shelf, but in an oral community, the "book" is always open; the "computer" is always on.[7]

Perhaps this explains why names outnumber all other types of carvings by a wide margin. Guided by their cultural instinct, the herders recorded their own history on trees by giving preeminence to themselves, to each individual, of which the name constituted the ultimate expression. Proof of such appreciation for individuality is the pride Basques take in their surnames, which also explains why the Basque Country is famous for the number of its surnames.[8] No other place its size has distributed so many surnames around the world. Garcia (Gartzia) is said to be "the third most common surname in the world."[9]

This sense of individual worth made every herder the narrator of his own experience in the United States. Thanks to this historical device, we are assured of an unfiltered interpretation of the events, of what the herder thought and felt. Other histories might be written by a trained professional with no personal connection whatsoever with the subject or people involved, but the sheepherders entered history through a special door—their own—which they opened and closed behind them.

The Basque keenness about keeping track of everyone made a lot of sense in the village environment where the majority of the herders were born and raised. This "village" effort was based primarily on the institution of *auzolan*, a communal effort to repair roads, bridges, and other structures, in which every farmstead was required to participate by providing its share of able-bodied (male) workers.[10] But when the herders carved their names in Nevada they were no longer representing their farmsteads. In this foreign land the idea transcended to representation of their hometowns or regions—which explains the dozens of "hurrahs" carved to boost Banka, Aldude, Nabarra/Nafarroa, and Bizkaia.

Not every arborglyph can be explained by resorting to traditional Basque culture. A number of them reflect the American milieu and life in Nevada. What is the significance of the tree carvings within the context of U.S. history? At first glance the arborglyphs seem alien and archaic to Americans. Who uses trees as a writing vehicle today? Aspen carvings can be cataloged as western American folklore; and although they were produced by Europeans, they exhibit certain affinities with the petroglyphs produced by Native Americans.

The carved information does not mirror the standard Anglo-American literature on ethnic and immigrant groups, who were traditionally shunned, even by academia. Due to prejudice or a mistaken understanding of relevancy, the history of many groups went unrecorded until a few decades ago, when it became fashionable to study minorities. Such was the fate of most of the carvings, which disappeared because federal historians and archaeologists did not appreciate their historical value. Nevertheless, the Basques, through their carvings, escaped the historical anonymity others endured. In fact, if all the arborglyphs were recorded, no immigrant group would be as well documented individually as the Basques.

Mainstream American literature relishes stories of successful immigrants but had no interest in the nobodies. Again, the carvings, with their fixation on the individual, contrast profoundly with that practice. The real hero, the protagonist of the arborglyphic history, is John Doe. Too, traditional literature concentrates on counting the blessings of the New Country, the bright future in the United States, versus the drab existence in the Old. Some of these topics are directly and enthusiastically covered in a number of the carvings—particularly those hailing America—however, the general tone, while one of affirmation and optimism, is somewhat guarded and restrained. My impression is that most of the carvings were inscribed by herders who found the cultural gap too large to cross and eventually chose to return to Europe.

Ironically, the Basques should never have become sheepherders, a job best suited, it seems to me, for philosophers. Contrary to popular belief, the Pyreneans were not culturally well equipped to endure the rarified summer isolation in the high country. Basques are very social people who exceedingly enjoy company, as the carvings themselves indicate.[11] Yet, for a century they were identified primarily with that occupation.

A substantial number of the recorded carvings, specifically the graphics, deal explicitly with women and sex. The carvings have been called pornographic, but I question that designation because their creators never intended to sell or otherwise purvey them. Should people be offended by the arborglyphs? Are we offended by crude ancient art? Aspen art existed in the mountains for a hundred years without offending anyone, but now groups of people visit the forests to view it, and suddenly it has become pornography.

Some of those who view the carvings may be scandalized or get the wrong idea about sheepherders—and all Basques—but this would be misinterpreting historical material taken out of context. In any case, the carvings are rather innocuous vis-à-vis contemporary mainstream American literature. But they contrast with Basque literature even more: The sheepherders' fantasies must be recognized as the first erotic literature produced by Basques on either side of the Atlantic. While many cultures regard the depiction of detailed sexual scenes as offensive or wrong, the herders, immersed in the animal world—with its endless cycle of procreation—were induced to view human sexual activity in the same light.

If we consider only Nevada, with its tradition of mining and the fluid, unstable, and transient society produced by that industry, we realize that the herders did not fit that norm. The sheep industry (except for the 1930s) was much more stable than mining, with its periodic cycles of boom and bust.[12] Nevertheless, the same herders' names occur in widely disparate areas of Nevada and California. That could only mean that a number of herders switched outfits, infected with the same restlessness that affected the majority of the people in the American West.[13] The carvings, year after year, record these whereabouts of hundreds of sheepherders.

Since the sheepherders' job kept them away from urban areas, their acculturation process took a different path from that followed by the majority of immigrants. The Basques took longer to assimilate than most other immigrant groups of their size. If an individual stayed all his life with the sheep or lived on a sheep ranch surrounded by fellow Basques, regardless of his aptitude for languages, there was little chance for him to master English. I have known a good number of men and women, especially women, who lived seventy or more years in the United States and still spoke little English.[14] On the other hand, many herders carved their names in English, and some even tried to write in English, phonetically, proudly carving "siper" (sheepherder) to name their profession.[15] A number of herders who were barely literate nevertheless engaged in tree carving. If they had remained in the Basque Country, they almost certainly would have written less.

The arborglyphs may contain an important lesson for Basque historiography as well. They portray the unadulterated pre-1960 *baserri* culture in written form. This culture undoubtedly possesses its own canons, imbedded in the language, which has always been the only native system of cultural transmission in Euskal Herria.[16] The minds of the uneducated Euskaldun herders were freer from alien cultures and standards. In that sense, the arborglyphs may contain precious cultural nuggets. And this in turn creates yet another paradox: the most genuinely Basque literature, the one that conforms best to the rural Basque cosmology and mentality, may be found only in the United States.

We should not expect the tree carvings to feature sophisticated debates over complex matters. That would be ridiculous. The intended viewers of the inscribed messages were the sheepherders themselves, and they saw no point in discussing matters unrelated or irrelevant to their lives. In that sense they behaved like the rest of us. It remains to be seen if studying Plato's philosophy is intrinsically a higher form of mental exercise than a herder carving that he is horny.

There is certainly a temptation to view arborglyphs as a curiosity with marginal significance. After all, what can primitive carvings by uneducated sheepherders offer to today's laptop-toting, Internet-surfing academic? But if we gaze at history like the sheepherders watched the valley from their lofty mountain posts, a different panorama emerges. Sheepherding may be an occupation with zero social prestige today, but it is one of the oldest human occupations.

Over the past several thousand years kings and generals who bloodied battlefields captured the attention of historians. No one paid any attention to herders, who made little noise, caused no urban riots, and did not write bestsellers. But once upon a time, not just queens but goddesses were chasing sheepherders. The Sumerian goddess Inanna, for example, preferred the sheepherder Dumuzi over other possible husbands.[17] The Basque term for "wealthy," *aberats*, refers to the ownership of sheep, cows, horses, goats, etc. The old Romans used the same principle to count money, using the term *pecunia*, which refers to animals.

These considerations can help us understand the

arborglyphs in context and on their own terms. While the sheepherders must be given credit for the carvings, we must not forget that the phenomenon was the result of a happy coincidence: the meeting of lonely young men with smooth-barked aspens. The herders were the active components, the aspens the passive. It is doubtful that the herders would have left comparable information on rocks or paper. The ease of carving, coupled with the artistically attractive results, nurtured the activity year after year. These trees were very important to the herders. They were both the confidants and the communicators of the herders' innermost feelings and secrets. No wonder we have herders talking to the creeks, the Sierra, even the aspens and the groves.

The arborglyphs supply minutely detailed facts regarding some aspects of sheepherder life, but very little about others. On the other hand, they can be great facilitators of information. By this I mean that they help others to remember past events and people. I have witnessed this again and again. Suppose you ask an old sheepherder to tell all he knows about sheep history in the Gerlach, Nevada, area. If he is endowed with an excellent memory, he may provide a few names or even relate an episode or two he heard about or witnessed. But the data gathered from the groves in the subject area greatly enhance his memory. Furthermore, if you play him a video with the actual tree markings, new memories are likely to flood his mind: "That fellow, well, I knew him personally. In fact I remember one time at Homestead Camp . . ."[18] And that is how their stories often unfold.

I would argue that tree carvings are superior to paper as information bearers. Compared with carving, writing with a pen is effortless. More significant, paper can be moved from its original historical environment, can be easily taken out of context, but the aspen stands witness to that one day when the herder walked up to it and carved his message. Arborglyphs offer other benefits: how can you compare the pedestrian exercise of reading a piece of paper to a vigorous hike in the high Sierra Nevada trails, admiring the wildflowers, enjoying the wildlife, pumping fresh air into your lungs, while being awestruck by the panoramic view of Lake Tahoe?

Most aspen art has a common characteristic. Some people call it "Picassoesque," but we might better settle for "crude and sketchy." The images created by the sheepherders fall perhaps within the domain of folk or primitive art because of their rural character and general lack of sophistication. By its own definition, art tends to be an elaborate affair, and the arborglyphs are anything but that. Judging art is always difficult because it is a personal matter. And as one observer rightly said, it is not easy to determine where popular art begins and where it ends.[19] Realistically, sheepherders' tree art is as unique as Euskara; there is little to compare it with, and certainly nothing identical with it.

Today the carvings have emigrated to town, where slides of typical images are shown to urban audiences for whom sheepherding may be as alien as the Basque language. In front of such people, sitting in comfortable chairs, the images surely "feel" foreign and out of place.[20] Certainly the sheepherders who carved them never thought that someday sophisticated urbanites might actually look at full-color images of their glyphs. For a Nevada sheepherder in 1925, such a scenario was as inconceivable as a moon landing. Had they suspected it, they would never have inscribed many of the carvings we have surveyed—I am sure of that.

My point is that the sheep ranges are the proper environment for tree carvings—especially the erotic ones—and they should be judged within that context. The underlying message of the nude women in aspen art is a desperate desire for female companionship expressed through sexual fantasies. I met many Basques who felt somewhat uncomfortable discussing their erotic artwork, but I did not find a single one who was apologetic. I think that is quite a statement.

What is missing from the arborglyphs may sometimes be as important as what is represented. There are no sheep, for instance. But why would a herder carve an animal he saw day in and day out? Nor are there references to quarrels with cattlemen or complaints of discrimination by Anglo-Americans, although the literature tells us that such things were not uncommon.[21] Writers are often selective. The herders remember numerous times when hungry cowboys dropped by camp for a quick bite and a shot of wine, but such events were not written about. Epic quarrels make much more interesting reading.[22] On the other hand, the sheepherders may have been just as discriminating as other writers and avoided carving about certain issues altogether—although fear of retaliation by cowboys was definitely not a concern (they could not understand Euskara, or even, in many cases, Spanish). The carvings are cluttered with diatribes against the camptenders, and a number of them even contain accusations against the sheep bosses.

The carvings were instrumental for a number of herders in redefining their Basque identity and rediscovering their patriotism, as well as fostering solidarity with other Basques—though, to be sure, there were limits to solidarity. The hundreds of carvings that herders saw and read during the summer months forced them back to their own Old World culture, to dwell upon, interpret, reevaluate, and reduce it to encapsulated statements about truths and other elements. Some were inspired by earlier messages and art.[23] The aspens were "tabloids," a forum where problems relating to their work could be aired and discussed, and jokes exchanged. Arborglyphs were the sheepherders' media, their radio and television, long before these were even invented.

The tree carvings were magnets drawing strangers together and helping them to establish a new social fabric.[24] Seeing on trees the names and stories of many herders who had preceded them persuaded the herders that they belonged to the great family of "Basques on the Range." As a result, they learned each other's dialects, and their ethnic and group consciousness evolved and rose to a higher level.

Names taken from the tree carvings can be tabulated and compared with the U.S. censuses and other official records to arrive at a more exact head count of Basque immigrants. Without doubt, the arborglyphs helped rescue many Nevadans and Californians from historical oblivion by preserving their names, their whereabouts, and the duration of their careers.

The Future of the Sheepherder Legacy

Little of more than a century of Basque sheepherding in the American West remains today. Many of the cabins that marked the sheep camps have been abandoned and burned. The mountain archives—the arborglyphs—are subject to slow, gradual decay, and sometimes, sudden removal. Eventually every one of the aspens must fall, disintegrate, and return to humus. Once the tree has fallen, the inscriptions are quickly rendered unreadable, thereby putting an end to the "growing" of history.

Faced with this certainty, we should comb the forests and record at least the better specimens (see appendix 5 for recording techniques). Recording is a form of preserving, and the Basques themselves should take the initiative in recording their heritage. It can be a lot of fun (I certainly enjoyed doing the research), and we need not labor alone. In the last ten years or so the media in northern Nevada and northeastern California have taken an interest in the arborglyphs and sheep camps. Consequently, growing numbers of enlightened individuals are becoming actively involved in the preservation efforts. The North American Basque Organization's leadership could prove decisive in this respect if it continued to foster cooperation between its members and federal archaeologists. In the last few decades, federal agencies have realized the importance of preserving sheepherder history. Tahoe National Forest has initiated efforts to recover Basque history in the Sierra Nevada and has taken the lead in preserving bread ovens and sheep camps, as have Humboldt and Toiyabe National Forests. But much remains to be done by other districts.

No one thinks that all the trees should be cut, stored, and preserved in museums. It follows, then, that they are destined to disappear; but in the meantime they must be managed. Federal agencies place a great deal of emphasis on management, and there are specific rules governing historic resources. At first, BLM and Forest Service officials and historians were of the opinion that the arborglyphs should be hidden from the public. Thus, for decades, tree carvings were a ghost resource. But of what value are they if no one visits the groves to contemplate the tree art?

Federal agencies today are more comfortable with an open approach. They are particularly pleased to know that the public forests contain one more resource. Within a short time, some federal agencies have gone from ignoring the arborglyphs to exploiting them. When President Clinton and high government officials converged on Lake Tahoe in the summer of 1997, top Washington administrators were given a tour of the aspen groves.[25] Ranger districts in California have begun printing brochures about tree carvings for the general public, although other outdoor maps that they sell to hikers still do not mention them.

Museums throughout the American West could also become players in the preservation effort, as they are ideal repositories for exceptional carvings that might be found in their respective areas. The Basque Museum in Boise, Idaho, could very well specialize in aspen art, and could easily assemble the best collection of tree trunks in the American West. The Northeast Nevada Museum in Elko and the Douglas County Museum in Gardnerville, Nevada, already possess a few specimens of carved aspen trunks, as does the University of Nevada's Basque Studies Program in Reno.

Two Sheepherders Sing

Lest the reader feel that the sheepherders have not spoken loud enough, *bertsolariak* and ex-sheepherders Jesus Goñi of Reno and Martin Goicoechea of Rock Springs, Wyoming, now do the honors for them.[26] Today these two, along with Jesus Arriada and Jean Gurutchet of San Francisco, travel throughout the American West delighting their Basque audiences with the improvised verses they sing to popular tunes.[27]

I will begin by transcribing two *bertsoak* from an audiocassette that Goicoechea recorded.[28] They describe his feelings when he bade farewell to his family and came to Wyoming:

Pena ateratzea leku xamurretik
apartera banua holako lurretik
ezin ahaztua daukat alako agurreki
etzaitut ikusiko ama beldur beti
gaizkiturik jarria bi astez aurretik.

What a pain getting away from such a loving place
I am going far away from such a land
I can never forget those last good-byes
"I will never see you again," Mom was always afraid
She took ill two weeks before I came.

Han utzia ote nun nik paradisua
penak ezin gordea gurasoan sua
nunahi zerala guri eman abisua
utzitzen zaitugu semetxo gaixua
itsatsia daukat nik orduko musua.

I wonder if I left paradise there
the grief I could not hide, and my parents' energy
"Wherever you are, make sure you let us know,
now we bid good-bye to you, our dearest son."
That kiss is stuck in my brain.

I asked Goñi to compose a few verses that touched on the highlights of his life as a sheepherder. He created eight *bertsoak* in which the reader will surely appreciate the issues that were important to him:

Amerikara etorrita hau da bizimodua
dolarraren tirriak gizona hola bortxatua
bertzek egin nahi etzuten lana guretzako autatua
zenbat aldiz damutu zait utzirik Euskal Herri gaixua.

Arriving in America, what a lifestyle we lead!
The thirst for the dollar compels a man in such a way
We have chosen for ourselves a job no one else wanted
Many times I felt sorry for having left the beloved
Basque Country.

Ardi bildotsak harturik baguaz mendirat goiti
bide arrotzak ezagutzera geure asto eta guzi
beti pentsatzen noiz behar dugun herrira itzuli
etorriko ginan baina ardiak bakarrik ezin utzi.

With our sheep and the lambs, we are headed for the
high country
to get acquainted with strange roads in the company of
the donkey
always thinking about our next return to town
We would have come gladly, but we could not abandon
the sheep.

Mendi gainera iganta, ai hura gauza tristia!
gau ilunian hartzak orroka ta koiotean karresia
deskuidatuz hil dezakete artaldearen erdia
arma hartuta egin behar guk gaba guzian goardia.

After ascending the mountain, wow, what a depressing
place!
In the dark night the bear is roaring and the coyote
yelping
If we are not careful, they can kill half of the sheep herd
With the rifle in our hands, we must keep watch all
night long.

Zortzi egunetikan baten agertzen da kanperua
beti aitzakia ekartzen du, ta gezurra segurua
bere eskutan egoten baita artzain guzien korreua
kartarik ekartzen badu, gure bihotza poztua.

Once every eight days, the camptender shows up,
He is always full of excuses and he lies like the devil,
because in his hands rests all the herders' mail
If by chance he brings a letter, what a joy in our hearts!

Letra egiten hasi eta, beti Eskual Herria goguan
ia itzultzen geraden diru puxkat egindakoan
ongi dakigu geure aita ta amak han daude esperuan
naiz batzuk gelditu diran ezin yoanak goguan.

As we begin to write a letter, the Basque Country is
always on our minds
Hoping to return someday, after making a little money
we are well aware that our fathers and mothers are there
waiting,
although many, wanting to return, have been unable to.

Esperantzarik ezbada itzultzeko atzera
Amerikako zuhaitzetan izainda zerbait diona
artzain berriak hor ikusiko du, zaharraren izena
horrek erranen dizu lehendik hor ibili den euskalduna.

If there is no hope to return back home
there will be something left on the trees of America
The new herder will see there the name of the old one,
the trees will tell you of the Basques who were here before.

Ai, artzaina, ahal bada ez dirutan loriatu
naiz urruti etorrigatik zoriona ezin harpatu
gaztetako sasoi ederrak mendian bakarrik pasatu
denbora pastuta gero, diruak bat'e balio ezpaitu.

Sigh, sheepherder, if possible, do not celebrate money
Even after coming this far, happiness is not easily achieved
The healthy vigor of the youth we have wasted alone in the mountains
Because when you have no more time left, money is worthless.

Horra zortzi bertso hemen artzainentzat ezarriak
gauz hoken berri ondo dakite Ameriketan ibiliak
kontuak ondo ateratzeko etorri nahi duten guziak
galdu ez daitzen betiko Eskual Herriko semiak.[29]

There you have eight verses composed regarding sheepherders
those (Basques) who have been to America are well aware of these things
to know the truth, come on over all of you who want to
so that the sons of Euskal Herria may not disappear (from here) forever.

APPENDIX I

List of Herders by Year(s) and by Geographical Location

CALIFORNIA AND NEVADA

Gray Creek, Nevada, and Juniper Creek, California

1910–1920 D. S., N. Y., Santiago, Santos, Martin Banca, C. Paul, Santso Martinez, Jose Setuain, Eustaquyo Aboytyz

1920–1930 Luciano Pipaon, A. D., Jose Ricabi, J. M. U., Enld (?), Martin Banca, Luciano Borda, Ibiricu, A. C. O., Mr. Pete Elgart, Fabiano (?) Olano, Javier, Navarrete Arasonis, Ramon Erasu, Jose Errecalde, Martin Urrutia, Frank Rodriguez, J. J. Badostain, Eustaquio Bernedo

1930–1940 D. A. (same as A. D.?), P. A., Martin Banca, R. N., Pite (Pete) Etcheberri, Phanou, V. A., Rodriguez, Joe (Jose) Setuain, Erramouspe, Jacinto Diaz, Ignacio Badostain, Pete Ardohain, Pete Elgart, H. Miller,[1] Cruz Ariztia

1940–1960 Bernardo, Marticorena, Eustaquio, Visconte, Valerio (Zubiri), Joseph Arnauld (Ezponde?), Santiago Presto, Jose Almirante, Jean Bortari, Joe Goicoechea, Vicente

Post-1960 Jose Maguregui, Luis Ojer, Fausto Lavari, Lopez, Carmelo Castejon, Elpidio Muro, Leon Riaño, Alberto Viteri, Jesus Arriaga, Eugenio Sarratea, Fulgencio Sarratea, Juan Zabalegui, Antonio, Joe Setuain, Melchor Aznar

CALIFORNIA

Monitor Pass Area

1910 Gazton Arichet, Echenique (?)

1911 Ricardo, Vicente Ardayz, Pedro Yriart

1913 Candido Urrizadi, Juan Vergara, Beremundo Urryziain, Miguel Yturralde

1914 Jess (Jesus?), Jose Vizcay

1917 Juanes Uhalde

1918 Yuanes Archtato ("Little Bear"), Teodoro Calvo, Laurent Incaby (Intzabi)

1919 Martin Errota (?), Vicente Yrazoqui

Tahoe National Forest

1900–1910 Rousson (?) (1900), J. Elcano (1901), Edis (1902), Jose Manha (1903), Bautista Maya (1906), Alphonse Bolnerot (1909; recorded by Tahoe NF, 1971), Azonio (1906; recorded by Tahoe NF, 1979; could be "Antonio")

1910–1920 Hernandez (?), Mateo Aranguren, Villanueva, Laurent, Martin Banca, Martin Larralde, Pascual, Joe Marseille, Jean Barbieri, B. A., Victoriano Azaolaza (or Apaolaza), Urbays, Bernardino Erro, Beremundo Urriza(ga?), Domingo Oroz (Oros?), Charlie Paul, Bones, J. L. Pasqual, Elcura (could be a nickname), Jose Urrutia, Juan Urzuegui, Pablo Navarre(te), Jose Leache, J. Inda, Jose Zubiru (or Zubiry)

1920–1930 Anjel Apaolaza, Balerio Leache, Ibarrola, Domingo Arburuas, Asensio Berystain, Juan Gregorio (or Gragoyo?), Asuanz, Martin Jauregui, Lazaro Ermeta, D. Borel, Jose Carraldo, Candido Arancibia (or Aransibia), Isidoro Barrenechea, Jose Zubyry, Lartirein (or Lartirain), Baptista Etchebarren, Jenaro, Bernardo, Arnauld Uhalde, Antonio, Rodriguez, C. Paul, Lorenzo Oroz, Valerio "Cascabel" (Zubiri), Bernedo, Pete Arretch, Eusebio, Felix, Vicente Arrobarrena, T. Larrola, Gregoryo Aldizta, J. Larralde, Mateo Aranguren, Domingo Alvarez, Juan Amostegui, Garcia, Paul Lopategui, Jose Inda

1930–1940 Bautista (Etchebarren?), Martin Banca, Jose Goicoechea, Makaya (this is a place-name, could be someone's nickname), Jo Labore, Jose Elorga, Juan Aleman, Ibbi (probably short for Ibiricu), M. Villanueva, Frank Rodriguez, Bernar Luro, A. Arocena, Steve Landa, Pierre Ardohain, Jose Iribarren, Juan Monasterio, Zesario Unzue, Raymond Etchevers, Joseph Arnaud Ezponde, Martin Aldax, Cruz Delgado, Roman Etulain, Aytegui (?), P. Uhalde

1940–1950 Ignacio Gallues, Charlie Paul, Serapio Chamilotena, Cruz Ariztia, Tomas Uranga, Zesario Unzue, Asencio Berystayn, Ramon Urrizola, Pablo Navarrete, Esteban Urtasun (not a complete list)

1950–1960 Pete Bengoa, Charlie Lanathoua, Fermin Yrigoyen, Barbier, Domingo E., Pierret Amestoy, Pascual, Pete Saroberry, A. Hiriarte, Amado, M. Arbilla

Post-1960 Antonio Ojer, Jose Otazu, Alfredo Asurmendi, Tony Otazu, Jose Anton, E. Mainz, Perfecto Naredo, Nikasio Karrikaburu, Severo Hernandez, Angelo Martinez, Berroza, Abel Mendeguia, Rodolfo Hernandez, Bartolo Hernandez

Toiyabe National Forest, Bridgeport

1910–1920 Roman Azparren, Nick Saryo, Michel Karrika, Rufino Aranguena, Sabala, Domingo Segura Dolcomendy
1920–1930 Charles Urrea, Daniel Garamendi, Pablo Navarrete, Jose Jaunsaras,[2] Pedro Errea, Peiyo Uranga, Martin Larzabal, Pete Ernaga
1930–1940 Roy Zorrilla, Santos Bilbao, Jose Azcarraga, Vidal Fernandez, Tomas Barbosa, Jean Etchart, Arnaud Salaberry
1940–1950 Jean Dendarieta, J. B. Etchamendy, Michel Iriart, Frank L. Garde
1950–1960 Michel Saldubehere, Tristan Arrosteguy, Arnaud Etchamendy, Nicasio Lorea, Miguel Erro, Laurent Munho, Pierre Heleta, Mateo Barrena, Jose Azcarraga
1960–1970 Eustaquio Uriarte, Jose Zayanz, Miguel Perez, G. Goñi, Julio Gorriz, Vicente Yrigoyen, Gabino Goñi
1970–1980 Redin, Julio Gorriz, Javier Asua
1980– Julyan Presto
No date Iribarne, Sampius

Warner Mountains, Modoc National Forest

1890–1900 Y .O., C. B., E. S. (Eugene Salet), F. O., E. C., B. P. (Batita Poko), Ben
1900–1910 Pierre Lase, F. C., Piarre (?)
1910–1920 Johanes (?) Oguilategy, R. F., Jose Urquiola (?), Modesto Elgutio
1920–1930 Samiel Urien, P. A., Dionisio Echeverria, Malion (Mauleon?), Valentin Ungusa
1930–1940 Michel Karrika, Manuel Ynda, E. Fernandez, Ularru Eruto, Jake Deshazea
1940–1950 Joseph Bidart, Michel Carrika, Manuel Ynda, Frank Satica, Mitchell Carrica (same as Carrika)
1950–1960 Ylarriet, C. M. B., Jean Harriet, Juan Pedro Goyeneche, Michel Carrica, Jonpit (John Pete) Vicondoa, Jonpit Sosenea Vicondoa
1960–1970 Jose Artazcoz, Candido Olano
Post-1970 Juan Urreaga, Guisasola tar Yoseva, Dick Sallaberry
No date M. Errea, Ayarbe, Johnnie Ayarbe, Juanita Vicondoa, J. Olaechea, Mich Etcheverry, Berro, Jose Imaz, Michel Bicondoa

NEVADA

Berry Creek, Humboldt National Forest

1900–1910 G. M., Martin Minagui, Ramon Belozagui
1910–1920 Gueria
1920–1930 Isidro Urruzola, Frank Carnita
1930–1940 Joe Elolz (Elorz?), Frank Saraitas (?), Jose Echegaray, Benicio Arystu, Pancho Ruiz
1940–1950 J. J. Adams, Miguel Alluntiz, Martin Rayola, Uriarte, Pedro Urrizaga
1950–1960 Angel Mendeguia, Felipe Echarte, J. Baptiste Guiroz
1960–1970 Ignacio Zubiria, German Lasarte, Anselmo Garcia, Francisco Azpiroz
1970– Luciano Echegaray, Ricky Echegaray, Daniel Uriarte, Lourt (Louis?) Uriarte
No date Josefina Uriarte, Venancio Urrutia, Tomas Arregui

Cave Lake, White Pine County

1900–1910 Little Indian George, Omartin
1910–1920 Little Indian George, Samuel Gonzalez, Keesing, Samuel Gonzalez, Grochas
1920–1930 Jean Etcheverry, John Etcheberry, R.W., Felix Ezpilanda, Martin Arteta, Pete Etcheverry
1930–1940 Pete Haristeguy, Jean Paul Sorhuet, Jean Goyenetche, Peter Goyenetche, Pete Harizpuru
1940–1950 Lee Whitlock (his name is in several sheep areas)
1950–1960 Jesus Celaya
1960–1970 Graciano Echaide, Domingo Iturralde, Angel Zubi

Cherry Creek

1910–1920 D. Gaona, Domingo Ynchauspe, Domingo Yribarren, Francisco Yribarren, Urara (?), Juan Yrazoqui, Pedro Yribarne, Pedro Oroz, N. B., N. B. K., Nick Kykatar (a nickname?), Pedro Ordoqui, Juan Jaureguito, Jose Manterola, J. B., Juan Manterola, Frank Careaga, Francisco Yrazoqui, Jose Dominguez, Jose Gonmes, Pedro John Dominguez, Dominique Etchebarne, Domingo, Gabriel Pena (?), Juan Urryzaga, Ramon, Pete (?) Mure
1920–1930 Palacios, Mike Bayones, J. L. I. (unclear), Pedro Iturralde, Pedro Aguren, Racios (?), Jose Auon (unclear), J. Sall (Salla?), Wm. Baerb, R. R. Ross, E. G. C., Fidel Zubeldia, F. J., Manes Auzqui, Martin Saldube, Gregorio Yzaguirre, Pete Uspartorea (unclear), Gratien Etchepare, Pete Etchepare, Juan Manterola, Luis Magas, Pedro Palacios, Jean Tipy, Baptiste Castorena, Jean Sorhuet, Howard Jackes, Jose, Francisco Lazago, Buztanguren, Pete Sero, Luis Careaga, Pete Etcheberry, Pete Eicazaga, Agustin Badostain, P. P. A., J., Juan Echegoenaga, Pete Etchart, Echeta, Ithurbide, John Bonetpels, G. E. (Gratien Etcheberry or Etchepare?), Serapio Chanchotena, John Pete Goyenetche, Felix Louis, Pedro Ordoqui, Florentino Lar-

rañeta, Gratien Etcheberry, Gerret, Martin Darir (Sarry?), Paul Sorhuet, Leopold Sorhuet, J. Jerre Lassi, Frank Yrazoqui

1930–1940 Frank Acha, Jean Bidee, Jean Bidart, R. B., J. P. Etcheberry, R. F., Paul Sorhuet, Jaureguiberry, M. S., Gratien Etchepare, Michel Franchichena, J. P. F., Molin, Frank Ygoa, Pierre Ygoa, Eri Motza, San Babro, Pete Etchart, Jean Lourbide, Gret Etcheverry, Aguirre, John Urizar, Piti Etcheverry, Emma Ordoqui, Jean Louis Bidee, Manes Bidart (same as Jean Bidart?), E. Arriz, Ramon, Jean Tipy, Pete Ordoqui, Raymond Eyherart (or Eyherard), Joe Selo, Jerome Green, Pete Dunat, John Ayarbe, Gilen Goyenetche, Juan Bayones, Pascual Tipy, Antonio Aranguren, Anza, Jean Baptiste Hachquet, A. B., G. O., Agustin V. Badostain, Gibert Bruchet, Leon Zarie, Martin Sarry, R. Luis Careaga, Pete Hachquet, Pete Etchepare, Epifanio Rivera, Gilen Yrigoyen

Genoa Peak Area

1900–1910 Emilio Alvarez, Pablo Elcano,[3] George N., Francisco Errola, Antoine (or Anton) Inda, Martinez, Ria, J. J. D., Louie Johnson, Er(r)amun Esterençuby, J. Azt (?), R. Q. M., John Egoscue

1910–1920 Emilio Alvarez, Francisco Ezcurdia, J. Manterola, Ross, Melchor Gariador (?), Anton Urrutia, Sorondo, Eizejianto (Eztereçuby?), E. S., A. J., Mr. Jesus, J. M., Pit (Pete) Aldabe

1920–1930 O. A., Agustin Odriozola (same as O. A.?), Carlos Ydolesuy, Francisco Ezcurdia, Jose Vilain, J. P. Borderre, Manuel Marco, Batista Ameztoy, Bertrand Borda, Cruz Ariztia, Behorleguy, Cruz Delgado, Arnot Urruty, Luis Aguirre, Mr. Segundo Epal (?), J. M., Jose Esain

1930–1940 Jose Esain, Batista Borda, J. F., E. F., Martin Borda, L. A., Merle Hachquet, Jose Mitjana, N. H., Ernest Marquez (?), Seguntio, Felipe Neri, N. U., Ernest Duly, Narciso Villanueva (same as N. U.?), Cadet Arrosalucia, Paul Etcheguy, Etienne Maizcorena, Martin Lanathoua, Arnot Urruty, Marcelino Urrutia, J. P. Laxalt, Jean Azconaga, Helen Hachquet, J. P. B., Emile Iratçabal, Araio Arostegui, Albert Grenade, Bill Teslia (?), Jean Leon Borderre, Tardets (?), Gregorio Arno, B. Harcundick[4]

Granite Mountain, Gerlach

1920–1930 J. A.

1930–1940 L. Urruty, B. A. P., M. Pasce

1940–1950 Elizzetche

1950–1960 Juan Aroz, Segundo Oroz, Bernard Desperben, Ayarbe, Martin Larraneta (dated from 1952 to 1961), Jose Larramendi, J. J.

1960–1980 Juan Urreaga, Jose Artazcoz, Larraneta, Fortunato Arbeloa, Jesus Artazcoz, Martin Larreaga, Miguel Eguzquiza, Larramendi, Frank Erro, John Espil, Eusebio Erquiaga, Bengoechea, Jose Ayarbe

No date Patchy Ayarbe, John Ayarbe, Leandro Arbeloa, Severo

Mount Rose

1940–1950 Dominique Ibarloucia, Leandro Urrutia, Jose Almirante, Manuel Inda, Francisco Redin, Bautista Saralegui, John Laborda, D. Harlouchet, Alfonso Sario, Santiago Presto, J. M. Carrico, Anton Paris, R. Beyron, Joe Arnaud (Ezponde, probably)

Pine Forest, Humboldt County

1910–1920 Joe Arizguren, Artiaga, Angel Cartago, Joe Ibarren, Martin Dufurrena, Yribarren, John Lacave (?), Antonio Sarratea, Angel Arrilaga, Santi Garcia, Jose Gaztelu, Esteban Hirisarri, Serigia (?), Arechuleta

1921 Bernardo Alvarez, Nicolas Echeverria, Dionisio

1922 E. Z., Jean Aro (unclear), Cosme, Jenaro, Roman Arostegui, Salbador Goñi

1923 Mariano Ordorica, Jose Garcia, Laureano Sorasu, Elcano, Joe Gastelu, Nicolas Lacarra, Bernardo Yruleguy, Francisco Arruarena, Martin Sarratea, Pablo, Manuel Echarte, Nicolas Echeberria, Pete Etchat, Ramon Montero, Jose Aguirre, Fortunato Arbeloa Garijo, John Goyheneche, Esponda

1924 Bernat, Santiago Legarza, Esteban Insarri (Hirisarri?), John Alzuyet, Yribarren, Alvarez, Antonio Larranaga (Larrinaga?), Lacarra, G. Gueresana (?), Dionisio Edeve (Echave?)

1926 Blanco, J. Alzuyet, Jean Etchemendy, Esteban Arostiain (?), Bidart, Blanca (Blanco?) Zarra (the latter could be a nickname)

1927 Jose Ariztia, P. Alvarez, Ramon Montero

1928 Valerio Pedroarena

1930–1940 John Inchauspe,[5] Pierre Gortari, Pablo, Jose Arizguren, Beiti, F. Bidart, Pedro Aznarez, Placido Aristu, Louie Bidart, John Begue, Phaulo Zabala, Martin Pete, Gregorio Elizondo, Martin Petoteguy

Pine Nut Range

1910–1920 Juan Saroiberri, Martin Banca, Beltran Bizcay (unclear), Alejandro, Ramon (Borda?), M. D. E., B(orda?), Nazario Hachquet, D. L. Leon, M. B., P. Laxalt, Daniel Peyron, Bert Buffin, J. P. Laxalt, Apino Panierda, Juan Garaber, Juan Bergara, Miguel Gaigas (or Mike Gayaga), Miguel Elizagoen, J. G., B. Y. ("Spanish Boy"), Martin Larralde, Arnaud Jaureguito, Arbaduey (unclear)

1930–1940 Bertrand Borda, German Lenoux, J. S. (?), Felix

Arrospide, Martin Larzabal, Domingo Gonzalez, J. A., E. J., Jean Lespahe, Domingo Pirata, Eugenio Sanchez, Alejo Yrastorza, Pedro Gurutsagui, S. D. M., Pablo Perez, Pedro Royz, Jose Mitjana, Viktor Casto, Gavino Ruyz, Martin Larzabal, Bidart, Michel, Bill Geddes, Omcoura, Bengoechea, C. Q. Kiser (?), Perry, B. V. M., Esteban Sanchez, Jose Mitjana, Mikel Leon, Bengoechea, Michel Jean, Tomas Ubarry, Bertran Gallegos, Leon Ernaut[6]

1940–1950 Mateo Pedroarena, Miguel Rekarte, Tomas Ubarry, Pierre Lanathoua, M. Reco, Pablo Inda, Aramendy-Arreta, Francois Basco, Frank Rodriguez, Jose Gonzalez, Felipe Errea, Frank Arizmendi, Leon Ernaut, Vicente Elizaga, Gregorio Bengoechea, Angel Gavica

1960s Toribio Gallues, Jesus M. Pedroarena, Jose Zapirain, Alberto Alonso, Antonio Alonso, Jose Fernandez, Modesto Gonzalez, Jean Etchart

Plumas (California) National Forest

1887 W. H.

1889 P. M.

1900–1910 Bautista Maya, B. M. (could be Maya)

1910–1920 E., Raymond Mocho, Jose Legarrea, Sal(e?)t, Benito Yroz, Pierre Errea

1920–1930 Telesforo Larra, Daniel, Santiago, Jean Yrac(abal), Alex, Varoules

1930–1940 Jean Ascue, Alex Riart, Bonars, Henri Peyron, Martin Banca, L. Dean, Rumbo, Bony Amostegui, Arnauld Carrika, Alexandre, Joe Ithurralde, Espil, Fausto Lavari, Vigil, Pierre Gortari, John Erramon Matnolu (?), Jean Erramouspe, Jose Zuvia

1940–1950 Ithurralde, Marco Urruty, Marcelino Urrutia, Stuart Ardura, Jean Baptiste Espil, Angel Dufur

1950–1960 Julyan Presto, Jesus Yrasabal, Elissetche, Santiago Presto, Manuel Ezque(rra?), Serapio Chanchotena, Espil

1960–1970 Mario Urrutia, Domingo Ajuria, Ricardo Uscola, Jesus Mallea, Antonio Ojer, Pete Larrea, Alfredo Asurmendi, Agustin Garde, Fernando Oroz, Perez, Fermin Oteiza, Ramon Zabalegui, Pedro Lugea, Jean Bidart

No date Jose Garra (elsewhere dated in the 1930s), Vicente Ureta

Summit Lake Area, Humboldt County[7]

1900–1910 Ynasio, Ramon Urrutibeascoa, Y. N., Oroz, B. Uriarte, Francisco Uriarte, J. M., Usano (unclear), Legarza, Ynasio Bengoechea, Ocafrain, Nabarniz (could be someone's last name or birthplace), Bitoriano Setien, George Itçaina, Manes, Inacio Elizondo, Leon, B. Urrutia, Lauzirica, Baleyanpor (?), Belro (Pedro?) Ibarriet, Felix Arospide, Pedro Uriarte, Ysidro Cenoz, Juan Goytia, Ramon Goytia, Anitia, Pete Uharriet, P. E., Tomas Uriarte, P. E. . . . (unclear)

1910–1920 Sabino Indiain, Jose Mariately, Juan Uriarte, Telesforo Gonzalez, Casildo Uriarte, Inasio Bengoechea, Jose Mulrripnial (unclear), Tomas Uriarte, Lequerica, Francisco Uriarte, Jose Irurita, Jose Bilbao, Felix Arospide, Jose Yturriza, Pastor[8] Uriarte, Arnaud Etchemendy, Yturria, Jose Urriza, Gregorio Elizondo, Pedro Uriarte, Leon Inda, Jose Arizguren, Jose Burugio, Frank Aldunate, Ysidro Ybero, Klemente Gonzalez, Jose de Goicoechea, Emilio Arista (unclear)

Telegraph Mountain, White Pine County

1882 E. . . . Miles

1887 Frank, F. (could be Frank again) B.

1889 M. Od . . . Lodeu, H. Coumt (unclear), D. I. (or T. I.)

1896 (Figure of a vulva, a tree, and several other unidentified carvings)

1900–1910 Vicente Astorquia, Diziroial (?), Sabary, Domingo (Y . . .)ena, Jose Irisarri, Johnny Sala, D. Iribarne

1910–1920 Pete Etchecopar, Enrique Goldaraz, Santiago Irazoqui

1920–1930 Iribarren, Toribio Zeta, Isidro Sala, Jean Bidart, Jose Garro, Jose Solas

1930–1940 Olechea, Serapio Chanchotena, Mike Aristegui, Baptiste Dugud(. . .), Pete Dunat

1940–1950 J. J. Adams, Manuel Lasarte, Gilen Irigoyen, Peter Yrigoyen, Bert Paris, Juan Iroz

1950–1960 Walter Peanum, Agustin Garcia, Juanito Lecumberri

1960– J. J. Adams, Santos Perez, Basilio Casares

No date Gabriel Iturria, Ayarbe, Domingo Iasa, Marilux Lapeyrade

Timber Creek, Humboldt National Forest

1899 Jean J.

1900–1910 B. L., Austis Petierr (unclear), Bautista Larraburu, Jean Baptiste Etcheberri

1910–1920 Miguel Urrutia, Venancio Urrutia, Serapio Urrutia, Juan Sala, Domingo Soto, Francisco Alzugarai, Choperena, Julian Barbero, Ernest Randall, Martin Elo(r)gui, Blanchard

1920–1930 Jose Arriz, Jacinto Garralda, Domingo Soto Rio, Juan Sala, Salbador Loarpe, F. Ch.

1930–1940 Daniel Beason, Etchelar Belxa

1940–1950 Walter Peanum

1950–1960 Dan Sanchez, Mendeguia, Jean Baptiste Guiroz, Cesario Ibarra, Bernardo Andueza, Airaco Poteros, Anselmo Garcia

1960–1970 Ignacio Zubiria, Jose Maguregui

No date Arviso, Joseph Myrua, Joe Ayarbe, Marie Laxague, Pedro Elia

Toiyabe National Forest, central Nevada

1900–1910 H. P., E. U., H. Marotin, M. H. Azcarraga

1910–1920 B. Elarchte (?), Frank Martiao, Arnauld Lascarot, Raymond Mocho, Elarchiepe, Justo Larrea, Ramon Uriona Garnicaguena, Ramon Amudia, Manterola, Pierre Haran

1920–1930 Justo Larrea, Pierre Arrat, Gregorio Elizondo, Justo Sabala, Yratçabal, Gorroño, Martin Lartitegui, Rodriguez, Pierre Haran, Castor Larre, Jean Baptiste, J. A. Bakio

1930–1940 Pierre Arrat, Martin Gania

1940–1950 Francois Gariador, Jean Gariador, Demetrio, Jean Pierre Carrica, Antonio Jauregui, Pete Louisena, John Vicondo, Antonio Vicondo

1950–1960 Jean Pierre Carrica, Frank Garat, Francois Gariador, Dermody, Pierre Garcia, Ciriaco Osa, Guillermo Yrigoyen, Jose Azcarraga, Gonzalt Gurutchet Haichet, Antonio Jauregui, Jean Jaureguy

1960–1970 Fidel Karrikaburu Zaldua, Juan Coscarat, Juan Garat, Jean Gariador, Salbador Oteiza, Bernard Tocoua, Jesus Ibarlucea

Post-1970 Juan Gariador, Fidel Karrikaburu, Eustakio Mutuberria, Frank Gariador, Tomas Sukunza

No date Joe Azkona Garate, Bellur, Segundo Arosteguy, Pedro Elias, Victor Oama, J. B. Sarasola, Jaio, Artola, Echapere Harria

Toiyabe National Forest, Table Mountain

1495 Chris Columbus[9]

1900–1910 P. E. (Pete Elgart), R. E.

1910–1920 Felix Mendiola, E. M., L. Z., Pet (Pete)

1920–1930 Maximo, O. V. Daniels, Arbunies, Mr. George, John Iracabal

1930–1940 F. C., Bernard Segura, Carraras (unclear), Jean Etchegaray (normally he spelled it Echnegaray), Antonio Surano, Gregorio, Diez, Jose Gogenola, Mr. George Ydoeta

1940–1950 J. J. Adams, Mr. Pete Elgart, Miguel Gallegos, Bertrand, Joe Onaindia

1950–1960 Benedicto Urralburu, Miguel Gallegos, Angel Arrinda

No date Souza, Don Dufurrena, Santy Gogenola

APPENDIX II

Sheepherders on Peavine Mountain,
Washoe County, Nevada

A question mark (?) means the carving is unclear.
Probable readings are given in parentheses.

1901	C. Paul
1902	Valentin Leiva (?)
1907	Luis Aliaga
1909	Luis Aliaga
1911	C. Paul
	Frank Moutrousteguy
1915	Basqo
	M. Jose
	Franc Elizalde
1917	C. Paul
	S. E. espanol vasco
1918	S. E. (?)
1919	C. Paul
1920	Lorenzo Maritorena
1921	C. Paul
1922	Antonio Maritorena
1923	Mar—cillo Joe
1924	Jean Othats
1925	C. Paul
	Antonio Vasco
	J. N. Marseille (?)
1926	Jenaro Inza
1927	John Etchemendy
	Jesus Marticorena
	Martin Urrutia
1928	Frank Rodriguez
1929	C. Paul
	D. Luro
	Ambrose Arla
1930	John Etchart
	Batista Etchebarren
	G. Paul
	John Elizzetch
1931	Alerude (Aldude?)
	Malen Zuberogoitia
	Frank Rodriguez
1932	A. Arla
	Frank Rodriguez
	Luis
	Tomas Alkayaga (?)
	Magencio Valencia[1]
1933	C. Paul
	A. O. G. (?)
	M. V. (Magencio Valencia?)
	Setuain
	Maria Pier (Marie Pierre?)
1934	A. Arla
	G. L. A.
	John Elizzetch
1935	C. Paul
1936	John Etchart
	Jim Mahoney
1937	Juan Barnech
1938	Ignacio Badostain
	Esteban Irisarri
	C. Paul
1939	Alerude (Aldude?)
	Joe Lacroix
	Paquito (F. Rodriguez)
	Al Varela
	Charles Aldabe
1940	C. Paul
1941	C. Paul
1942	C. Paul
1943	Joe Lacroix
1945	Ernaut
	Joe Lacroix
1946	Luis Segurola (?)
1947	Luis
1949	C. Paul (1947?)
194-[2]	Esteban Urtasun
1950	Jean Pierre Lanathoua
	Pablo Urrutia
1951	Jean Biscay
	Charles Lanathoua
1952	J. P. Lanathoua
	Charles Lanathoua
1953	Juaquin Urreaga
	Al Varela
	Urrutia

	Salvaletea (Zabaleta?)
1954	Jean Biscay
	John Guidu
	Juan Urreaga[3]
	J. Urreaga
	M. H.
	Jean Othats
1955	J. P. Lanathoua
	Eugenio Vidaurreta
	J. Urreaga
1956	Juan Marticorena
	J. Urreaga
1957	Jose Goicoechea
	J. Urreaga
1958	Magencio Valencia
	P. Garrohry (?)
1959	Jose Javier Otazu
	M. Valencia
	Ignacio Arrupe Acorda
	Jean Biscay
	J. Pierre Phanou
	J. Urreaga
	Jose Leoz
	Frank
	Steve Landa
1960	Y. Arrupe
	Juan Zabalegui
	Jose Eguiroz
	M. Valencia
	J. S. (Justo Sarria)
MCMLX	J. E. (Jose Eguiroz?)
1961	Y. Arrupe
	Jose Leoz (also, Juan Jose Leoz)
	M. Valencia
	J. Zabalegui
	J. S. (J. Sarria)
1962	Justo Sarria
	Y. Arrupe
	Manuel Inda
1963	Justo Sarria
	Fidel Aspi (Espi?)
	Luis Ojer[4]
1964	Nikolas Ibarra
	Y. Arrupe
	J. Zabalegui
1965	Jose Leoz
	Antonio Ojer
1967	Isidro Martinez
	P. J.
1968	B. Pyer (?)
1969	Pete Borda
	Isidro Martinez
	Eugenio Vidaurreta
	J. Zabalegui
1970	A. Arla[5]
	Valentin Uriarte
	Xent (?) Iriarte (same as Uriarte?)
1971	Lauria (?)
	Larraneta
1972	Domingo Ajuria
	T. S.
1973	D. Ajuria
	Juan Aizbaso
	Trinidad Suqylvide
1974	D. Ajuria
1975	D. Ajuria
	M. B.
1976	Miguel Lorca (Fresno)
1977	D. Ajuria
	E. E.
1978	D. Ajuria
	J. Zabalegui
1980	A. Leon
1981	Lauro (?)
	Amostegui
1982	Nicolas
1987	Esteban Alvarez
1988	E. Alvarez
1989	Felipe Rojas
No date	Martin Tipy Arrosa[6]
	Sabala
	Urreltz
	Avila
	Isasi
	Fermin Oteiza
	Pit Salvaje[7]
	Elorriaga
	L. Marquez
	Ramon Jayo
	Joe Barbara
	Eugenio Goni
	Frank Garcia
	Jose Lasa[8]

I know of other Basques who herded on Peavine, though their names are absent from the trees: Alfonso Sario and Jean Uhart, for example, who were sheep owners; Sario from Carson City; and Uhart from Silver Springs. One of Sario's herders, John Goyenetch ("Manex"), worked on Peavine in 1959–60.[9] Surely there were others.

APPENDIX III

Names in Copper Basin,
Elko County, Nevada

BASQUE NAMES, OCCURRENCES, AND DATES[1]

Name	Occurrences	Year
Abdias (various forms)		
Abdias	6	(19)70, (19)71
Abdias Comu?	1	—
Abdias Namas?	1	(19)72
J. J. Adams	3	1948
Adelita (woman's name)	1	—
Jose Alberto Al . . . ?	1	(19)79
Alcorta, Alkorta (various forms)		
Frank Alkorta	1	—
Fransysco Alcorta	1	—
Juan Aleman	3	1914
Arano Alkueta	1	1928
John Ampo	1	1911
Pio R. Aquino	1	(19)96
Aramburu (three brothers)		
Aramburu	1	1923 (1933?)
Julian	1	1934
Julian Aramburu Bros.	1	1935
Manuel Aramburu	14	1932, 1933, 1934, 1935, 1936, 1938
Manuel	1	—
Manuel A.	1	—
Teles	1	—
Telesforo Aramburu	2	1934, 1935
Albe Aramouk	1	1963
Armolea, Armaolea (probably relatives)		
Antonio Armolea	2	—
Jesus Armaolea	1	(19)74
Arrate (various combinations)		
Edorta Arrate Tar	1	—
Eduardo Arrate	4	—
Arretche (various renditions)		

Name	Occurrences	Year
Arno Arretche	2	1933
Enrique Arretche	1	1915
Julian Arrillaga	3	1956
Pedro Asparne	2	1919
Ax (various comb.)		
Ax	1	(19)84
Manuel Ax	1	1936
Juan Badiola (Gabiola?)	5	1911, 1926
Barajas	1	(19)89, (19)90
Manuel Barrena	1	1932
Barruetabena (usually written with "ñ"; various forms)		
G. B.	6	1911, 1915, 1948, 1949, 1954
Gregorio Barruetabena	4	1915, 1948
Gregorio J. B.	4	1911, 1915, 1948, 1951
Jose Basterrechea	5	(19)50, (19)53, (19)54, (19)72
Tedy Bear	1	1929
Ysidro Bengochea	4	1957, 1976
Beria	1	—
Bilbao (various comb.)		
Tomas	18	(19)54, 1957, 1958, (19)59, 1959, 1960
Tomas B.	2	1959, 1985
Tomas Bilbao	22	1950, 1955, 1957, 1958, 1959, (19)60, (19)68
Brust (first name in three languages)		
J. Brust	1	1930
Jean Brust	5	1930, 1931, 1932
Juan Brust	6	(19)29 1929
Mr. John Brust	1	—
Pete Brust	3	1924
Jean Burr?	1	(19)29
Calbo	1	1974
Carlos	1	—
Castillo	50	(19)49, (19)50, (19)52, (19)53, (19)54, 1953, (19)55, (19)65, (19)71, (19)72, (19)73, (19)74, (19)75
Castro	1	(19)72
Celaya, Zelaya (various comb.)		
Felix Celaya	1	1920
Felix Zelaya	1	1917
Leon Y. Chitana (Chitana probably a dog's name)	1	—
Cia (Cea?)	1	1903
Calvo Cia	1	1907
Norberto Cia	3	1910, 1913, 1914
Coeran	1	(19)71
Colqui (various comb.)		
Jaime	1	1961

Name	Occurrences	Year
Jaime Colqui	3	1970 (1990?)
Jaime Colqui Ventura	1	—
Jaime C. V.	4	—
Jaime Q.	1	—
J. C. V.	1	(19)91
Corta	1	1953
Luis Cresio (Crespo?)	2	1972, (19)74
Demetrio	2	1947, 1954
Heguy Dendary	1	1952
Diaz (various combinations)		
Adolfo	4	1971
Adolfo Diaz	5	1970
Adolfo Diaz Simi?	1	—
J. Diaz	1	1926
Norberto Dominguez	1	1941
Juan Duaroto	1	—
Felix Echave	3	1969
Elza (woman's name)	1	1905
Julio Erreal	1	1916
Juan Espinoza	4	(19)94, (19)95
Estudillo	1	1974
Etcheverria	1	(19)38
Etcheverry (probably relatives)		
Etcheverry	1	1952
Mike Etcheverry	3	—
Pete Etcheverry	1	(19)52, 1952, 1953
Fagoaga	2	(19)53
Pete Forua	2	1938
Gabica, Gavica (various comb.)		
Gabica	1	1941
Inacio Gavica	1	(19)57
Pablo Gavica	3	(19)57, (19)58
Jual Gabiola (*see also* Badiola)	2	1926
Pete Gamio	6	952, 1953, 1957, 1958
Celestino Garamendi	4	1943
Garcia (various comb.)		
Garcia	1	—
Miguel Garcia	2	1935
Steve Goicoeche	1	192(5)6, 1952
Heguy (relatives)		
Bob Heguy	1	1973
Heguy	2	(19)63, (19)70
Joe Heguy	1	(19)83
Lori Heguy	9	(19)75, 1975, (19)76, (19)77, (19)78, (19)79, (19)80
Mitch Heguy	2	(19)70

Name	Occurrences	Year
Hipolito	1	—
Juan I. (?)		
Ibarluzia	1	—
Fausto Ilado	1	1924
Inda Pirpi (Pirpi is probably a nickname)	1	1916
Inza (various)		
Jenaro Inza	1	1934
Martin Inza	2	1916
Esteban Iryarte	1	1923
Jas ?Apant	1	(19)90
Jennins	1	1933
Juan (from Chile)	1	—
Juan Juaristi	1	1967
Julia (girl's name)	1	1973
Julian (*see* Arrillaga)	2	1973, 1976
Ken	1	1955
Kopentipy (various comb.)[2]		
Kopentipy	4	1920, 1924, 1929, 1934
Michel Kopentipy	1	—
Pete Labat	2	1928, 1929
Martin Lartitegui	1	1944
Mateo Legiso? (uao?)	1	—
Leiva (Lleva?)	1	1974
Leniz (various comb. relatives)		
Jorge Lenix	1	(19)63
Anton Leniz	2	(19)67
Antony Leniz	1	—
Juan Leniz	1	(19)75
Tony Leniz	1	—
Lusarreta (various comb.)		
Angel Lusarreta	1	1928
Manuel Lusarreta	1	1927
Madarieta (various comb.)		
Madarieta	3	1960, 1961, (19)64
Joe Mike Madarieta	1	1965
Mariana (Lugea) (probably Jess Goicoechea's wife)	1	—
Marie (girl's name)	3	—
Mayi (girl's name)	1	—
Mendeguia (various comb.)		
Serapio	4	1944, 1947
Serapio Mendeguia	2	(19)44, 1947
Ray Mendive	1	(19)85
Steve Moiola	1	1980

Name	Occurrences	Year
Moncerio, Monohario (various comb.)		
Jose Moncerio	2	1954
Jose Monohario	2	1954
Pablo Nick	1	—
Juan Nolido	1	1916
Mo—y Numi-uh (?)	1	1931
Frank Olague	4	1920, 1921, 1922, 1923
Jim Onote	1	1936? or 1938?
Jack Oyharzabal	1	1957
Pello	1	(19)91
Maximiliano Pena	2	1988
Quinones (Quiñones)	1	1983
Manuel Rambo	1	1933
Miguel Rapeain	1	1927
Juan Reyes	1	—
Riveira	1	1971
Roberto	4	1929, 1959
Hortencia Romani (girl's name)	1	1982
Rubio	1	—
Ruiz	1	(19)75
Sampiar	1	1919
Pendes Santander (place-names)	1	1947
Satur Santesteban	3	1954
Martin Santos	2	1977
Carlos Sarrabe	1	1941
Sebastian	1	1991, (19)92
Carlo Sha-be (*see* Sarrabe)	1	1949
Juan Sumbilla	1	—
Leon Tellerya	1	—
R. Tobin	1	1979
Uman, Human (various comb.)		
Human	1	(19)89
Miguel Human	2	1980, 1982, 1983, (19)84, 1986, 1987
Migel (misspelled)	3	(19)94
Migueel (misspelled)	1	(19)85
Miguel	20	(19)84, (19)84–88, (19)85, (19)89
Miguel U.	1	—
Miguel Uman	6	(19)84, (19)85
Migul Uman (misspelled)	1	(19)85
M. U. S.	14	(19)81, (19)84, (19)85, (19)90
Uman	5	(19)81, (19)84, (19)85, 1985, (19)88
U. M. S. (misspelled)	1	(19)88
Jenaro Uria	1	(19)70
Uriona (various comb.)		
Juan Uriona	2	1959, 1963

Name	Occurrences	Year
Uriona	1	—
Urruti	1	—
P. S. V.	1	1913
Vaquero, Baquero (various comb.)		
Baquero	1	—
Vaquero	2	(19)85
Ventura (various comb.)		
Alfonso	2	1973
Alfonso Ventura	20	1972, 1973, 1974, 1980, 1981, 1982
A. V.	1	—
A. Ventura	1	1980
Simeon	2	1984, 1988
Simeon Ventura	3	1982, 1984
Fred Verfiedo	1	1959
Ander Vicrat	1	1940
Pedro Ybarra	1	1923
Ybarruri (place-name)	1	1916
Martin Yparraguirre	1	1911
Pete Ythurralde	1	1932
Jose Yturri	1	—
Zabala	1	—
Zara	1	—
Pedro Zuca	2	1917, 192?

APPENDIX IV

A Sample of Sheepherders' Birthplaces and Ethnic Identities

In the following list regionalism and ethnic identity are variously expressed. They are transcribed as they appear on the trees, except that the year is given first and a comma is supplied after the name in order to differentiate the surnames from the place-names.

1901 Vicente Astorquia, Nachitua
1907 Alfonso Zuricaran, vyzcaino Espana
1909 P. Laxalt, Bazco D-r(r)—n Raza
1911 Pedro Yriart, Apino, Aldude July 30
1912 Felix Mendiola, Marquina
 Juan Saroiberri, 5 June Basco Francis
1914 Joe Manterola, Navarra
 Juan Uriarte, Abadiano
1915 Juan Manterola, Yanci
 Santiago, basco espanol
1916 Gregorio Elizondo, navarro
1918 Esteban Goicoechea, natural de Mendeja Vizcaya Sept. 10
1919 Jose de Goycoechea, natural de Mungia, dya 15 de agosto
1920 Vicente Arrobarrena, Bazco Navarro Espana (the last word is crossed out)
1924 Feliz Ezpilanda, espanol bekayno (bizkaino) bazko
1925 Antonio, vasco
 Ruperto Gantzo, Biba Luzaide
 Jose Errecalde, espanol navarro
1926 Batiste Castorene (Castorena?), Bidarray
 Jean Tipy, June 3 Bidarray
1928 John Pete Goyenetche, Urepel Franci
1929 Manez eri motza, Sept. 1st Vive Francia
 Martin Larzabal, July 11, Hendaya
 Mr. George (Lasuen), Marquina Vizcaya
1930 Batiste Iribarne, Basco Fracais
 Dominique Landabure, Makaya
 Bernard Segura, luzaidarra
 Gratien Etchepare, Frances Basco Anhaux B.P.
1931 Pite Etcheberri, Yholdi France
 Pierre Ygoa, 14 de agosto Basco frances Urepel Basses Pyrenees
1932 Erramouspe, Soule
 Santos Ibar, Bidarrai
1933 Antonio Garcia, Arrieta valle de Arce Navarra julio 13
 J. P. Etcheverry, Bidarray France
1934 Jean Pierre Etchepare, julio Urepel
1936 Pierre Gortari, Basco Fracais
1937 Pete Dunat, luastarra Lapurdi France
1938 Jean Goyhenetche, French Basko
1943 John Dendarieta, aldudarra France
1947 Fermin, vasco navarro, euscalduna
 Serapio Chamilotena, Bazko julio 26
1949 J.B. Urricariet, Urepel
 Gilen Yrigoyen, Arneguy Basses Pyrenees
1950 Pete Lanathoua, Basque Frances Ahaxe
1951 Fermin Gamio, Baigorry
 Felipe Errea, Mezquiriz, vasco navarro espanol
1952 Esteban Ampo, Valcarlos
1953 Santiago Presto, Linzuain Navarra
 Gratien Ocafrain, Banca
1954 Domingo Olaciregui, junio Hendaye Biba Hendaye
1955 Jean Pierre Saroiberry, urepeldarra Vive la France
1956 Jose Armaolea, Elanchove Vizcaya
1957 J. Urreaga, Iruritacoa Navarra
 Inasio Isasti, de nasionalidad de Rigoitia (Rigoitia [Bizkaia] national)
 Juan Pedro Iroz, Valcarlos
1958 Inazio Arrupe Acorda, Ibarranguelua Vizcaya
 Miguel Recarte, 30 June, vasco Arizcun

1961 J. P. Begue, Biba Francia
1962 J. Sarria, Lekeitio euskotarra
Javier Llona, vizcaino
1963 Bernardo Tocoua, Hasparne
1964 E. Egana, Ereno dia 21 junio
1965 Pedro Sarratea, Almandoz Valle Esteribar Navarra
1966 Toribio Gallues, Viva Rocaforte Viva Navarra
1967 Jesus Arriaga, Marquina Vizcaya
1968 Luis Lesaca, Navarra
Nicolas Perez, Uztarroz Navarra
1969 Eustakio Mutuberria, Zubieta Espana
1970 Fidel Carricaburu, Arizcun Navarra Espana
1972 Salbador Oteiza, vasko
Julian Urien, Barrio Sagasta Abadiano Vizcaya, años 26
1979 Jose Oleaga, Gernica Vizcaya Espana

APPENDIX V

Methodology for Recording Arborglyphs

Scholarly research requires systematically finding, recording, reading, interpreting, and cataloging the arborglyphs. Below I describe the basics of the methodology I use.

PROBLEMS

The main challenge the researcher faces is time; there is probably not enough time to record all the carvings. The next most difficult problem is reading badly scarred trees. These are normally the oldest aspens; their bark is often very rough, cracked, and split, and the letters are likely to be very distorted. Several techniques may be helpful. First, videotape the carving *very* slowly from about three feet away. Then, repeat the procedure close up using a macro lens.

Video catches details that the eye sometimes does not, and close examination of the videotape may help you to decipher some of the letters. Try backing away from the trunk and looking at the whole carving rather than at each letter. Another technique that sometimes works is to sketch the carving, "repairing" the cracks as best you can.

Other difficulties involve the carver's deficiencies. Many words are slang, and others are misspelled. A high degree of familiarity with the culture of Euskal Herria is a must to undertake this work. At times only a few letters remain and the researcher must, as a last resort, make an educated guess based on that knowledge. For example, if only the letters ET-H-V—-Y can be read, you can conjecture that they are part of the surname Etcheverry. A minor problem is the translation of highly colloquial words into English.

RESEARCH TOOLS

A camcorder is the most useful and economical tool to record aspen carvings. I started, as instructed, with a regular 35mm camera and a small plain-paper notebook, and then discovered video. Its versatility and advantages over still photos were indisputable, but I had to argue with the state Office of Historic Preservation and federal archaeologists, who favored the more traditional tools.

I began with a regular VHS camcorder but soon graduated to an SVHS camcorder,[1] which produced broadcast-quality images but was also bulkier and heavier, a serious drawback. No one seems to know how long the videotapes will last, but according to vendors, the shelf life of an SVHS tape should be more than twenty years.

Then the small size and picture quality of 8mm camcorders got my attention. I now use a Hi8mm camcorder equipped with the "steady-shot" feature. Although it is slightly larger and more expensive than an 8mm camcorder (the tapes cost almost twice as much), it possesses superior dubbing capabilities. Of course, you can also use professional camcorders costing thousands of dollars, but they are much too heavy and awkward to operate in aspen groves. The newer digital camcorders with the 3-CCC system are the best alternative because they offer professional quality in a small, light package that can be downloaded directly into a computer or the Internet, and copies can be made without generational loss.

Video became such an integral part of the research that I stopped taking notes and drawing sketches in the field—a time-consuming task—and started transferring data from film to notebook at home, a perfect winter activity. That also allowed me to use my field time exclusively for videotaping.

In conjunction with the video I am building a computer database.[2] I started using dBase IV but switched to Alpha Four. Both are fairly compatible with other database software programs.

In the latest development, I now transfer data directly from the camcorder to the computer. I have thus eliminated sketching altogether—anathema for most traditional archaeologists. Sketching the arborglyphs takes a lot of time, and even the most accurate sketch cannot match the fidelity of video. When I need a hard copy of an arborglyph, I print one directly from video.[3]

The arborglyph research involves many thousands of images, but it is possible to streamline data handling by feeding the video directly into the computer, freezing the best images (using a video-grab program), and storing them. They may be manipulated—for example, to get rid of unwanted scars—before being stored.[4] As technology stands today, DVD seems to be the most suitable medium to store and disseminate the glyphs. Digital and computer images have the advantage of lasting virtually forever. They do not fade like black-and-white photographs—still the medium of choice for many museums—nor do they age like videotape.

I use scanners and image-manipulation software to overcome the ever-present problem of bark distortion.[5] Sometimes it has been possible to decipher a message by inverting the digital image. Inversion can help detect certain shades in the groove, which in turn enables the viewer to distinguish the original incision from other grooves caused by natural splits in the bark. In 1990 I started using a Macintosh computer and "Digital Darkroom" to remove many distortions from the arborglyph images and to obtain a more faithful approximation of the original carving executed by the sheepherder. Today this type of technology is much more sophisticated and accessible.

MAKING THE MOST OF THE VIDEO

Video has another clear advantage over the traditional tools: the camera can be "walked" around the tree. Often the carvings gird the aspen and cannot be photographed from a single position. When two or more photographs of the same inscription are taken, matching them sometimes becomes a chore. Videotaping solves this problem. More important, the reading of the arborglyphs, their translation, and any comments can be made on camera. Thus you have voice and image together.

When videotaping, use manual focus whenever possible. Autofocus tends to flicker back and forth as it gets "distracted" by the "quaking" leaves in the background. Manual focus can be sharper, and it saves the battery as well, a good thing when you are hours away from the closest electrical outlet. I highly recommend purchasing chargers that work off the cigarette lighter in the dashboard. Take extra batteries to the field, and keep them warm. Tripods are burdensome to carry and a bother to operate in the forest, but they are indispensable. A video taken without a tripod is extremely tiring to watch, even when the "sure shot" feature is used.

Shadows are the enemy of video and still cameras. The combination of sun, bright surfaces (aspen trunks), and shadows has a particularly negative effect on a Hi8 camcorder, which works best in the shade. I use a neutral medium dark filter to minimize the imbalance. Another alternative is to remove branches that cast heavy shadows on the arborglyphs to be recorded. Or carry a cloth to block the sun. If you have no assistant to hold it for you, use a stick to hang the cloth; you can operate the camcorder with your free hand. Eliminating glaring contrasts makes all the difference in the quality of the video.

I continue to take still photographs of the most significant arborglyphs. Color slides, too, are very useful for presentations, but they are short-lived in comparison to black-and-white photographs, which last a hundred years or more.

When taking still pictures, one exposure may not be enough to cover the whole carving. In that case one of two procedures may be followed: either try to include as much as possible with one shot, or take several exposures. In the latter case, be sure to overlap them slightly, and always stand the same distance from the tree.

If you decide to limit photographing to one exposure, make sure that you see the basics of the message in the viewfinder. In a common inscription consisting of first name, last name, and date, the crucial elements are the last name and the year. They outweigh the first name and the month, for example.

GETTING READY FOR THE FIELD

Since there are so many unrecorded groves, how should priorities be established? The most heavily carved mountain ranges should be recorded first. Aspen groves known to have been the site of sheep camps are sure to contain arborglyphs. Waste no time searching through small or young groves. Record first the most extensive groves with large trees, as they normally contain older and more significant carvings.

Some herders were compulsive carvers who inscribed numerous repetitious glyphs in the same grove. It isn't necessary to record all of these. For example, if the carver inscribed his name and the same date in four adjacent trees, recording the clearest one will be sufficient. Ten percent or more of the arborglyphs found so far may be in this category. If the dates are different, however, videotape all the inscriptions.

Many carvings have a definite format consisting of first name, last name, date, month, and year. Their variations, though numerous, contain repetitions. Many of the first names, for example, are Jean, Pierre, Baptiste, Batita, Martin, Francisco, Pedro, Juan, etc. This knowledge can be helpful in reading carvings that are distorted or lack one or several letters. The same method does not work as well with the last names, which vary much more.

A "Tree Carving Dictionary" of the carvings is being developed. Eventually it will contain every word found on the trees. When tree distortion prevents reading a whole word or message, the dictionary can be consulted for the closest word known to have been carved by the herders. The name of the mountain range or forest where a particular word occurs could be added to the dictionary entries as a helpful feature.

MAPPING THE SITE

Elaborate site maps have been an essential part of archaeology for decades, but painstakingly accurate mapping is unnecessary for studying arborglyphs. Quite simply, if you see aspens, you may have arborglyphs; if there are no aspens, there is no research site. Some archaeologists pinpoint and number each carved aspen. I think this fastidious approach is both unnecessary and unwarranted when there remain so many unrecorded trees. It is sufficient to outline the parameters of the groves on the map. In fact, most aspens are found alongside creeks and canyons or near other bodies of water. If the recording methodology described below is observed, relocating each tree, if it should be necessary, should not be a difficult task.

At the beginning of each research day, record the date and the time on the tape. Next, before leaving home, spread the map of the state on a flat surface, or hang it on the wall, and record the area to be researched. Replace the map with a 7.5-minute topographic map of the area in question showing the quadrant, section, and township numbers, and finish by zooming in on the creek or canyon you are going to record. Use a colored pencil to delineate the periphery, if known. Provide the official names of the creek or camp, if known. Sometimes local people have different names for them.

As you drive to the field, you can stop and videotape paved road numbers that lead to the mountain as well as gravel road signs, cattle guards, or other salient topographic features such as the entrance of the canyon, the creek, and so forth. More detailed information is needed for the remotest areas, less for the more accessible groves. When you reach the grove, provide a detailed description of your position by using a Geographic Positioning System (GPS) device.

COLLECTING DATA

As most groves are surrounded by higher mountains, the first thing I do at a research site is climb up and take a panoramic video of the grove in question, indicating the roads and salient geographic breaks.

Always begin at the bottom of the grove; that is, work uphill. This should be simple because many groves are located along creek beds and in canyons. Use the same principle for recording intermittent groves in wet meadows. The herders entered the groves at their lowest point and from there proceeded to higher country. It makes sense to retrace their steps in the same fashion.

Walk from one tree to the next nearest one. Walk around every tree to avoid missing what is carved on the other side. When proceeding from one tree to another, always give the necessary details that will help others locate the aspen. For example: "This tree is about twenty feet north of the last recorded aspen."

The creek is an excellent marker. Note every time you cross it. If a particular carved aspen stands near the water, I always try to get the creek in the shot. If someday you need to return to the grove, this procedure will facilitate finding the arborglyph.

As a rule, take a minimum of two videos: one of the whole trunk and its surroundings from about ten to fifteen feet away, the other a close-up of the carving itself. If the tree is carved all around, "walk" the camcorder around it. Depending on the message and the type of carving, a very tight shot may be needed.[6]

As you videotape, first read the message aloud in its original language, then translate it into English. If the letters are not clear (which is not uncommon), move close to the tree and use your index finger or a pencil to point out the individual letters, spelling words as much as you can.

Make the necessary comments explaining the significance of the message by addressing such details as:

Is the carver's name new or a repeat?

Where are his other carvings located?

Is he an artist deserving special mention?

Every comment is important because this is the information that you will depend on for subsequent analyses.

Often the same tree has two or more carvings by different people, sometimes dated decades apart. This should be made clear by dealing with each one separately, but without neglecting to comment on their connection, if any.

An assistant is invaluable for this research because four eyes are better than two. One can take photographs while the other uses the video.

ORAL INFORMATION

The nuggets of information retrieved from trees can be amply supplemented by interviewing old herders of the area. In fact, I strongly recommend it. Often they can provide valuable details about carvers who left scant information about themselves. But conducting interviews is not always easy. In my experience, most herders are tight-lipped or prudent and will not tell a lot of what they know. I have a technique that in many cases works pretty well.

I run the video of their former sheepherding grounds along with their own carvings or those of their former companions. Their reaction is usually immediate. They loosen up and start "remembering" a great deal more. Allow them to comment at their own pace on the various carvings. Images of their old mountain ranges, the groves, and the old sheep camps evoke strong emotions and past memories.

INTERMOUNTAIN ANTIQUITIES COMPUTER SYSTEM FORMS

Using a camcorder eliminates the need for the elaborate sketches that federal archaeologists usually file with every IMACS or similar site form. Recorders have spent a great deal of time making these sketches. I know; I used to be one of them. But we must free ourselves from old habits. Sketches are not needed if we have videos of the inscriptions and VCRs or monitors to watch the videos.

The main problem with existing IMACS forms is that they lack guidelines to properly evaluate arborglyphs in their own historical context. It is no wonder that all the previously filed forms I have seen answer question number 23 of part A in the same way: "Not eligible for the National Register." Of the hundreds of groves in eleven states of the western United States containing tens of thousands of inscriptions, is not even ONE grove eligible? In my view, all of them are eligible.

How do we evaluate the significance of the arborglyphs? The same way other historical data are evaluated. Basically, groves that offer the most information on the topics listed in the IMACS Supplement Form should rank as more valuable. A grove that addresses all or most of the themes might be outright eligible for the National Register.

Other considerations may add to the merit of a grove: the originality of the artistic figures, the quantity and quality of the information supplied by the statements, and the state of preservation of the arborglyphs. Some outstanding carvers deserve to be singled out as artists or informants. Statements in the Basque language deserve special attention. The herder who carved in Basque was experimenting and using his knowledge of French and Spanish orthography to spell in Euskara. Another consideration is how many different names of herders appear in the grove. The more names, the richer the grove may be.

It is more difficult to assess the value of isolated groves. Herders moved from range to range, from winter ranges in the lowlands to the high country in the summer. Perhaps evaluations should be based on allotments. The whole summer range allotted by the government to each sheep company could be considered as a unit. The arborglyphs found in such an arca tell thc story of that particular sheep outfit.

IMACS SUPPLEMENT FOR RECORDING ARBORGLYPHS

1. Aspen Grove
   - Large (more than 100 inscriptions)
   - Medium (25–100)
   - Small (fewer than 25)
   - Actual number of inscriptions
   - How many readable/understood
   - How many unreadable/partly understood
   - Number of textual messages
   - Number of graphics
   - Oldest date
   - Most recent date
2. Human Group Identity
   - Basque (from France and Spain)
   - Hispanic (Mexican, Chilean, Peruvian, etc.)
   - Spanish
   - Other (American)
3. Historic/Cultural Themes
   - Number of different carvers (counting initials)
   - Number of years covered
   - How many more than 50 years old
   - Language (Spanish, Basque, English, etc.)
   - Country/region of origin
   - Patriotic (Gora Euzkadi, Biba Amerika, etc.)
   - News on sheepherding
   - News on sheep bosses/wages
   - News on pasture/weather
   - Personal statement (loneliness, etc.)
   - Female/erotic/sexual-related messages
   - textual
   - graphic
   - Old Country fiancée
   - American women (prostitutes)
   - Self-portraits (graphic)
   - Interpersonal matters among herders
   - Animal figures
   - Duration of the job
   - Humor, swear words
   - Symbols (crosses, etc.)
   - Old Country (houses, etc.)
   - stars
   - The "good-bye" ritual
4. General Assessment of the Grove
   - Extremely important
   - Important/significant
   - Average
   - Poor

Contains
sheep camp
outdoor bread oven
harri mutil (stone marker) in the vicinity

5. Management
Interpretative sign suggested
Incriptions to be curated
Vandalism observed
Accessibility

6. Eligible for the National Register
Yes
No justification

NOTES

Introduction

1. In 1997 I noted dozens of carvings on the aspens lining the road to Jarbidge, but I do not remember seeing any in 1968. But they were there, no doubt.

2. In July 1997 I hiked to the saddle directly south of the top of Copper Mountain (northern Elko County) where I remembered seeing the first aspen carvings. Although I came across some interesting glyphs, my recollections did not match the real carvings that remained on the trees.

3. Many of the articles in the Bibliography are based on data gathered by other people.

4. I spent the 1972 Fourth of July holiday up on Little Bighorn Mountain in Wyoming visiting with several sheepherders. I expected to see the range; it snowed and instead I spent most of the time inside sheep wagons. Although I had no plans to record arborglyphs, the nearby aspen groves surely contained them, and I probably would have seen some of them if it had not snowed.

5. "The Lonely Sheepherder," *American West* 1, no. 2 (1964): 2, 36–45.

6. Frances Wallace and Hans Reiss called their study an "extensive search," but they mentioned just one short sentence ("Troke 12 mi") in the two pages of text. Furthermore, the accompanying figures are more fantastic looking than they ought to be because apparently the authors did not differentiate the carver's work from the natural distortions of the tree—or at least did not alert the reader about it. See Frances Wallace and Hans Reiss, "Basquos," *Basque Studies Program Newsletter* (University of Nevada, Reno), no. 5 (1971): 3–5. Also see Philip I. Earl, "Carving the Basque Experience: The Etchings of Sheepherders Become Legitimate Culture," *Reno Magazine* 2, no. 4 (1980): 13; "Basque Folk Art in Nevada: Aspen Tree Carvings Are Called Mountain Picassos," *Las Vegas Sun*, 14 November 1982; "Lonely Sheepherders Carved up Aspens," *Reno Gazette-Journal*, 19–25 October, 3.

7. They are curated at Special Collections, University of Nevada Archives, Reno.

8. Jan Harold Brunvand and John Abramson, "Aspen Tree Doodlings in the Wasatch Mountains: A Preliminary Survey of Traditional Tree Carvings," in *Forms upon the Frontier: Folklife and Folk Arts in the United States*, ed. Austin Fife, Alta Fife, and Henry H. Glassie. Monograph Series 16, no. 2 (Logan: Utah University Press, 1969), 89–102.

9. According to the newsletter of the Folklore Program of Utah State University, August 1992.

10. James B. DeKorne, *Aspen Art in the New Mexico Highlands* (Santa Fe: Museum of New Mexico Press, 1970), 11–12. The book is mostly pictorial and was not intended to be a real research study on the carvings' content.

11. From a 1989 telephone conversation with an NFS archaeologist in New Mexico. Regrettably, I do not remember her name or her address. We will return to the issue of the "pornographic" nature of tree carvings.

12. Richard Lane, "Basque Tree Carvings," *Northeastern Nevada Historical Society Quarterly* (winter 1971): 1–7.

13. See David Beesley and Michael Claytor, "Adios, California: Basque Tree Art of the Northern Sierra Nevada," *Basque Studies Program Newsletter*, no. 19 (November 1978): 3–5, 8; an expanded version appears in "Aspen Art and the Sheep Industry in Nevada and Adjoining Counties," *Nevada County Historical Society Bulletin* 33, no. 4 (1979): 25–31.

14. "Aspen Art and the Sheep Industry in Nevada and Adjoining Counties," 25–31; David Beesley and Michael Claytor, "The Basque and Their Carvings," *Sierra Heritage* (June 1982): 18–21.

15. Brunvand and Abramson published a figure of fish in "Aspen Tree Doodlings," 101. In "Adios, California," 4, Beesley and Claytor stated that fish are rare, but in "Aspen Art and the Sheep Industry in Nevada and Adjoining Counties" they said otherwise. Regarding dogs and coyotes, it is not always easy to distinguish them.

16. Beesley and Claytor, "Adios, California," 8.

17. Ibid.

18. Piles of stones the herders built usually as range markers; these are discussed in chapter 5.

19. It is probable that more articles than those cited here or listed in the Bibliography have appeared in local newspapers and magazines of the western states, but the job of locating them is enormous.

20. The three crew members were Inaki Bizkarra, producer; Auxtin Goenaga, cameraman; and Lourdes Guridi, audio.

21. Actually, they spent only summers at this high elevation, but that was something the poet from Europe had no way of knowing.

22. This volume is an expanded version of that Basque manuscript.

23. It was basically the same type of agreement as the original one with Toiyabe National Forest.

24. English-speaking readers interested in Basques may consult the following sources: Roger Collins, *The Basques* (New York: Basil Blackwell, 1986), which deals predominantly with the early history; William A. Douglass and Jon Bilbao, *Amerikanuak: Basques in the New World* (Reno: University of Nevada Press, 1975), an excellent overview of the subject; and Rodney Gallop, *A Book of the Basques* (Reno: University of Nevada Press, 1970), a good source for Old Country rural culture. Several good books are available on modern Basque political history, among them Robert P. Clark, *The Basques: The Franco Years and Beyond* (Reno: University of Nevada Press, 1980). For a current discussion on the Basques, see Mark Kurlansky, *The Basque History of the World* (New York: Walker, 1999).

25. In 56 B.C. on the northern side of the Pyrenees the Romans encountered a tribe called the Ausci, a term that some believe is reminiscent of Eusk.

26. Decades ago, Basque ethnologists and anthropologists, such as Barandiaran'dar J. M., author of *Euskalerri'ko leen gizona* (Donostia: Benat Idaztiak, 1934), argued that the proto-Basques were a Cro-Magnon type of people. As of today, that is still the most plausible theory.

27. Interestingly, the idea of life in the caves has not been completely erased from popular memory. According to oral traditions that survived to the present century, the Great Goddess Witch of the Basques—Mayi/Maya or Mari—and other *lami* (pre-Christian witches) and *sorgin* (typical witches) dwelled predominantly in caves.

28. Linguistic "family" here makes reference to Euskara and perhaps to certain Caucasian languages, which according to some linguists are related. See L. Luca Cavalli-Sforza, Paolo Menozzi, and Alberto Piazza, *The History and Geography of Human Genes* (Princeton: Princeton University Press, 1994), 276, 300.

29. Thomas J. Abercrombie, "Europe's First Family: The Basques," *National Geographic* 188, no. 5 (November 1995): 74–97.

30. Mark Kurlansky, *Cod: A Biography of the Fish that Changed the World* (1997; reprint, New York: Penguin, 1998), 29.

31. Peter Boyd-Bowman, *Indice geobiográfico de pobladores españoles de America en el siglo XVI*, vol. 1: 1493–1519 (Bogotá: Instituto Caro y Cuervo, 1964), xxx.

32. Michael C. Meyer and William L. Sherman, *The Course of Mexican History* (New York: Oxford University Press, 1979), 162; according to another source, Santa Fe was founded in 1605. See Ralph Emerson Twitchell, *The Leading Facts of New Mexican History*, vol. 1 (Albuquerque: Horn and Wallace, 1963), 333.

33. See A. Martínez Salazar, *Diego de Borica y Retegui 1742–1800. Gobernador de California* (Vitoria-Gasteiz: Diputación Foral de Alava/Arabako Foru Aldundia, 1992), 179, 189; see also Virginia Paul, *This Was Sheep Ranching: Yesterday and Today* (Seattle: Superior Publishing, 1976), 27.

34. Some of them were acting governors; after Micheltorena, the rebel Pío Pico took over the governorship. See *The Works of Hubert H. Bancroft: California*, vol. 2 (San Francisco: A. I. Bancroft, 1885), passim.

35. See Frances F. Guest, *Fermin Francisco de Lasuen (1736–1803), a Biography* (Washington, D.C.: Academy of American Franciscan History, 1973). Sebastian Vizcaino, Quadra Bodega, and Bruno Hezeta are but a few of the names associated with American history; see Sylvia L. Hilton, *La Alta California española. Colección España y Estados Unidos* (Madrid: Editorial Mapfre, 1992). In the interior, many of the commanders—such as Juan Bautista Anza, instrumental in the founding of San Francisco—were Basque; see Herbert Eugene Bolton, *Anza's California Expeditions*, 5 vols. (New York: Russell and Russell, 1966). Furthermore, according to Douglass and Bilbao, the Spanish settlement of Alta California was greatly influenced by the writings and arguments of the Basque Juan Echeveste. See *Amerikanuak*, chap. 4.

36. The Bizkaian Guido Labesarri or Lebesares (his name is written various ways) and his German partners associated with Cromberger in Seville, Spain, imported *ingenios* (machines) into Mexico in 1536 to start mining operations. See J. Mallea-Olaetxe, "Mexicoko lehen euskaldun meatzariak," in *Origen de la Comunidad Vasca en México*, ed. Koldo San Sebastián (Getxo-Gernika: Harriluze, 1993), 83; Alexander von Humboldt, *Political Essays on the Kingdom of New Spain*, 4 vols. (London: Longman, 1822), 3:554. Fausto Elhuyar Lubice, of Iparralde parentage, was hired and sent to Mexico in 1788 to revitalize the mining technology. He founded El Colegio de Minería, but the latest European technology imported from Germany proved inferior to the old *patio* system of silver mining. See Clement G. Motten, *Mexican Silver and Enlightenment* (New York: Octagon Books, 1972), 10–11, 55–65; Angel Martínez Salazar and Koldo San Sebastián, *Los Vascos en México, estudio biográfico, histórico y bibliográfico*

(Vitoria-Gasteiz: Eusko Jaurlaritza-Gobierno Vasco, 1992), 168–69. During the gold rush the basic mining technology was imported from Mexico into California and the Comstock. This included the *arrastre* method, a term probably derived from the Basque *arri* (stone) and *auste* (breaking or grinding action).

37. In 1931 in northern Humboldt County, Nevada, two Basques (Tomas Alcorta and Juan Ondarza) and an Aragonese (Eusebio Azenares) staked claims to what would become the Cordero Mine (now the McDermitt Mine), which has been producing mercury since 1941. Alcorta discovered the cinnabar while herding his sheep on the sage-covered hillside (videotaped conversation with Alcorta's daughter Tomasa, who lives in Winnemucca, Nevada, 1991). I saw part of an article on the history of Cordero Mine titled "Quicksilver . . . ," dated 6 February 1965. It appeared to have been published in Burns, Oregon, and included a photograph of Alcorta, who at the time lived in McDermitt, Nevada. See also the McDermitt Mine brochure (no publisher, no date). Aguer(r)eberry spent a lifetime working his claims and became a well-known figure in the Mojave Desert, where Aguereberry Point is named after him. See George C. Pipkin, *Pete Aguereberry, Death Valley Prospector and Gold Miner* (Trona, Calif.: Murchison Publications, 1971).

38. Even those who did not succeed in digging for gold in California or sheepherding in Nevada had at least traveled and seen some of the world. See J. S. Holliday, *The World Rushed In: The California Gold Rush Experience* (New York: Simon and Shuster, 1981), 95.

39. I was reminded of this by a personal communication from Jesusa Mallea-Yzaguirre of Bend, Oregon. Monolingual Basques tend to substitute *f* for p, thus Galipornia. The substitution of k for *g* is rarer. The h of Idaho is silent in Bizkaia; thus it sounded like "Idao" (E-dah-o). After spending years in the United States many ex-herders returned home still pronouncing it "Idao."

40. Robert Laxalt, "Basque Sheepherders, Lonely Sentinels of the American West," *National Geographic* (June 1966): 870–88.

41. William Douglass, "Lonely Lives under the Big Sky," *Natural History* 82, no. 3 (1973): 28–39.

42. Michele Strutin, "Lords of the Range," *Rocky Mountain Magazine*, 28–35 (no volume number or year found).

43. Robert Glass Cleland, *Cattle on a Thousand Hills, 1850–1880* (San Marino, Calif.: Huntington Library, 1964), 140ff.; Paul, *This Was Sheep Ranching*, 30. For more details on the early Basques in California, see Douglass and Bilbao, *Amerikanuak*, 212–38.

44. Beesley and Claytor, "Aspen Art and the Sheep Industry in Nevada and Adjoining Counties."

45. J. F. Bannon, ed., *Bolton and the Spanish Borderlands* (Norman: University of Oklahoma Press, 1964), 73.

46. The Basques in the Los Angeles area are still prominent in the dairy business. Albert Erreca of Chino, California, told me that the Basques are second only to the Dutch. Interview, September 1998.

47. Byrd Wall Sawyer, *Nevada Nomads: A Story of the Sheep Industry* (San Jose, Calif.: Harlan-Young Press, 1971), 17.

48. As far as sheepmen are concerned, Altube was a latecomer in Nevada (he devoted most of his efforts to cattle, anyway). For example, BLM archaeologist Roberta L. McGonagle, in "Frank Rodriguez, Sheepherder and Artist," *Women in Natural Resources* 12, no. 1 (1990): 40–41, stated that there were sheep on Bates Mountain, south of Battle Mountain, by the 1850s. Sheep were present in the Toiyabe Range of central Nevada by about that time as well, according to forest rangers in Tonopah (personal communication, 1990).

49. *Humboldt Star*, 21 May 1923, 1:1.

50. "An Interview with Frank Yparraguirre," conducted by R. T. King, 22 May 1984 (University of Nevada, Reno, Oral History Program, 1984), 31.

51. This seems to indicate that Spanish was still the majoritarian language. See Ferol Egan, *Frémont: Explorer for a Restless Nation* (Reno: University of Nevada Press, 1985), chapters 19 ff.

52. For information on Garat and Indart I am indebted to Mimi (Garat) Rodden, who provided me with family papers, photographs, and a manuscript by Margaret Brenner Garat (Oakdale, California, 1989).

53. The Spanish Ranch, today part of the Ellison Company, includes almost 50,000 deeded acres around the old ranch and more than 50,000 additional acres elsewhere in Nevada. In total, the company runs 2.2 million acres. Ellison is also the largest sheep operation in the state today with 9,500 ewes (personal communication, Aulene Evans, Spanish Ranch, Elko, 1998). One other ranch in Nevada, Walker Winecup and Gamble, owns more deeded land—some 248,000 acres—but overall is only half as large as the Spanish Ranch (personal communication, Charles Steiner, Sierra Pacific Power Company, Reno, 1998).

54. Spelling change from "Zavala."

55. Mimi (née Garat) Rodden could not find any information in the family archives about this Ramon Garat. When a party of researchers came to the area in the summer of 1997 they could not find this arborglyph, which the museum in Elko sought to curate. The bark had crumbled or was missing.

56. Craig Campbell, "The Basque-American Ethnic Area: Geographical Perspectives on Migration, Population, and Settlement," *Journal of Basque Studies* 6 (1985): 83–89.

57. See Jeronima Echeverria, *Home Away from Home: A History of Basque Boardinghouses* (Reno: University of Nevada Press, 1999).

58. Typically, herds in the Basque Country are small, averaging a hundred head or less, and tending them is a much more confined affair.

59. J. C. of Ely, Nevada, personal communication, 1991.

60. Clel Georgetta, *Golden Fleece in Nevada* (Reno: Venture Publishing, 1972), 348.

61. This is the longest, and a rather pompous, inscription, from which someone later erased the word "gentleman"; see the complete text in *El Morro Trails*, El Morro National Monument, New Mexico, 18th ed. (Tucson: Southwest Parks and Monuments Association, 1987), 9. Eulate was probably from Nafarroa.

62. None of these inscriptions is accented. See John Slater, *El Morro: Inscription Rock, New Mexico* (Los Angeles: Plantin Press, 1961).

63. Ibid.

64. In 1734 in Mexico City, Elizacoechea laid the cornerstone of the famous Colegio de San Ignacio, popularly known as the Colegio de las Vizcaínas (College of Basque Women); see *Los Vascos en México y su Colegio de las Vizcaínas* (Mexico, D.F.: Instituto de Investigaciones Históricas, UNAM, 1987). For Ibarra, see J. Lloyd Mecham, *Francisco Ibarra and Nueva Vizcaya* (Durham, N.C.: Duke University Press, 1927); Vicente Zavala, *Francisco de Ibarra* (Bilbao: Ediciones Mensajero, 1988); and Oakah L. Jones, *Nueva Vizcaya Heartland of the Spanish Frontier* (Albuquerque: University of New Mexico Press, 1988).

65. *Aspen: Ecology and Management in the Western United States*, ed. Norbert V. DeByle and Robert P. Winokur (USDA Forest Service Report RM-119, Fort Collins, Colo., n.d.), 7.

66. E. L. Little Jr., *Atlas of United States Trees*, vol. 1: *Conifers and Important Hardwoods* (USDA Forest Service Publication 1146, 1971), no page numbers.

67. Mark Shaffer, "Arizona's Aspens: White Trees Disappearing from Majestic Landscapes," *Arizona Republic*, 1 August 1999, A1, A18.

68. I have received literature or personal communications from at least four different Forest Service districts in California and a fifth one in eastern Nevada where programs of aspen regeneration are planned or under way.

69. Personal communication, archaeologist Michael Baldrica, July 1999.

70. Little, *Atlas*, 19.

71. I have seen a handful of carved saplings (see Douglass, "Lonely Lives under the Big Sky," 38), but those were exceptions. Saplings are too small to draw on, and the growth of the tree will probably deform the arborglyph beyond recognition.

72. According to Matilda Rogers, *A First Book of Tree Identification* (New York: Random House, 1951), hackberry trees were the first choice of the carvers.

73. Personal communication with the author, 1990.

74. Personal communication with the author, 1989.

75. Information contributed by a herder from Tolosa, Gipuzkoa, in the late 1980s. I cannot remember his name.

76. Mrs. Murphy of Verdi, Nevada, provided the information about this rock.

77. See *Aspen: Ecology and Management in the Western United States*, 96–106. Forestry people periodically burn, cut, or clean up cankerous or diseased aspen groves, and valuable carvings may have been lost in the past during these operations.

78. Videotaped interview, KNPB, 1991.

79. "Aspen Tree Doodlings," 95.

80. Interview with Leon Legorburu, Chino, California, 1998.

81. In Robert C. Lamm, *The Humanities in Western Culture: A Search for Human Values*, 4th ed. (Madison: Brown and Benchmark, 1996), 296.

82. I am indebted to William A. Douglass for this remark; personal communication, October 1998.

83. But this line of thinking can be misleading because chances are very good that these herders made similar comments on the Spanish Civil War in more than one grove. The problem is finding them.

84. Most people agree that Basque peasants are more thoroughly and culturally Basque than their brethren in the cities, who might be bilingual or speak only Spanish or French.

85. Gallop, *Book of the Basques*, 56.

86. For example, the information handed down by the African *dyely* or *griots* may be as reliable as anything written on paper. See *The Norton Anthology of World Masterpieces*, vol. 1, exp. ed., gen. ed. Maynard Mack (New York: W. W. Norton, 1995), 2335–45.

87. Jose Miguel Barandiaran, aided by his companions Telesforo Aranzadi and Enrique Eiguren, excavated dozens of caves and formulated this theory. See *El hombre primitivo en el País Vasco* (Buenos Aires: Ekin, 1953); see also Bosch Gimpera, *Etnología de la Península Ibérica* (Barcelona: Editorial Alpha, 1932).

88. The Pyrenean region, as evidenced by the Paleolithic cave paintings, is home to one of the earliest artistic traditions in Europe. The Basque Country is situated near the geographic center of many of these great prehistoric sites, which lie on both sides of the Pyrenees.

89. Egloga V. See Paul Alpers, *The Singer of the Eglogues: A Study of Virgilian Pastoral* (Berkeley: University of California Press, 1979), 33.

90. Cited by Brunvand and Abramson, "Aspen Tree Doodlings," 91–92.

91. Kit Flannery, "A Dissertation on Arboreal Anaglyphs or Tree Graffiti," *Holiday* 55, no. 4 (1974): 49, 66. See also John Bakeless, *Daniel Boone* (Harrisburg, Pa.: Stackpole, 1939); and Brunvand and Abramson, "Aspen Tree Doodlings," 89–102. By the way, an aspen near Reno proclaims: "Daniel Boone kilt a bhar here, 1776."

92. *Australian Dreaming: 40,000 Years of Aboriginal History*, comp. and ed. Jennifer Isaacs (1980; reprint, Sydney: Ure Smith Press, 1992), 239–60; George Milpurrurru, *Kangaroo*, Ochers on Bark (Sydney: Art Gallery of New South Wales, 1988), poster no. 00094.

93. My thanks to the Txasio family of Markina, Bizkaia, for combing the Basque Country for tree carvings. They found them in beech groves located in summits and other high places where people congregate. Other informants noted the existence of many carvings by hikers and visitors in the mountains above the Sanctuary of Arantzazu, Oñati.

94. Lane, "Basque Tree Carvings."

95. Flannery, "Dissertation on Arboreal Anaglyphs," 49, 66.

96. Shoup cut down the aspen in 1948. It was hollow but still had some green branches. Its rings were counted at the University of Idaho in 1976, yielding sixty-one plus or minus two rings. The core was forty-nine rings, plus or minus one. Higher up in Agency Canyon on the walls of a place called Birch Creek Canyon, Shoup found and read other Indian pictographs, which he interpreted as referring to Josh Jones's party. They said: "Eight persons in companionship with 12 other persons, all being well with them and they are going upon the mountain and travel eastward." Shoup claimed that Josh Jones was an inspector of furs at the port of St. Louis in 1808. He wrote the Missouri Historical Society about his findings, but the society was unable to confirm any of the stories. See "Quaken Aspen Carries Tale of Josh Jones, Early Trapper," *Idaho Recorder Herald* (Salmon), 1 June 1950. See also Merrill D. Beal, *History of Idaho: Personal and Family History*, vol. 3 (New York: Lewis Historical Publishing, 1959), 94–99. I want to thank the archaeology department of Lemhi National Forest for sharing this information with me.

97. I am indebted to James Snyder for sending me copies of the recorded blazes and sharing his insightful observations regarding the use of the land by sheepherders and others.

98. According to Snyder, Yosemite was a backwater for herders; Emigrants Pass, north of the park, and Mammoth Pass or the headwaters of San Joaquin River were much more important. During World War I cattle were permitted in Yosemite to support the war effort, Snyder said.

99. See Jeronima Echeverria, "Basque Tramp Herders on Forbidden Ground: Early Grazing Controversies in California's National Reserves," *Locus* 4, no. 1 (1991): 41–57. Some of the herders in Yosemite moved around a lot and carved their names on aspens in different areas of the eastern Sierra. Antoine or Anton Inda, for example, herded in the mountains over Carson City before going to Yosemite. In 1903 he was cited in Stanislaus National Forest for grazing illegally on reserve land.

100. Fermin Leizaola, *Euskalerriko artzaiak* [The sheepherders of the Basque Country] (Donostia: Etor, 1977), section 8.

101. Luis Pedro Peña-Santiago, *Arte popular vasco*, 5th ed. (San Sebastián: Editorial Txertoa, 1985), 160–62; see also *Artzaintza. Cultura pastoril* (Bilbao: Euskal Arkeolojia, Etnografia eta Kondaira Museoa, 1988), 40–41.

102. As told by Jose Mallea, Munitibar, Bizkaia. Mallea said that the original read: "Me marcho para Argentina," but he did not remember Zabala's first name or the exact year of the carving.

103. Some years ago I wrote: "Aspen art, as found in the American West, is nonexistent in the Basque Country" (in "History that Grows on Trees," *Nevada Historical Society Quarterly* 35 [spring 1992]: 23), but the statement may have been premature.

104. See Sally J. Cole, *Legacy on Stone: Rock Art of the Colorado Plateau and Four Corners Region* (Boulder, Colo.: Johnson Publishing, 1992, 2d printing), passim.

105. Arnie Lynn Cunningham, "Analysis of the Interrelationship of Archeological Surface Assemblages: An Approach to Petroglyphs" (master's thesis, University of Nevada, Las Vegas, 1978).

106. *The Men of the Old Stone Age*, trans. from the French by B. B. Rafter (London: George G. Harrap, 1965), 175. See also Anatoly I. Martynov, *The Ancient Art of Northern Asia*, trans. Dimitri B. Shimkin (Urbana: University of Illinois Press, 1991); Gómez-Tabaneras, *La caza en la prehistoria (Asturias, Cantabria, Euskal-Herria* (Madrid: Colegio Universitario de Ediciones Istmo, 1980), 313.

107. Bear-track carvings are found in bear country, but they are not signed or explained, and it is thus impossible to ascertain whether the herder meant to represent bear tracks or something else.

108. See Jarl Nordbladh, "Some Problems concerning the Relation between Rock Art, Religion and Society," in *Acts of*

the International Symposium on Rock Art, ed. Sverre Marstrander (Oslo: Universitets-forlaget, 1972), 185–210; Campbell Grant, *Rock Art of the American Indian* (New York: Thomas Y. Crowell, 1967), 151.

109. His name was Francisco Goitiandia, a Bizkaian (personal communication from William A. Douglass, 1999).

110. Leizaola, *Euskalerriko artzaiak*, 107.

111. Dale W. Ritter and Eric W. Ritter, "Medicine Men and Spirit Animals in Rock Art in Western North America," in *Acts of the International Symposium on Rock Art*, 97–125; Grant, *Rock Art of the American Indian*, 28–29. In Nevada many petroglyphs have been associated with hunting because they are located near water or in canyons where game passed; see Robert F. Heizer and Martin A. Baumhoff, *Prehistoric Rock Art of Nevada and Eastern California* (Berkeley: University of California Press, 1962), passim.

112. This took place in Humboldt County, Nevada, in July 1991.

113. Grant, *Rock Art of the American Indian*, 38–39.

114. See Donald E. Weaver Jr., *Images on Stone: The Prehistoric Rock Art of the Colorado Plateau* (Flagstaff, Ariz.: Museum of Northern Arizona, 1984).

115. Dekorne, *Aspen Art in the New Mexico Highlands*.

116. Sawyer, *Nevada Nomads*, 165.

117. In the Lakeview, Oregon, area Irish sheepherders outnumbered the Basques, but they seem to have carved few trees. Fremont NF archaeologist John Kaiser led two PIT projects in the summers of 1997 and 1998, and according to reports I received from volunteers, relatively few carvings were found.

118. Yvonne Michie Horne, "Treeffiti," *Westways* 70, no. 11 (1978): 28–29.

119. Ernest L. Abel and Barbara E. Buckley, *The Handwriting on the Wall: Toward a Sociology and Psychology of Graffiti* (Westport, Conn.: Greenwood Press, 1977), 143.

120. But we would not say that about our political leaders, even though their names may be written all over public spaces; and we do not label a 2500 B.C. Egyptian hieroglyph a graffito, although it may well have been one.

121. Robert Reisner, *Graffiti: Two Thousand Years of Wall Writing* (New York: Cowles Book Company, 1971), 203.

122. Ibid., 23.

123. Abel and Buckley, *The Handwriting on the Wall*, 41–42.

124. Alfred C. Kinsey, *Sex in the Human Female* (Philadelphia: W. B. Saunders, 1953), 675.

125. See Alan Dundes, *Life Is a Chicken Coop Ladder: A Portrait of German Culture through Folklore* (New York: Columbia University Press, 1984), passim.

126. Reginald Reynold, *Cleanliness and Godliness* (London: George Allen and Unwin, 1943), 33.

127. Gallop, *Book of the Basques*, chap. 4. Are the Basques always reserved? Of course not; it depends on the audience.

128. This same idea was expressed by Eugenio Sarratea of Reno in a 1989 interview, and Beesley and Claytor quoted a Bizkaian herder who gave the same explanation in "Aspen Art and the Sheep Industry of Nevada and Adjoining Counties," 30.

129. In daily conversation Basques use *ez* (no) at least several times more often than *bai* (yes). In contrast, I would say that in English one hears *yes* more often than *no*.

130. Salvador de Madariaga, *Anglais, Français, Espagnols*, 12th ed. (Paris: Gallimard, 1930), passim.

131. C. Zilboorg, "Loneliness," *Atlantic Monthly* (January 1938): 40.

132. As told by a bartender who heard it in Munitibar, Bizkaia; personal communication, 1995.

133. Robert Laxalt, *Sweet Promised Land* (1957; reprint, Reno: University of Nevada Press, 1986), 1.

134. J. Mallea-Olaetxe, "History that Grows on Trees: Basque Aspen Carving in Nevada," *Nevada Historical Society Quarterly* 35, no. 1 (spring 1992), 33.

135. This aspect of the arborglyphs reminds me of the crowded figures on Paleolithic cave walls, where many engravings, but fewer paintings, are superimposed or even retouched. Annette Laming and André Leroi-Gourhan consider them a conscious association. Magín Berenguer disagrees (see *Prehistoric Man and His Art*, trans. Michael Heron [Park Ridge, N.J.: Noyes Press, 1973, 19–20], 85–86). Superimposition, retouching, adding, and even scratching over happens on trees also, and, I guess, on ghetto walls in the cities.

136. Marija Gimbutas, *The Gods and Goddesses of Old Europe, 7000 to 3500 B.C.: Myths, Legends and Cult Images* (Berkeley: University of California Press, 1974), 148, for example.

137. Based on conversations with two sheepherders, Reno, 1995.

138. From Forest Service IMACS forms.

139. See J. P. Mallory, *In Search of the Indo-Europeans: Language, Archaeology and Myth* (London: Thames and Hudson, 1989), 182–85, 242–43.

140. T. J. of Elko, Nevada, interview with the author, 1990.

141. For example, when Basques notice in the audience a person with a big belly, no one will say, "He has a big belly," because that is obvious. Rather, the statement will run something like, "Han zilbor ttipi bat," which literally means, "There [is] a belly little one [guy]" (or "There is a fellow there with a little belly").

142. For related topics, see Douglass and Bilbao, *Amerikanuak*, 265–97.

143. Georgetta, *Golden Fleece in Nevada*, 320.

144. Conversation with Joe Ciscar, Ely, Nevada, 1992.

145. Arriandiaga's Memoirs, in the Basque Studies Program archive, UNR.

146. This is the modern spelling for the carved form, Vicente Astorquia.

147. Most Basques took trains to reach their destination out west. I have been told many stories about these trips. Unable to say one word in English, many did not eat at all for days. Others were more imaginative. Benturo Bengoechea of Ispazter, Bizkaia, came through New York City in 1902 and took the train to Salt Lake. In New York City he was given a piece of bread and told that whenever he was hungry, he should go to the restaurant car and mimic eating by putting the bread in his mouth. It worked. From Pete Bengoechea (his son), interview with the author, Winnemucca, Nevada, 1991.

148. This figure is based on many conversations with sheepherders. Jean Lekumberry told me that "90 percent" of herders carved (videotaped interview, 1990).

149. The original is uniquely written: "Haqi estubo Felix Mendiola, 1912."

150. *Code of Federal Regulations*, Parks, Forests, and Public Property, 36, pts. 1–1991, rev. 1 July 1992 (Washington, D.C.: Government Printing Office, 1992), 248.

151. According to some reports, aspens are declining rapidly in the American West. If the trend continues, half of them will be gone in four hundred years. See David Tippets, "Acid Test for New Perspectives," *Forestry Research West* (USDA Forest Service, Fort Collins, Colo.) (November 1991): 1–11.

152. See Brunvand and Abramson, "Aspen Tree Doodlings." I spoke with Polly Hammer in 1990.

153. Mark Shaffer, "Trees Are Living History: Basques Made Their Marks 100 Years Ago," *Arizona Republic* (Flagstaff), 1 August 1999.

Chapter One. A Forest of Names: Peavine Mountain and Copper Basin

1. David Beesley and Michael Claytor found "praying hands" in Tahoe National Forest; see their "Adios, California: Basque Tree Art of the Northern Sierra Nevada," *Basque Studies Program Newsletter*, no. 19 (November 1978): 3–5, 8. But all the hands I recorded conveyed the idea of identity.

2. The word is not clear; "corrals" is my best guess.

3. James Snyder, Yosemite historian, letter of 2 July 1996.

4. James B. Snyder, with James B. Murphy and Robert W. Barrett, "Wilderness Historic Resources Survey. 1988 Season Report" (Studies in Yosemite History 1, Yosemite Research Library, Yosemite National Park, California), 27; see also Synder et al., "Wilderness Historic Resources Survey. 1989 Season Report" (Studies in Yosemite History 2, Yosemite Research Library, Yosemite National Park, California), 10.

5. William Shepherd, *Prairie Experiences in Handling Cattle and Sheep* (London: Chapman and Hall, 1884), 186–87. In June 1890, as soon as the bridge over Yosemite was repaired, herders driving twenty thousand sheep crossed it (Snyder, "1988 Season Report," 21, 35).

6. Jerónima Echeverría, "Basque Tramp Herders on Forbidden Ground: Early Grazing Controversies in California's National Reserves," *Locus* 4, no. 1 (1991): 41–57.

7. In December 1997 the Borda Sheep Company donated 1,880 acres in Kings Canyon near Carson City to Toiyabe National Forest. Raymond "Dutch" Borda, the last of the Borda sheepmen (his brother Pete Borda had recently died at Kings Canyon), when asked to comment on the event, said: "We sheepherders don't have much to say. But we love this place." See Tim Anderson, "Deal Protects Carson-Area Land from Development," *Reno Gazette-Journal*, 19 December 1997, D1, D6.

8. "Pinot vale verga, Mateo Pedroarena 1953, 1954."

9. Conversation with Eileen Green, BLM, Carson City, Nevada.

10. Segundo Mugarra, South Warner Mountains, California.

11. Twelve Indas arrived in the West through New York City between 1897 and 1902 from Aldude, Benafarroa. Anton came earlier, in 1895, to Plumas County, California, when he was twenty-one years old.

12. Beesley and Claytor, "Adios, California," 3; Frances Wallace and Hans Reiss, "Basquos," *Basque Studies Program Newsletter* (University of Nevada, Reno), no. 5 (1971): 3–5. I searched the Marlette area but did not find the 1895 tree.

13. Personal communication, 1992.

14. I am grateful to Robert Laxalt for this information.

15. My thanks to Hal Klieforth from the Desert Research Institute for informing me about Little Valley and for taking me into it.

16. Incidentally, the name Sario means "sheep corrals" in Basque.

17. "Ecology" understood as an urban phenomenon.

18. Stephen F. Bishop, district ranger, Tahoe National Forest, Sierraville, California, letter to the author dated 12 April 1990.

19. According to the documents delivered to me.

20. Byrd Wall Sawyer, *Nevada Nomads: A Story of the Sheep Industry* (San Jose, Calif.: Harlan-Young Press, 1971), 8–9.

21. According to Tahoe NF archaeological site records, FS

05-17-57-292, p. 2. Evratchu appears to be Basque, but the spelling may be off.

22. Personal communication, 1992.

23. My thanks to Mr. Foster for providing copies of the arborglyph sites in Plumas.

24. Personal letter from his daughter Josefina Urrutia of Reno dated 12 April 1998.

25. The aspen on Lake Tahoe dated in the 1850s does not have the old authentic look that this one in Plumas has.

26. *Licensing News* (Sacramento, Calif.) 18, no. 1 (1999). My thanks to Mr. Sendek for forwarding a copy of the newsletter.

27. The Espil Sheep Company sold out in 1997 or 1998, although the Espil brothers still manage it for the new owner. Wesley Cook, of Cedarville, California, is the other sheep operator.

28. See Frank Bergon, *Shoshone Mike* (Reno: University of Nevada Press, 1994).

29. See Dayton O. Hyde, *The Last Free Man: The True Story behind the Massacre of Shoshone Mike and His Band of Indians in 1911* (New York: Dial Press, 1973), 175.

30. Personal communication, 1991. His views were also broadcast on KOLO-TV Channel 8 of Reno on 8 February 1991. Some people in the Cedarville, California, area were not happy with the TV station's program. See also J. Mallea-Olaetxe, "Indioek ala bakeroek hil zituzten Nevadako hiru euskaldunak? Erahiltza eta Lertxun Marrak" [Who killed the Basques in Nevada, Indians or Cowboys? The murders and the aspen carvings], *Argia*, no. 1337 (May 5, 1991): 22–25.

31. J. Mallea-Olaetxe, "History that Grows on Trees: The Aspen Carvings of Basque Sheepherders," *Nevada Historical Society Quarterly* 35 (spring 1992): 21–39.

32. It was supposed to be Marie Jean, she said.

33. This is not a name in the Basque Country or in Spain, but Madera is Spanish, and he and Refugio are listed as being from Spain; the others are all Iparralde surnames.

34. Personal communication from Jean (Duque) Wright, 1990.

35. Clel Georgetta, *Golden Fleece in Nevada* (Reno: Venture Publishing, 1972), 351. Later Flanigan went broke; in 1914 he bought the Garaventa Ranch in Weeks, Nevada; ibid., 348–58.

36. Personal communication, 1992.

37. I am indebted to Marcelino Ugalde of Reno for much of the news on the Gerlach Basques.

38. According to Teresa Ugalde, Marcelino Ugalde's grandmother, personal communication, 1998.

39. Georgetta, *Golden Fleece in Nevada*, 339.

40. The area surrounding the lake itself is Indian land, which I did not research.

41. Personal communication, 1990.

42. Francisco Uriarte was twenty-six when he came to Winnemucca in 1898. Juan Pedro came in the same year; he was only fifteen. Juan Jose, thirty-five, was married and came two years later. It appears that Juan Jose and Juan Pedro shortened their names at carving time.

43. Effie Mona Mack, *The Indian Massacre of 1911 at Little High Rock Canyon of Nevada* (Sparks, Nev.: Western Printing and Publishing, 1968), 110–11.

44. Around 1980 Frenchie Montero bought out the Bidarts, his partners. Conversation with Buster Dufurrena, Denio, Nevada, 1990.

45. Interview with Lawrence "Frenchie" Montero, 1991; see also the report on this one-room school in *Reno Gazette-Journal*, 23 October 1994.

46. I often had to interpret the sketches made by BLM recorders, who sometimes were seasonal assistant archaeologists ill prepared for reading the languages used by the herders. Nevertheless, the tally of names is impressive, and I am very grateful to Regina Smith and the archaeology department.

47. Interview with the author, November 1990.

48. This arborglyph was read in 1959 by Jess Lopategui, who herded sheep in the Columbia Basin for several years.

49. Juanita Vicondoa, interview with the author, Eagleville, California, 1995.

50. Information from J. Lopategui, Elko, Nevada, and others. I asked several chemists about this, but none could explain why water would not cook beans at four thousand feet. John Clevenger, of Truckee Meadows Community College in Reno, suggested that the herders quit using the water in Pumpernickel because of its bad taste, and eventually the story changed to "you cannot cook beans with it," and finally to "beans don't cook" (personal communication, 1998). I asked another individual who had done some sheepherding in Pumpernickel; at first he had no recollection of the water problem, but later he seemed to agree.

51. Cruz Bilbao of Elko still remembers life on Altube's Spanish Ranch, where he worked from 1912. Today another Basque, Frank Arregui of Elko, owns Altube's branding iron. Conversation with Cruz Bilbao and Frank Arregui and members of their families. In January 1999 I spoke with Mr. Allustiarte of Stockton, California, whose father, a native of Gernika, Bizkaia, herded sheep in the Independence Mountains in 1906 for a man named Stewart.

52. One such area was Burns Flat (or Basin), according to David Frippo; personal communication, Reno, 1991.

53. According to the U.S. Census of 1900, Pedro Elgart, a herder, arrived in San Diego County, California, in 1890. An Elgart, perhaps the same person, later carved more trees in

at least two other mountain ranges of Nevada, hundreds of miles away.

54. J. L. M., personal communication, Elko, 1990.

55. Krenka is a descendant of Pete Itçaina.

56. See Siege Dechaud, "Basque Sheepherder," *Range Magazine* (Reno) (spring 1991): 16–19.

57. Conversations with Forest Service personnel during the trip to Table Mountain, July 1991.

58. Taped conversation, July 1993. The cash was a significant detail. During a number of conversations I have had with children of Basque immigrants, they often provided that specific information with pride. The conveyed message was that their ancestor was not only a hard worker but also frugal, and while others bought sheep with borrowed money—the easy way—their grandfather, father, or uncle paid with cash.

59. Jess Goicoechea knew "Amoto Txiki" (Litte Amoroto, Onaindia's nickname). One time Onaindia fell off the wagon and broke his leg. There was no one around, so he had to climb back up on the wagon and drive it to town, which was miles away. He made it. "He was a tough guy," Goicoechea said.

60. My thanks to Joe Ciscar and Sharon Etcheberry for their assistance while my son, Erik, and I recorded the arborglyphs in White Pine County, Nevada. Sharon photographed John's Wash for me. They not only volunteered to be our chaperons and guides, but also graciously took us in as if we were family—and I had never met them before. My teenage son is no stranger to Basque culture, but he was nevertheless impressed by that welcome. "Basques are incredible," he said, beaming.

61. Mañiz (Jim) Ithurralde, interview with the author, 1991.

62. Jerry Etcheverry, personal communication, Winnemucca, Nevada, 1995.

63. Today there is a Swallow Ranch west of these mountains.

64. He was related to the Bidarts, part owners of the Leonard Creek Ranch in northern Humboldt County; personal communication, Frenchie Montero, 1990.

65. For further details, see William A. Douglass, "The Vanishing Basque Sheepherder," *American West* 17, no. 4 (1980): 30–31, 59–61.

66. We reached the high meadow from the back side—the road came in from the other side—and it appeared that wildlife was not used to seeing people arrive from that direction.

67. See Beltran Paris, as told to William A. Douglass, *Beltran: Sheepman of the West* (Reno: University of Nevada Press, 1979).

68. Information from Nikolas Olaziregi, personal communication, 1996. Some of the information regarding sheep activity in Cherry Creek was gathered from an interview with Al Arnold, owner of the Currie Store, in Currie, Nevada, September 1990.

69. Based on carving activity, the high periods of sheep industry in Nevada were—notwithstanding the Depression—1925–39 and 1953–71. Not enough early arborglyphs survive to allow drawing further conclusions.

70. Some sheepman told me that these wagons are easiest to tow on the highway, surpassing even modern trailers, which wander from side to side behind the truck. Blanton Owen of Reno was an expert in sheep wagons.

71. Partially recorded by Esther Dew of Modoc National Forest.

72. See Marie-Pierre Arrizabalaga, "A Statistical Study of Basque Immigration into California, Nevada, Idaho and Wyoming between 1900 and 1910" (master's thesis, University of Nevada, Reno, 1986), 40, 72–73.

73. Ibid., 82.

74. I have seen maps indicating 8,260 and 8,270 feet as well.

75. See Carmelo Urza, *Solitude: Art and Symbolism in the National Basque Monument* (Reno: University of Nevada Press, 1993), 91.

76. Ibid., 20–28. Elsewhere we will have the opportunity to compare the romantic notions of urbanites, as symbolized in this monument, with the real life of the sheepherders.

77. It occurs to me that if those involved in the monument project had been more cognizant of tree art, some aspects of it could have been incorporated into Basterretxea's piece. On the other hand, we wouldn't have the present controversial sculpture, and controversy is an essential part of Basque culture.

78. The announcement of the unveiling in that day's *Reno Gazette-Journal* was decorated with a Basque flag and colors; see p. 3-A.

79. The arborglyphs in Dog Valley and Long Valley were carved by the same herders, but they are not included in the count.

80. To be precise, the date is a toss-up between 1949 and 1947.

81. An almost verbatim transcription from taped conversation, July 1994.

82. In the 1990s this area was totally transformed by unrelenting development.

83. Personal communication, June 1998.

84. Conversations with Jeannette Landa of Reno, 1993; and Candido Olano of Cedarville, California, 1994.

85. Videotaped conversation, October 1992.

86. Reginald Meaker, *Nevada Desert Sheepman*, as told to Sessions S. Wheeler and Gerald Meaker (Sparks, Nev.: Western Printing and Publishing, 1981), 104, 123.

87. C. Elizabeth Raymond, *George Wingfield: Owner and Operator of Nevada* (Reno: University of Nevada Press, 1993), chaps. 9 and 10.

88. Her daughter Jeannette married "March" Landa.

89. Interview with the author, Reno, 1991.

90. Thirty dollars was the standard wage, according to Reno sheepman Abel Mendeguia; personal communication, 1994.

91. "Viva la depression de 1932" in the original.

92. Parts of the data from Peavine were first published in UNR's *Basque Studies Program Newsletter*, no. 43 (April 1991): 3–6.

93. In the early 1960s Sarria used to go up to visit Juan Zabalegui and Antonio Ojer, two herders from Nafarroa whose names I recorded. On one such weekend, Sarria told me, a military airplane crashed in a nearby Peavine peak, scaring the daylights out of them (personal communication, October 1998). It might be objected that my conclusions could be flawed if some of the names on the trees were carved by Basques living in town, but the methodology I used has built-in safeguards against such pitfalls.

94. Videotaped interview, Gardnerville, Nevada, winter 1991.

95. Personal communication from Juan Zabalegui of Reno, nephew of the carver in question, 1992. Many people know Juan, or "Johnny," who tends bar at Louis's Basque Corner in downtown Reno.

96. As a general rule, the carvings will be translated and quoted in English with the understanding that the original is in Castilian Spanish, although in special cases I include the original as well. I consider all the messages in Basque significant, and many of them are included verbatim along with the translation.

97. The political platform and war cry of those who held to the Carlist ideology, which was particularly strong in Nafarroa, Zabalegui's homeland.

98. In the Basque community one hears of extreme cases, like the herder who never got paid. He worked for his uncle, a big sheepman, and during forty or more years as sheepherder he never saw a penny of his pay. One day he retired and settled his account—with interest, of course. Everyone agreed that he could never spend all of it.

99. The message in Basque says: "Hemen hitza hunak," written in the dialect of Iparralde, but the carver was Bizkaian. This mixing of the dialects on the range is discussed in chapter 2.

100. The Basque original says: "Gosho da itzalian."

101. As told by his cousin P. L. of Bakersfield, California; personal communication, 1993.

102. *Chingada*, a favorite Mexican swear word, normally refers to the act of fornication, but according to *Cassell's Spanish Dictionary* (1982), *chingar* also means "to get drunk," and in South America "to shit." There is little doubt of which of the three Rodriguez meant.

103. According to Reginald Meaker there were two sheep owners named Padera (should be Pradera): Martin, who owned eight thousand head, and Jim, who owned two thousand. Meaker himself warned about possible misspellings of Basque names; see *Nevada Desert Sheepman*, 146–47.

104. Some people took on the task of "cleaning up" the erotic material. Brunvand and Abramson reported a lot of four-letter-word "sanitation" in Utah's aspen groves as well; see Jan Harold Brunvand and John Abramson, "Aspen Tree Doodlings in the Wasatch Mountains: A Preliminary Survey of Traditional Tree Carvings," in *Forms upon the Frontier: Folklife and Folk Arts in the United States*, ed. Austin Fife, Alta Fife, and Henry H. Glassie. Monograph Series 16, no. 2 (Logan: Utah University Press, 1969).

105. A. A., personal communication, 1994.

106. High and Low Nafarroan dialects are spoken in geographically adjacent areas. The former is from the mountains of northern Nafarroa; the latter is spoken in Benafarroa, in France.

107. In Basque it says: "Nire maite Marie hemen banuse, nik chikua harekin." Educated people would say that that is grammatically incomplete, but sheepherders would agree that it is sharp, clear, and to the point. "Banuse" should probably be "banizu."

108. I assume that Linney is a woman, though I have no proof. The name might be misspelled.

109. The message is for Jose Otazu and Juan Zabalegui. They and Mac were born in the same town.

110. All the messages are in Castilian—these herders spoke little Basque—with some English words mixed in, an indication that the carvers were not recent arrivals.

111. Data gathered on Peavine suggest that, with few exceptions, herders did not last more than six or seven years on the job. What is more, no fewer than fifty-seven of the names appear during only one year.

112. On the Altubes and Spanish Ranch, see Inaki Zumalde, "Un pionero vasco en el oeste americano," *Aranzazu* 51 (1971): 22–23. As for the Garats, I have in my possession "Family Archives: Garat and Indart, Brenner and Eastham," a manuscript written by Margaret Brenner Garat (Oakdale, California, 1989). For a general history on early Elko

County, see Edna B. Patterson, Louise A. Ulph, and Victor Goodwin, *Nevada's Northeast Frontier* (Sparks, Nev.: Western Printing and Publishing, 1969); see also "Pedro Altube, Spanish Ranch Founder, Honored," *Elko Free Press*, 13 February 1960, 3.

113. Georgetta, *Golden Fleece in Nevada*, 117–20.

114. Richard Harris Lane, "The Cultural Ecology of Sheep Nomadism: Northeastern Nevada, 1870–1972" (Ph.D. diss., Yale University, 1974), 26–27.

115. *Twelfth Census Report Taken in the Year 1900*, vol. 5: *Agriculture* (Washington, D.C.: U.S. Census Office, 1902), 461. It must be noted that just ten years earlier the figure was 43,655 sheep, after the catastrophic losses of the 1889 winter; see also Richard Lane, "The Cultural Ecology of Sheep Nomadism," 52–56.

116. *Census of the United States Taken in the Year 1910*, vol. 7: *Agriculture, 1909 and 1910* (Washington, D.C.: Government Printing Office, 1913), 72, 80.

117. *1992 United States Census of Agriculture*, vol. 1, pt. 28: *Nevada* (Washington, D.C.: Government Printing Office, 1994), appendix C, C-17.

118. J. G. of Elko, Nevada, videotaped conversation with the author, 1990.

119. See Arrizabalaga, "Statistical Study of Basque Immigration into California, Nevada, Idaho, and Wyoming," 73. That is considerably fewer than Humboldt County's 84 individuals in 1900 and 275 in 1910. However, the greatest concentrations of Basque immigrants were not in Nevada but in California and in Boise, Idaho.

120. E-mail correspondence from archaeologist Fred Frampton, Humboldt National Forest, Elko, October 1997.

121. Georgetta, *Golden Fleece in Nevada*, 362 ff.

122. Beltran Paris eventually became a prosperous sheepman and rancher himself in White Pine County, Nevada. See *Paris Beltran: Basque Sheepman of the American West*, 39–42.

123. Georgetta, *Golden Fleece of Nevada*, 122–24.

124. Goicoechea is a biologist in the Reno BLM office; personal communication, February 1998.

125. Incidentally, PIT volunteers found the remains of a previously unknown log cabin at the bottom of the basin, and Forest Service historians and archaeologists are looking into the possibility that it is Goicoechea's original homestead. Basque immigrants tend to build with stone rather than logs, but it is also true that a log house goes up a lot faster than a stone one.

126. Georgetta, *Golden Fleece in Nevada*, 380.

127. Humboldt-Toiyabe National Forest Historic Range Files, Elko District, Nevada. I am indebted to Fred Frampton for supplying the maps and other information pertinent to Copper Basin.

128. Fred Frampton, Humboldt-Toiyabe National Forest, Elko, personal communication, 1997.

129. According to Jeannette Landa, the main reason herders switched outfits was food and drink. They were always looking for better fare.

130. This situation could be peculiar to the Ellison Company.

131. The PIT project lasted a week, and by its end we had recorded more than twelve hundred carvings. We divided the volunteers into teams of four or five and assigned each individual a task. One handled video, another the still camera, a third sketched the glyphs and filled out the IMACS form, and the other scouted ahead in search of arborglyphs, which were flagged with a ribbon. The ribbon was removed after the carving was recorded.

132. Personal communication, Reno, February 1998.

133. There was an earlier Arretche in Copper Basin, Enrique, dated 1915. He may have been Arno's relative, and perhaps facilitated his sheepherding job in Nevada. The Arretches are numerous in the Surprise and San Joaquin Valleys of California.

134. Laxalt, *Sweet Promised Land* (Reno: University of Nevada Press, 1957), 82–92.

135. There are no written accents in Euskara.

136. Basque women have always enjoyed a great deal of respect. In fact, some anthropologists have argued that Basque society is or was matriarchal in nature.

137. Oyharcabal should be written "Oyharçabal," and Lasaka is probably "Lesaka," a town of northern Nafarroa, not in France. But in this case France denotes the carver's origin.

138. Susie's is an establishment in the red-light district of Elko.

139. It sounds better in the original: "Locas paciones [*sic*] y andar solitario."

140. The original says: "Amor fugaz dinero y perberso."

141. The original—"Amor maldito, porqué busco seres tormento"—is ungrammatical, like the previous quotation.

142. For example, the Peruvian herder we talked to in Copper Basin was married, as was his coworker, the old Mexican he mentioned.

143. "Mueran las cabezas rojas." Although in English this might be taken as a reference to a woman, the Spanish grammar indicates otherwise.

144. The Castilian grammar in this message is so butchered that it guarantees the Basque origin of its author: "No pregunte a que patria haser por ella."

Chapter Two. Culture in the Carvings: Euskara, Homeland, Politics, and English

1. Confessions of Santi Basterretxea, of Reno, translated from Basque. Basterretxea and I were classmates in primary school in Munitibar, Bizkaia, where the "wallet" he mentions became a nightmare for the majority of children who spoke only Basque.

2. The Spanish Fascist Party, the only legal political party.

3. For example, the artwork of Goiko (short for Goikoetxea), who was popular in the Basque Country in the 1950s, often portrayed *baserritarrak* as slow or ridiculous.

4. And there was the awful thing called "the wallet," which was given to the unfortunate student caught speaking Basque. At the end of the week, whoever had the wallet had to stuff in it one *peseta* and hand it to the teacher. The government directives were not applied with the same rigor in all the Basque towns. Much depended on the local political environment, and especially on the teacher's personal convictions.

5. Eugen Weber, *Peasants into Frenchmen: The Modernization of Rural France*, 1870–1914 (Stanford: Stanford University Press, 1976).

6. See James E. Jacob, *Hills of Conflict: Basque Nationalism in France* (Reno: University of Nevada Press, 1994), chaps. 1 and 2.

7. Ibid., 55–57.

8. A good example is this carving found in the Lake Tahoe area: "JULLEITE LE 29 POUR LA NEE 1937 MONSIEUR ARNAUD JAUREGUY JE SUIS FRANCAIS Y QUERO SER TANBIEN GOOD BY SIR VERY Y FRIENDS."

9. Joseba Intxausti, *Euskara, euskaldunon hizkuntza* (Vitoria-Gasteiz: Eusko Jaurlaritza, Hizkuntza Politikarako Idazkaritza Nagusia, 1990), 69–74.

10. Xabier Quintana, "Literatura euskerica de los siglos XVI y XVII" (paper presented at Conferencia en el Ateneo de Madrid, 27 January 1976).

11. Although *burua* (the head) is the common form, in some Iparralde dialects it sounds like *buria*.

12. According to a 1995 poll, 1 percent of the people were monolingual Basques, 21 percent were bilingual with 6.7 percent being more fluent in Euskara and 8.4 in Erdara (French or Spanish), and 69 percent spoke only Spanish or French. See *Euskararen jarraipena. La continuidad del Euskera. La continuité de la langue basque* (Vitoria-Gasteiz: Eusko Jaurlaritza-Gobierno Vasco, 1995), 62–63.

13. Only recently have great strides been made in educating the Basques in their own language. In 1998, 29.5 percent of the elementary school children in Euskadi (Bizkaia, Araba, and Gipuzkoa) studied in Spanish, 26.9 percent attended bilingual schools (50:50 Spanish and Basque), and 43.6 percent received all instruction in the Basque language. See *Noticias de Euskadi*, Boletín de Información sobre la Comunidad Autónoma del País Vasco, no. 16 (Septiembre 1998): n.p.

14. John P. McKay, Bennett D. Hill, and John Buckler, *A History of Western Society: From Absolutism to the Present*, 5th ed., vol. 2 (Boston: Houghton Mifflin, 1995), 667, 853–54.

15. Claudio Lozano, *La educación en los siglos XIX y XX* (Madrid: Editorial Síntesis, 1994), 120–22.

16. Includes Euskadi and Nafarroa; see *Noticias de Euskadi*, no. 16. The government of Euskadi commits 25 percent of its total budget to education (ibid.).

17. In the 1540s Bishop Zumarraga sent textbooks from Mexico to Durango (Bizkaia), his hometown, for the public school run by a priest. See J. Mallea-Olaetxe, "Juan Zumarraga, Bishop of Mexico, and the Basques: The Ethnic Connection" (Ph.D. diss., University of Nevada, Reno, 1988), 338; Euskaltzaindia, *Euskararen liburu zuria* (Bilbao: Euskaltzaindia, 1978), 350. The word *men* here must be taken literally, as few women probably attended school.

18. Mikel Zalbide, "Euskal eskola, asmo zahar bide berri," in *Euskal eskola publikoaren lehen kongresua. Primer congreso de la escuela vasca*, vol. 1 (Vitoria-Gasteiz: Eusko Jaurlaritza-Gobierno Vasco, 1990), 211–71.

19. *Eusqueraren berri onac, eta ondo escribitceco, ondo iracurteco, ta ondo itzegiteco Erreglac* (Iruñea: A. Castilla, 1761), 16–17.

20. Etxeberri wrote (modern spelling):

> Burlatzen naiz Garibaiez
> bai halaber Etxabez
> zeñak mintzatu baitire
> erdaraz Eskaldunez.
> Ezen zirenaz geroztiz
> Eskaldunak hek biak
> Eskaraz behar zituzten
> egin bere historiak.

From *Eliçara erabiltçeco liburua* (Bordelen, 1636). Esteban Garibay published extensively about Basque history and culture (*Compendio historial*, 1571), and Baltasar Etxabe, while residing in Mexico, composed an apology of the Basque language (*Discursos de la antiguedad de la lengua cántabra*, 1607), but both men wrote in Spanish.

21. Only selected carvings in Basque are considered in this chapter. Many others are discussed elsewhere in this volume.

22. Sonia Nanclares, "El batua se inventó en América antes que aquí, asegura Mallea. Investiga las huellas de los vascos en EE.UU.," *Correo Español–El Pueblo Vasco* (Bilbao) (5 July 1992): Cultura, 59.

23. Linguists call their dialects High and Low Nafarroan, respectively.

24. Videotaped conversation with J. C. of Ely, Nevada, 1992.

25. Personal communication, Reno, 1991.

26. No one has conducted an in-depth study of the Basque language as it is spoken in the American West. Gorka Aulestia lived and worked in Nevada while working on his *Basque-English Dictionary* (Reno: University of Nevada Press, 1989) but did not touch on the topic. When I suggested to him the merit of incorporating some of the American Basque vocabulary, he answered that such a dialect did not exist. More recently, Estibaliz Amorrortu studied the same subject in Elko and concluded that Euskara in the United States is a myth. She was really too late. The study should have been done in the 1960s at the latest. See "Retention and Accommodation in the Basque of Elko, Nevada" (master's thesis, University of Nevada, Reno, 1994).

27. Conversation with J. I., June 1991.

28. Interview with J. Lopategui, 1991.

29. Pete Bengoechea and Mary Mentaberry of Winnemucca, Nevada, personal communication, 1991.

30. Mary Irueta, personal communication, Winnemucca, Nevada, 1991.

31. Hank Gallues of Reno recalled that his father, an immigrant from Aibar, Nafarroa, used to say that Euskara lacked practicality, and he forbade its use at home. Hank did not learn Basque, but he and his brother Albert and sister Irene speak Spanish perfectly.

32. On-camera conversation with Mari Irueta, 1991.

33. Louise Shadduck, *Andy Little, Idaho Sheep King: An Anecdotal Biography of a Memorable Idaho Sheep Man* (Caldwell, Ida.: Caxton Printers, 1990), 132.

34. The company is no longer in operation.

35. The word is onomatopoeic and originally may have been used for horses.

36. This is the opinion of Bittor Kapanaga and others. Aitor Zuberogoitia, "Elkarrizketa: Bittor Kapanaga" [Conversation with Bittor Kapanaga], *Argia: Euskal Herriko Astekaria* 1, no. 685 (27 September 1998): 26–29. I am not talking about today's Euskaldunberri literature, a lot of which, to a native speaker, reads like a translation from Spanish.

37. Tirso de Molina, *La prudencia en la mujer*. Colección Austral, no. 369 (Madrid: Espasa-Calpe, 1964), 10–11; see also Miguel Herrero García, "Ideología española del siglo XVII. Concepto de los vascos," *Revista Internacional de Estudios Vascos* 18 (1927): 549–69; and Anselmo de Legarda, *Lo "vizcaíno" en la literatura castellana* (San Sebastián: Biblioteca vascongada de los Amigos del País, 1953), passim.

38. The *bertsolari* must improvise and sing rhymed verses on any given topic within a specified time. See Gorka Aulestia, *Improvisational Poetry from the Basque Country*, trans. Lisa Corcostegui and Linda White, foreword by William A. Douglass (Reno: University of Nevada Press, 1995).

39. There once were verse composers among the Hispanic sheepherders of New Mexico as well. Each chose a fancy pseudonym, such as "Rising Moon" or "Funny One," and they competed yearly. The last such meeting occurred in mid-century at El Cañon del Judío, New Mexico. One of their most famous versemakers, and a sheepherder, was a fellow named García. See Albert A. Baca, "El Pastor," *New Mexico Magazine* (November 1982): 47.

40. Clel Georgetta spelled it "Domengo Lauissena" in *Golden Fleece in Nevada* (Reno: Venture Publishing, 1972), 348.

41. Incidentally, in 1995 I met Mrs. Louisena in a retirement home in Cedarville, California.

42. Tentative translation.

43. Could be "hark" (he/she with ergative) or "gaur" (today).

44. In the dialect of Banka *erdi* could also mean "half" or "surprised," but I chose "to give birth," which seems to go along with "prostitute."

45. It is written "crande" but could be "grande," a word that is not used in Basque.

46. Quoted in James F. Varley, *Lola Montez: The California Adventures of Europe's Notorious Courtesan* (Spokane, Wash.: Arthur H. Clark, 1996), 167.

47. This carved aspen was cut and is now in a museum in the Basque Country.

48. A knot on the trunk of the tree is part of the word *egia*, which is debatable. This tree is preserved in the library of UNR's Basque Studies Program.

49. Some of the words are distorted, and the message is not 100 percent clear.

50. During this time many Bizkaian herders in the United States received *Zeutzat*, a mimeographed periodical published in Basque by several Bizkaian priests. It is not likely that this Nafarroan herder benefited from such a publication, and in fact, it does not appear that *Zeutzat* inspired many Bizkaian herders to carve in Euskara.

51. Jose M. Recondo, *San Francisco de Javier. Vida y obra* (Madrid: Biblioteca de Autores Cristianos, 1988), 1006.

52. See Robert Laxalt, "Land of the Ancient Basques," photographs by William Albert Allard, *National Geographic* (August 1968): 244.

53. Original Basque: "Ni nas Bernardo Tocua urte asko Américan eta beti pobre."

54. The American concept of the "black sheep" does not exist among the Basques. Herders, in fact, say that when a

ewe gives birth to twins, she will favor *buru beltza*, the black-headed lamb, over the white one.

55. Conversation with J. Y., 1994.

56. Original Basque: "Bizi tristia duk artzain bizia. . . . Fitezko behar diat utzi ofizio hau luzaz bizitzekotz."

57. Original Basque: "Arnot Ni hemen eneatia noizpa izaynda hementik yuaitia." Then Urruty switched gears to give himself a lift and ended the carving with: "Biba Fantzia 1926" (Up with France 1926).

58. Obvious indications are the place-names that dot the geography of the United States—familiar names such as New York, New Orleans, and New Mexico, and lesser-known ones such as New Ulm. Nevada has Minden, settled by Germans.

59. Some Germans in Nebraska whom I talked to confessed never "looking back" to their European homeland, but that feeling may be influenced by the fact that the United States fought two wars against Germany. See Russell R. Elliot, with William D. Rowley, *History of Nevada*, 2d rev. ed. (Lincoln: University of Nebraska Press, 1987), 254–55.

60. In the original: "De nasionalidad de Rigoitia."

61. The original says: "Ene sorlekia dut Erastan, Eskual (He)rrin." I have not been able to locate Erasta, but it must be a hamlet somewhere in Iparralde.

62. In the original: "Pete Ampo, a 16 D julio 1929 natural de balkarlos probinzia de nabarra casa alonbro arocor." The last word is strange, may be "aroco."

63. (Tranlation): July 1926 Gratian Laxague born (at?) Gorba in the Lower Pyrenees, commune of Baigorry, town of Aldude, house of Yrocho. On the 7 of July of the year 26 Gratian Laxague born (?) in 1901 in France in the Lower Pyrenees in Baigorry at the house of Pritcha Oila . . . (unreadable).

64. There is a duplication of the word *valley* here, as Esteribar means "Valley of Esteri."

65. He carved his name three different ways: Gilen (Basque), Gillaume (French), and William (English).

66. See, for example, J. Ortega y Gasset, *Obras completas*, 2 vols. (Madrid: Revista de Occidente, 1963), 1:187, 2:557–59.

67. This business of "national characters" is a slippery proposition at best and merited sarcastic remarks by Caro Baroja in his *El mito del charácter nacional. Meditaciones a contrapelo* (Madrid: Seminarios y Ediciones, 1970). But that didn't deter Salvador de Madariaga from trying in his *Englishmen, Frenchmen, Spaniards* (London: Oxford University Press, 1931).

68. E. Lafuente Ferrari, *La vida y arte de Ignacio Zuloaga* (San Sebastián: Editorial Internacional, 1950), 412.

69. Original: "Viva los bascos y Yanci Espana Juan Manterola."

70. Original: "1930 Biba Escual Herria Sorhuet Arneguy Vive la France."

71. Original: "1956 Viva Elanchove Vizcaya, Jose Armaolea."

72. Original: "Viva los vascos, Miguel Recarte Errazu Navarra Espana."

73. Original: "1959 Vive la France et Pays Basque Biba Bachenabarre Marc Gortari, August 27." The artist's rendition of this carving appeared on the cover of the brochure for the nineteenth annual meeting of the Western Society for French History, held 6–9 November 1991 in Reno.

74. Original: "1963 Luis Ojer, Viva Navarra siempre adelante."

75. Original: "1963 Viva Baztan Lecaroz Navarra 1964."

76. Original: "Gora Euzkadi B Azkoitia 1966."

77. Original: "Biba Fransia Martin Tocoua 1967."

78. Original: "Aupa Vizcaya."

79. Original: "Francisco Olea Gora Euzkadi gora jaungoicoa Ta lege Sarra." This is the political slogan of the Basque Nationalist Party, founded by the Bizkaian Arana-Goiri'tar Sabin in about 1895. Olea was Bizkaian.

80. Original: "Arriva el Pais Vasko" (no date, no name).

81. Not all Basques were on the same side; for a quick survey, see Robert P. Clark, *The Basques: The Franco Years and Beyond* (Reno: University of Nevada Press, 1980), 24–38, 59–76.

82. Original Basque "Alikoa" is not totally clear (should be "alakoa").

83. Correctly written it should say: "Eusko gudariak gara Euskadi askatzeko eta belarri motxak orpoz zapaltzeko."

84. Should be José Antonio Primo de Rivera, founder of the Falange, the Spanish Fascist Party.

85. Dave Goicoechea of Reno showed me a photograph of a carved *ikurrina* and underneath the slogan "Gora Euskadi Azkatuta" (Long live free Basque Country). He found the tree in White Pine County, Nevada; it is the best sample I have seen.

86. Incidentally, shooting accidents happened more than once on the range, and sometimes with dire consequences. The herder Martin Goikoetxea of Wyoming sang the following *bertso* to a fellow herder who almost bled to death after a bullet went through his torso:

> Zu zera gizon apala
> baina etzera makala
> gure artean zu izatea
> gu pozten gerala
> gorputzean pasa bala

ala ere bizkor zu hola
esan gentzake burni gogorrez
egina zerala.

You are a humble man
but not a weak one
that you are still with us
we certainly are happy about
the bullet went through your body
and yet you are still strong
we could say that
you are made of hard iron.

Azidente gogorra zinez
nola pasa dan jakin ez
ori pasata ardi txabolara
joan zinan oinez
odoletan zinan minez
otoitz aundiak eginez
gure artean nola zauden ni
sinestu ezinez.

In truth, a terrible accident
we do not know how it happened
after it took place, to the sheep cabin
you went on foot
you were in pain and bleeding
as you prayed fervently
I cannot believe how
you are still alive.

From "Martin Goikoetxea: Bertsolaria Ameriketan," audiocassette, San Francisco, California (1995).

87. The original combines Spanish and Basque: "Cuidado que te meto perro donivarra." *Donibarra* means "native of Donibane," but there are several Donibanes in the Basque Country, one called Lohitzun, the other Garazi. The carver was probably from Lohitzun (Lapurdi), located very close to Gipuzkoa, where ETA was very active.

88. No relation to the author.

89. During Garcia's lifetime there were in the Basque Country political and cultural organizations hailing "Irurac Bat" (the three are one—Bizkaia, Araba, and Gipuzkoa), "Laurac Bat" (the four are one—the former three, plus Nafarroa), and "Zazpiak Bat" (the seven are one—all the former, plus Lapurdi, Benafarroa, and Zuberoa).

90. I am grateful to Jim Potashin of Bishop, California, for taking me to this grove.

91. UNR history professor Francis X. Hartigan found this message intriguing. He said that in 1939 the French army was believed to be the strongest in Europe. Did the sheepherder's pity for his country indicate that he knew something the rest of the world did not? Not likely; and as it turned out, France's army proved no match for Germany's.

92. In the original: "Viva Hitler campeon de Francia."

93. The original says: "Viva America patron de todos."

94. For example, *1910 Nevada Census Index*, comp. Bryan Lee Dilts (Salt Lake City: Index Publishing, 1984), in which Antone G. AArutia of Paradise, Humboldt County, is also spelled Antone G AAnutia (p. 1); and Lucio Arechanala of S. Clair, Churchill County, is Lucio Arechauala (p. 4).

95. Told by Joe Eguen of Reno about a friend of his, 1996.

96. Told by Claudio Yzaguirre, personal communication, 1988.

97. According to Barandiaran'go Joxemiel, in the Basque Country the billy goat and goats in general were seen as protectors of the sheep. Fermin Leizaola, *Euskalerriko artzaiak* (Donostia: Etor, 1977), 11. In the late 1980s, a farmer in Bizkaia told me that billy goats have a smell that wards off diseases that can affect other farm animals and that is why he kept one in the stable.

98. Told by Hank Gallues, personal communication, 1991. Louise Shadduck related a similar incident in *Andy Little*, 132.

99. Told by J. T. Lekumberry, Gardnerville, Nevada, 1991.

Chapter Three. Sheepherding According to the Sheepherders

1. Robert Laxalt, *Nevada: A History* (Reno: Univerity of Nevada Press, 1991), 24.

2. Source: *Biztanleriaren inkesta ihardueraren arabera*, I-1988 (Vitoria-Gasteiz: Euskal Estatistika Erakundea, 1988), 93.

3. Fermin Leizaola, *Euskalerriko artzaiak*, 22.

4. Taped interview with eyewitness Joe Ciscar of Ely, Nevada, 1991.

5. Obviously this was carved in the United States (Tahoe National Forest), but it appears to refer to a herder in the Basque Country.

6. Bill O'Neal, *Cattlemen vs. Sheepherders: Five Decades of Violence in the West, 1880–1920* (Austin, Tex.: Eakin Press, 1989).

7. The carver forgot to include the "no." This arborglyph is not signed, but its author was from Santander, Spain.

8. A bumper sticker sold at Basque picnics, and intended for mixed marriages, proclaims: "Living with a Basque builds character."

9. In Bizkaian Basque, "Biergiñe zan ba!" and "Piñ utse zan," respectively.

10. In Argentina and parts of South America the expression "word of Basque" carries more weight than a written

document, and some of these qualities may be discernible in the United States as well. In 1992, for example, Ana Hachquet of Elko and her family went to a restaurant that accepted only cash. They asked to pay with a check, even though they had the cash. The waiter asked: "Are you Basque?" They answered, "Yes," and the waiter said: "OK, then." When I asked Ana if the establishment or waiter was Basque, she said no. Personal communication, July 1993, Elko.

11. Laxalt, *Nevada: A History*, 23.

12. J. Mallea-Olaetxe, "History that Grows on Trees: The Aspen Carving in Nevada," *Nevada Historical Society Quarterly* 35, no. 1 (spring 1992): 21–39.

13. Personal communication, March 1998.

14. This statement strikes an inner chord of Basque peasant culture, but non-Basques, too, had similar impressions. Reginald Meaker said of his Basque foreman, Martin Gallues, who worked for him for eighteen years: "Wonderful human being . . . I could not have done better if I had searched the entire United States."

15. Pete Mendiboure, Madeline, California, 1993.

16. Jean Lekumberry spoke to me in similar terms; personal communication, 1990.

17. If the cavemen of the Pyrenees operated anything like the sheepherders, the arborglyphs could shed light on their prehistoric paintings. Top historians struggle to understand them but generally agree that the many figures of animals painted in the caves were part of magic rituals. But what if they express nothing more than a desire to have meat? Or perhaps commemorate a good hunt? In other words, the paintings cannot be taken as evidence that these animals were part of the Magdalenian cave people's lives; it might be just the opposite.

18. Should have been B. F. M.

19. The Spanish original says: "Viva el vino y las mujeres en pelotas" and literally means "Hurrah for the wine and the women in balls."

20. Videotaped conversation with Frank Aramburu, 1990–91. In other parts of Nevada the herders drive the sheep up to the hilltop for the night. Others usually ate supper later than Aramburu did.

21. Laundry was a chore most herders hated. They usually heated water in the largest cooking pot, which also served as the sink where they washed their crusty Levis, socks, and everything else.

22. Robert Laxalt informed me in 1992 that his father, Dominique, used the area.

23. "Iturri artzainaren ur frescoa."

24. "Lo bat," a Basque expression for uninterrupted sound sleep.

25. In the original: "Emen au bier da, ostantzin gaztik aurri jatetsue, ba!"

26. Mendiola's Basque words have a ring of their own: "Mutillak, au dok gozotasune!"

27. Contributed by Joe Mallea, Elko.

28. J. L., northeastern California, personal communication, 1995.

29. Until the 1950s, among rural Basques, wheat was less plentiful than *arto* (corn), and *morokil* (corn-flour porridge) was an everyday staple. Wheat bread was more expensive than *arto* (cornbread), which unlike the U.S. variety had a solid consistency with a golden brown, hard crust. The Basques must know something American cooks don't because Joseph R. Conlin says that corn "refuses to make a decent crust however it is cooked." See Joseph R. Conlin, *Bacon, Beans, and Galantines: Food and Foodways on the Western Mining Frontier* (Reno: University of Nevada Press, 1986), 12.

30. Ignacio Urrutia—also known as "Mr. Idaho"—of Susanville, California, was raised on corn, beans, potatoes, milk, and fruit on a farm on the hills above Galdakao, Bizkaia. At age six, when he had his first meal in town, in 1919, he was served a steak and did not know what it was. Personal communication, 1999.

31. Eugen Weber, *Peasants into Frenchmen: The Modernization of Rural France, 1870–1914* (Stanford: Stanford University Press, 1976), 132–40.

32. Juan Antonio de Moguel, *Peru Abarka*, edición bilingue (1881; Bilbao: Editorial La Gran Enciclopedia Vasca, 1981), 29–30.

33. P. Z., interview with the author, Reno, 1992.

34. Byrd Wall Sawyer, *Nevada Nomads: A Story of the Sheep Industry* (San Jose, Calif.: Harlan-Young Press, 1971), xiii. Fred Fulstone, member of a pioneer ranching family of Smith Valley, Nevada, assured me that most of the work done by miners in the Silver State was fueled by high-protein mutton (interview, June 1996). As an insight, the first item featured on the menu of the upscale What Cheer Restaurant in nineteenth-century San Francisco was mutton. See Conlin, *Bacon, Beans, and Galantines*, 141.

35. This carving is a curious blend of languages and identity markers. Elissalde was from Iparralde, but he carved in Spanish; his first name, Nicolas, is also spelled in that language, while his last name remained unchanged. His choice of the term *piches* (phonetic spelling of *peaches*) indicates that he had been in the United States for some time.

36. Information from C. Y. of Winnemucca, Nevada, 1987.

37. Misspelling for Dangberg, the name of the sheep boss from Minden, Nevada.

38. In the original: "Biba artzainak eta sotoko oporra M. G. 1950."

39. Original Spanish: "Bautista Amestoy natural de la de Urdax español navarro borreguero de Ramon Borda of Garveville [*sic*]."

40. Information on sheepmen from Bishop provided by Richard Potashin of Bishop, California, 1997.

41. May have been carved by F. E.

42. The herders called mule's ears "tobacco," and some smoked it when they ran out of the regular stuff.

43. Information supplied by Patxi Txurruka of Reno, personal communication, 1987.

44. Personal communication, name withheld by request.

45. He probably meant to say "find" rather than "look for." The last words of the carving are unreadable.

46. This sheepherder must have been herding just the rams, because in the summertime the rams are usually kept apart from the ewes.

47. The herder's name appears to be Castilian, but the message is partly in Basque.

48. Meaker, *Nevada Desert Sheepman*, 91.

49. According to Eileen Borda of Carson City, 1991.

50. Interview with Pete Bengoechea, Winnemucca, Nevada, 1991; see also, Meaker, *Nevada Desert Sheepman*, 106–11.

51. This carving is located high in northwestern Humboldt County. "Frenchie" Montero of Leonard Creek Ranch is the official weather watcher of the area, and when I mentioned this carving he was surprised. According to his records, in that particular year more than eight inches of rain fell in the district in question—not a dry year by Nevada standards.

52. A large fish is carved next to the caption.

53. "An Interview with Frank Yparraguirre," conducted by R. T. King, 22 May 1984 (University of Nevada, Reno, Oral History Program, 1984), 110.

54. In the original: "Popatik artu daiela ardi danakin."

55. Beginning in the 1950s the Western Range Association imported many herders from Spain under a three-year contract. Once the contract had been fulfilled, the herder had to return to Europe and negotiate another contract if he wanted to return. Herders who wanted to remain in America felt that was unfair.

56. Humboldt National Forest, Elko County, Nevada.

57. Personal communication from Claudio Yzaguirre, Bend, Oregon, 1994.

58. Often Basques who tried to write in Spanish did so poorly. In the original, this carving reads: "Noviember 31 stan vando ace kalor." "Calor" here means "hot," which is unlikely if it was snowing. That is why I translated it as "warm."

59. Personal communication, 1992.

60. Personal communication, Reno, 1991.

61. The original reads: "Viva el rio y la redada." *Redada* means "a good catch with a fishing net."

62. Frank Aramburu, interview with the author, 1990–91.

63. Rufino Bena, interview with the author, Twin Bridges, Elko County, Nevada, 1968.

64. P. Z., interview with the author, 1990.

65. As told by Bert Paris, White Pine County, Nevada, 1990.

66. Information found among the field notes gathered—probably—by Martha Mitchell and deposited with the Basque Studies Program, University of Nevada, Reno.

67. Claudio Yzaguirre, personal communication, Bend, Oregon, 1994.

68. Interview by the author near Smoke Creek, Nevada, 1993.

69. Beesley and Claytor said that fish were "rarely carved" in California; see "Adios, California," *Basque Studies Program Newsletter* (University of Nevada, Reno), no. 19 (November 1978): 4.

70. See, for example, Adrian C. Louis, *Wild Indians and Other Creatures* (Reno: University of Nevada Press, 1996).

71. The Spanish *cabrón*—usually carved without the accent—literally means "cuckold."

72. The original is onomatopoeic: "empeso el ay ay con los coyotes."

73. P. Z., conversation videotaped by the author, Reno, 1990–91.

74. Personal communication from Jess Arriaga, Reno, 1998.

75. Robert S. Weiss and others, *Loneliness: The Experience of Emotional and Social Isolation* (Cambridge: MIT Press, 1973), 17.

76. It should be "vivan los borregueros," but the sheepherders never pluralized the verb *vivir*.

77. Probably the most celebrated writer in Basque literature, author of *Gero*, published in Bordeaux in 1643.

78. Original Basque: "Faustin ene lehen arsain laguna."

79. Frank Yparraguirre believed that the old herders were more dedicated. See "An Interview with Frank Yparraguirre," 112.

80. *Caca* was borrowed by the Spanish, but Basques today spell it *kaka*.

81. This is in Spanish, but the stronger native Basque language of the carver, J. U., comes out in right in the middle of the sentence in the "eta" he carved instead of "y" (and).

82. This is more than idle boasting. Ynchauspe is today a successful cattle rancher in Austin, Nevada.

83. Competitions in bread making, sheep hooking, weight carrying, and so on, are today an integral part of Basque picnics in the West. At first, Anglo-Americans did not participate in these games, but after some years they started competing in weight carrying and tug-of-war contests. Women may participate in bread baking.

84. It is not signed, but the dissimilar type of etching denotes a different carver.

85. In the original: "Kaka hire, zapoa, emakok ene ipurdiari musu polita diok, urde zikina," and "Tosia eta Arzamenta bi urde dira." These are probably nicknames.

86. *Gabacho* is a derogatory term applied to non-Basque French. See J. Mallea-Olaetxe, "History that Grows on Trees," 37.

87. The original Basque says: "Adio gure S kaka eguin behar du bere adiekin" (should be "ardiekin").

88. Mallea-Olaetxe, "History that Grows on Trees," 37.

89. Some of the words can be read, but the spelling and the colloquialism are such that the meaning is partly lost.

90. The original says: "Malacaca euscalduna."

91. According to the earliest written Roman sources, southern Nafarroa was the homeland of the "Vascones" or "Uascones." Historians agree that the Vascones were the largest Basque tribe, and later the whole country was named after them: Vasconia, Vascongadas (this applies only to written sources). The carver with anti-Basque sentiments came from southern Nafarroa, the original home of the Vascones, where two thousand years later, ironically, the people have lost their language and some of their ethnic identity.

92. It is not totally clear if he was disagreeing with the first carver—thus chastising him for keeping a lousy camp—or indicating where the bathroom was, or should be.

93. One author who touched on the subject wrote: "The shrewd camp tender treats his herders to a good deal of horseplay, jokes of a sexual nature, and consciously animated conversation." See William A. Douglass, "Lonely Lives under the Big Sky," *Natural History* 82, no. 3 (1973): 28–39.

94. The tree is near a couple of cabins that used to be called Sheep Camp, now Summer Camp.

95. Personal communication from retired sheepherder Frank Aramburu, 1991.

96. Videotaped interview with the author, Gardnerville, Nevada, KNPB, 1991.

97. Arriandiaga's Memoirs, 193.

98. Personal communication, August 1990. The original carving says: "Hotel Derrepente aun suele estar el año tres veces del año." He forgot "abierto" (open), but it is not necessary to understand the joke.

99. The Spanish *camelar* means "to seduce," but also "to deceive or screw," as in this case.

100. Told by Frank Gallues, Reno, 1991.

101. Personal communication with Claudio Yzaguirre, Bend, Oregon, 1996.

102. On loneliness, see also Douglass, "Lonely Lives under the Big Sky."

103. Personal communications, 1991.

104. The Forest Service people have not always seen eye-to-eye with sheepherders, but it is interesting to note that in the last couple of decades Anglo ranchers have broken ranks, have joined the Basques in the "Sagebrush Revolution," and have voiced strenuous complaints regarding the management of public lands.

105. Interview with the author, 6 April 1991, Paradise Valley, Nevada. Aramburu arrived in Nevada in the early 1920s and was acquainted with the period of "tramp" sheepmen.

106. Videotaped interview by the author, June 1991, Winnemucca, Nevada; taped interview with Frenchie Montero, 19 November 1980, Basque Studies Program, University of Nevada, Reno.

107. According to a flier I picked up in Jordan Valley in 1993, the first Basques in the area were Jose Navarro and Antonio Azcuenaga in 1889 (the information was probably obtained from William A. Douglass and Jon Bilbao, *Amerikanuak: Basques in the New World* [Reno: University of Nevada Press, 1975], 242). Agustin Azcuenaga came a year later, followed by Pedro Arritola, Luis Yturraspe, and Cipriano Anacabe. In 1915 they built a Basque handball court. The flier did not mention it, but the Jordan Valley Basques also helped build the Catholic church (ibid., 355–56, 392).

108. The last three words mean "Hello, come everyone." A photograph of the sign was published in *Voice of the Basques* 2, no. 12 (1976): 8–9.

109. Today we would write "Gure etxea, zurea da," but whoever in Jordan Valley was responsible for including it in the flier must be commended for the "gaure echie" (our house). Latin Americans say "Mi casa es su casa" (My house is your house), and it is "my house" in the United States, too, but that concept does not exist in rural Basque culture; rather, they say "our house." In the same vein they use "gure aita" or "gure ama" (our father/our mother) and hardly ever "my father/my mother." It reflects old *baserri* culture, which modern Basque speakers and writers lack, so they say and write "my house."

110. Original Spanish: "Estoy mas triste que un pinar al anochecer," unsigned carving in Elko County, Nevada.

111. The original Basque says: "Ampo hemen bici da unsa ala gaizki," White Pine County, Nevada.

112. Manex is pronounced "Mah-nesh."

113. The previous paragraphs offer a reconstruction based on what many herders told me.

114. Videotaped conversation with the author, 1991.

115. "Kazeno" must be a phonetic rendering of an English word. Leon Legorburu of Reno told me that he had carved *jota* (Basque folk dance) dancers at Kazeno in the late 1940s, but I did not find them.

116. One evening I got lucky and went "out to eat." Two quiet buckaroos who were spending a few days in a nearby cabin invited me to share their pot of beans. The two men were as authentic as anything I had ever seen or read about. As soon as I saw them I thought I had been transplanted to the last century. One was a Cherokee from Oklahoma, the other a Mexican. Their chaps were caked with grime, and their clothes were worn out and so filthy that I could not tell their original color. The Oklahoman was missing several teeth, and their faces seemed to be made of water-repellent material. I found out why they were so quiet: the Mexican spoke no English and the Oklahoman had mastered just a few Spanish words. I had a great time with them. Jim, the Cherokee, urged me to eat more beans, and I did my best—they were delicious. Next day they drove to Denio for *one* pack of cigarettes: thirty-five miles, one way, by mountain and gravel roads. Looking back, how I wish I had videotaped that evening!

117. Laxalt, *Nevada: A History*, 24.

118. Actually, *Amerikanua* is a recent term compared with the much older *Indianoa* (a returnee from the Indias or Americas), attested in Durango, Bizkaia, as early as the 1550s. See J. Mallea-Olaetxe, "Antso Gartzia Larrazabal (1489?–1554?) Euskal Herriko lehen 'Indianoa,'" *Swing Durango (Durangoko Aldizkaria)* 3, nos. 10–11 (1992): 26.

119. P. Z., videotaped interview with the author, Reno, 1990–91.

120. One day an old sheepherder nearing the end of his life returned to his native hamlet in Nafarroa after a thirty-year absence in the United States. Everyone believed the man was rich, but he had no close relatives and showed no interest in settling his affairs. The town doctor, the priest, and the secretary met to discuss the urgent matter and agreed that the priest should pay the feeble man a visit in order to bring up the subject of the will. (In Basque villages the mayor is mostly a figurehead. The secretary is a full-time employee and conducts all the routine business. He, the priest, and the doctor usually run the town's affairs.) To his surprise, the clergyman found the repatriate well disposed and receptive to the idea. They set the date, and a few days later the three town leaders went to the elderly fellow's home. When the sheepherder announced that he was prepared to make out the will, the secretary sat down ready with pen and paper. The ex-sheepherder declared rather solemnly: "My soul I entrust to God. My body I leave to the hard earth, my balls for the secretary, and my dick for the priest."

The will rhymes in the original Spanish:

Mi alma entrego a Dios,
mi cuerpo a la tierra dura,
mis huevos para el secretario
y la verga para el cura."

Told by F. G. of Reno, Nevada.

121. Notes and tape recording in the author's possession, 1971.

122. Laxalt, *Nevada: A History*, 24. Miners always made more money, especially those in the Comstock in the 1860s–1870s—up to four dollars a day—but the job was dangerous.

123. I heard this story often, but I do not remember the name of the herder.

124. Interview with the author, 1990.

125. In the original Basque: "Halo Maneuel uscic medioc baduc sosic biba ni eta arzainac."

126. Videotaped interview with B. R., Big Creek, Austin, Nevada, 1990; and ex-herder J. I. of Winnemucca, Nevada, 1991.

127. According to a conversation with Pete Paris, 1991, although afterward it was not totally clear to me if both of us were talking about the same carvings.

128. In the original: "Gaste ta pobre biba Fransia."

129. Personal communication with Pello Salaburu, who knows F. C., Bilbao, Bizkaia, 1993.

130. Interview with the author, July 1993, Elko.

131. One hears jokes in Nevada regarding sheepherder literacy. Once in Elko I was asked: "What is an 'XX' at the bottom of the check?" I said, "What?" And the man laughingly blurted: "An Italian co-signing for a Basque."

132. This conversation took place in Basque, in which the precise word *zenbaterako* (for how long a time) rhymes with *betirako*, which is stronger than *betiko* (forever). Information provided by Claudio Yzaguirre, of Bend, Oregon.

133. Told by Erremon (Ray) Jayo of Reno.

134. See Arnold van Gennep, *The Rite of Passage*, trans. Monika B. Vicedom and Gabrielle L. Caffee (Chicago: University of Chicago Press, 1960), chap. 1.

135. Interview with the author, 7 April 1991, Paradise Valley, Nevada.

136. Personal communication, Reno, Nevada, 1989.

137. This carving is located in the Gerlach, Nevada, area according to a contemporary herder from Tolosa, Gipuzkoa.

138. "Martin Goikoetxea: Bertsolaria Ameriketan," audiocassette, San Francisco, California (1995).

139. Information from James Snyder of Yosemite Research Museum and Library, 24 February 1994.

140. Interview with Emily (Zatica) Miller, Paradise Valley, Nevada, 1991.

141. Taped conversation by the author with Kozkorroza's daughter Delfina of Paradise Valley, Nevada, 1991.

142. Information from Reno's Abel Mendeguia and Ardans's nephew, Cody Krenka, Ruby Valley, Elko, 1997.

143. Ithurralde, interview with the author, Winnemucca, Nevada, 1991.

144. Personal communication, October 1998.

145. Laxalt, *Sweet Promised Land* (Reno: University of Nevada Press, 1986), 3.

146. *The Nevada Scene*, a documentary series produced by KNPB–Channel 5 of Reno, aired on 14 February 1990.

147. This is carved in a highly compressed fashion: "conse pal lesto" etc. (Un consejo para el que lee esto . . .)

148. The original Basque says: "Au da visiaren gogorra."

149. A misspelled reference to the hill where Jesus was crucified.

150. Virginia Paul, *This Was Sheep Ranching: Yesterday and Today* (Seattle: Superior Publishing, 1976), 84.

151. Victor Turner, *Dramas, Fields, and Metaphors: Symbolic Action in Human Society* (Ithaca: Cornell University Press, 1974), 243, 223.

152. Joseba Zulaika, *Basque Violence: Metaphor and Sacrament* (Reno: University of Nevada Press, 1988), 242–44.

153. The Basque term *kontrola* resembles somewhat the French pronunciation of Cointreau.

154. Examples can be found in Jeronima Echeverria, *Home Away from Home: A History of Basque Boardinghouses* (Reno: University of Nevada Press, 1999), and Gretchen Holbert, "Elko's Overland Hotel: A Family, a Culture . . . and a Memory," *Northeastern Nevada Historical Society Quarterly* 5, no. 3 (1975): 13–19.

155. Videotaped interview with the author, Gardnerville, Nevada, 1991.

156. In the original: "Adios ene maitia adios sekulako orai hemen guira bicico guziak."

157. The observant reader will notice that there is no 31 September in the calendar, a minor detail that did not bother the herder. Somewhere in Elko County there is a tree bearing on its trunk the "Sheepherder Calendar." Sometime in the 1960s a herder for the Allied Sheep Company kept track of his days in a particular range by carving the date every day on an aspen until he was finally relieved. When the new herder saw the calendar, he figured he would continue it and thus a new calendar was born: July 31, 32, 33, etc. Information from J. Lopategui, Elko, Nevada, 1990.

158. The original says: "Foken chik para los americanos." I do not know who this individual was.

159. The original says: "aurtengoz edo sekulakoz" and spells "Alpain" for Alpine.

160. The original is in "Spanbasque": "Adios erreserba kerida."

161. In the original: "Good by Pinut, asta la vista."

162. Transcribed literally. Pedroarena is a very active member of the Basque community in Minden-Gardnerville, Nevada.

163. Henri Breuil and Raymond Lantier, *The Men of the Old Stone Age*, trans. B. B. Rafter (London: George G. Harrap, 1965), 261–65.

164. One such tale comes immediately to mind: One day a man was walking in the forest and saw a wolf caught in a trap. Taking pity on the animal, he set him free. But the wolf was hungry and wanted to eat his benefactor. The man asked, "How could you do that?" The wolf saw that the man had a point and they decided to get counsel. They walked down the road until they found a donkey and asked him for his opinion. The donkey said that his ungrateful master often treated him badly, and the wolf was happy to hear that. They continued on the road and they found a fox. The fox indicated that before he gave his opinion he must first see how the wolf was found in the trap. So they went back and the man put the wolf in the trap again. The fox told the man to hit the wolf in the head with his hoe. This he did and the wolf died. The man thanked the fox and promised to give him some chickens; but instead of chickens he put dogs inside the sack. When the fox saw the sack he said: "Wow, these are some chickens!" "Yes, they are big indeed," the man answered, and he set the dogs free and they chased the fox. But the fox escaped and hid in his hole.

165. In the original: "Huna hemen Martin Larzabal asto pake jazerat."

166. Obviously these stories have a special flavor in the original language. I heard them in Burns, Oregon.

167. Lete is a Basque poet and singer known for his Dylanesque lyrics. His "Giza aberea" refers to factory workers, but I think it can be equally applied to this hybrid figure carved in the Sierra.

168. According to Basque folktales, the witches generally took female form, but these women had animal feet—duck, goat, etc.

169. Author's conversation with J. P., 1995. On communication with animals, see David Abram, *The Spell of the Sensuous: Perception and Language in a More-Than-Human World* (New York: Pantheon Books, 1996), 78–81, 151–52.

170. In the original Basque: "A zuan a! Astun esana ein

bier." I met this Bizkaian in a crowded bar in Elko in 1990, but he disappeared before I asked for his name.

171. This aspen was cut by Toiyabe National Forest personnel and is now in a museum in Bilbao, Bizkaia.

172. On another hilarious topic, each of the four *bertsolariak* assumed the role of a female lamb, a male lamb, the catcher, and the docker.

Chapter Four. Life Without Women

1. Rafael Castellano, *Erotismo vasco* (San Sebastián: L. Haranburu, 1977), text on the jacket; José M. Satrustegui, *El comportamiento sexual de los vascos* (San Sebastián: Editorial Txertoa, 1981), 9.

2. Rodney Gallop, *A Book of the Basques* (Reno: University of Nevada Press, 1970), 56–57.

3. V. W. Lehmann, *Forgotten Legions: Sheep in the Rio Grande Plain of Texas* (El Paso: Texas Western Press, University of Texas Press, 1969), 52–53.

4. Information by Jerry Etcheverry of Winnemucca, Nevada, 1998.

5. Videotaped interview with the author, 1992, Elko, Nevada.

6. E. O. James, *The Cult of the Mother-Goddess* (New York: Barnes and Noble, 1959), 23.

7. Robert C. Lamm, *The Humanities in Western Culture: A Search for Human Values*, 4th ed. (Madison: Brown and Benchmark, 1996), 10.

8. For figures of women as portrayed by Paleolithic artists, see Henri Breuil and Raymond Lantier, *The Men of the Old Stone Age*, trans. B. B. Rafter (London: George G. Harrap, 1965), 112–13; and Magín Berenguer, *Prehistoric Man and His Art*, trans. Michael Heron (Park Ridge, N.J.: Noyes Press, 1973), 19–20, 49–50. Scholars argue about the significance of headless animals in the cave paintings. Berenguer assigns them some unknown symbolism, 75.

9. Marija Gimbutas, *The Civilization of the Goddess: The World of Old Europe*, ed. Joan Marler (San Francisco: Harper, 1990), ix, 70, 83, 84, 91, 110–11, 261, 266, passim.

10. The Frenchman André Leroi-Gourhan is the author of a controversial theory on cave paintings, according to which everything revolves around sex. See Berenguer, *Prehistoric Man and His Art*, 73.

11. Robert S. Weiss and others, *Loneliness: The Experience of Emotional and Social Isolation* (Cambridge: MIT Press, 1973), 39.

12. In the original: "Cuando te mire benir le dije a mi corazon que bonita piedrezita para darme un trompezon."

13. The original Basque rhymes: "Ene maite pollita Lucita egingo daus(u)t fite bisita."

14. The herder called her "Lepo luzia" (long necked).

15. In the original: "Eusebian du andragaia ene picotsak pe saen dute 95 urte baac abrich."

16. "Chestnut" is a euphemism for pubic hair and female genitals.

17. This herder was from Benafarroa but spoke Spanish well enough to even rhyme the statement. The original says: "Tuve una novia y ya me perdido las relaciones con ella por cabron firmado y rubricado el primero qe lo lea el comete pecado, P. G. 1935." Second readers were welcome, I guess.

18. Based on conversations with herders.

19. The carving says: "El mal de amores es malo." The song says: "Amodioaren pena, oi pena kiratsa."

20. A conventional imagery of the aspen "literature." See J. Mallea-Olaetxe, "A History that Grows on Trees: Basque Aspen Carving in Nevada," *Nevada Historical Society Quarterly* 35, no. 1 (spring 1992), 35.

21. In the 1950s and 1960s many Galicians migrated to the Basque Country's Hegoalde as laborers.

22. Even today the nearest public telephone is thirty to forty miles away.

23. In the original: "JPI tiene la novia mui bonita pero puta jode con todo cristo." A popular Basque proverb says: "Andre ederra, etxean gerra" (A beautiful woman [causes] war at home).

24. Among the peasants of the Basque Country, Mari is a blanket name for women.

25. The Basque expression "jo eta kito" is not exactly "I am going to kill her" (*jo* means "strike"), but that is the meaning here.

26. In the original: "Amigos mios un favor voia pedirles a todos los que tieven mugeres no las dejen tato sola porce el diavlo no duerme amas ellas ke no son tontas se dan."

27. According to William A. Douglass, the man is thinking about being cuckolded; personal communication, 1999.

28. In case the reader is wondering, Basques hold this type of conversation only among familiars.

29. Aymeric Picaud, a twelfth-century Frankish traveler, wrote that the people of Nafarroa practiced zoophilia. He did not see it with his own eyes, however, because he wrote: "It is said that the Nafarroan hangs a padlock by the hind legs of his she-mule and his mare, so that only he and no one else gets close to them." Also he said that the Basque "lecherously kisses the genitalia of the woman and the mule." See *Liber Sancti Jacobi. Codex Calixtinus*, ed. A. Moralejo, C. Torres, and J. Feo (Santiago de Compostela, 1951), 521. Many Basque historians question the veracity of these statements.

30. The original Basque says: "Hemen zicotu dut asto urruz beroa" (Here I screwed a female donkey in heat).

31. One of the most celebrated writers was Aita Prai Bartolome Santa Teresa, *Euscal-errijetaco olgueeta, ta dantzeen neurrizco-gatz-ozpinduba* (Iruñea: Joaquin Domingo Nausijaren, eta Gaztiaren Liburuguillaan, 1816; facsimile ed., Donostia: Hordago, 1978).

32. The farmer as the prototypical Basque goes back at least to the eighteenth century, as seen in Mogel's *Peru Abarka* (first published in 1881), and later romanticized in *Garoa* (1912) and in the great poem *Euskaldunak* (1972). The concept was adopted by nationalist politicians as well.

33. Ramon Etxezarreta, *Hiztegi erotikoa* (Gasteiz: Hordago, 1983).

34. Ibid., 368–92.

35. For example, 1984 Basque Hall of Famer Domingo Ansotegui, born in McDermitt, Nevada, started sheepherding at age fourteen. See *Journal of the Society of Basque Studies in America*, vols. 11–12 (1991–92), 112.

36. J. A., personal communication, October 1998.

37. A Reno woman, after taking a tour of the aspen groves, remarked: "Well, if the ancestors of the Basques painted the walls of the caves, it does not surprise me that they also carved trees"; D. T., October 1998.

38. The oldest carved "puta," dated in the nineteenth century, is said to be in Buckeye Creek Canyon, Toiyabe National Forest, near Bridgeport, California; personal communication, James Snyder, February 1994.

39. My friend Eusebio Osa Unamuno, "Sakone," sent me this information. He said it is carved somewhere in Idaho.

40. They also use *eman* or *emon*. The Basques learned additional vocabulary in the United States from the Mexicans and South Americans, such as *chingar* and *coger* (to fuck). One hears *chingado* (bastard) almost as much as *puta*.

41. In the original: "Buztana azkar eta mocha falta nik banu aluia ziko bainio."

42. In the original: "Gaiso ene maite a ze buztana bezenequi."

43. In the late 1960s and early 1970s Mexicans working on the potato farms north of Winnemucca would come to the Winnemucca Basque Hotel wanting to break a ten-dollar bill into two fives. The brothels in Winnemucca are located nearby.

44. Translated from Basque, personal communication from J. G., Reno, 1993.

45. Conversation with John Altrocchi, professor of psychiatry and behavioral sciences at the University of Nevada, Reno, 1991. However, Altrocchi was not so sure about the meaning of other carvings that show women defecating or urinating.

46. D. T. of Reno told me that she had heard about a herder who came into town, cashed his check, and then had a prostitute come into his hotel room. After she left, the herder found that all the cash was gone; personal communication, October 1998.

47. Personal communication from D. T. of Reno, October 1998.

48. *The Old West: The Cowboys* (Alexandria, Va.: Time-Life Books, 1978), 145, 183–215. Admittedly, these incidents were part of popular culture more than history.

49. Drawn from conversations with a number of ex-sheepherders and Basque hotel operators.

50. The term *virgo* is usually applied to women as meaning "hymen." Here the herder is not suggesting that he had sex with a virgin, but that *he* was a virgin.

51. Difficult to read in part, but "wet it" may be a euphemism for intercourse.

52. Personal communication, October 1998.

53. In the original: "Dicen que en Francia no ay ni unas putaz ay az muchas."

54. The name is not clear.

55. Interview with Eugenia Mendiola, Winnemucca, 1991.

56. Interview with E. M., November 1990, Humboldt County, Nevada.

57. Interview, Elko, Nevada, August 1990. I was told of another incident that happened in Caldwell, Idaho: a prostitute showed up in the sheep camp ready for business, but the herders were too shy. She left, saying disdainfully, "The Basques are chickens." One herder vowed that next time she would not say that. When the woman returned, he immediately had sex with her in front of the other herders. S. R., personal communication, 1971.

58. According to one source, Mount Lola in Tahoe National Forest was named after Lola Montez, a "bad girl" who danced and entertained in California and lived in Grass Valley in the 1850s. See James F. Varley, *Lola Montez: The California Adventures of Europe's Notorious Courtesan* (Spokane, Wash.: Arthur H. Clark, 1996).

59. I did record one "Jani" (honey), but in this case I am using it in lieu of the real name.

60. All information retrieved from trees.

61. In the original: "Nunca no hizo devalde."

62. Not the real name.

63. Personal conversation, name withheld, 1998.

64. Jeanne Seagel, "Erotic/Fragment: Ariane Lopez-Huici's Tactile Photographs," *Arts* (November 1990): 98–100.

65. J. A. worked with Periko Zubiria for five years; personal communication, 1990.

66. I inquired about J. A. in Ely, and I talked with two people who said they knew him. Although the themes of his carvings are quite limited, this Native American must be considered an exceptional aspen carver.

67. J. M., Reno, January 1999.

Chapter Five. Artists, Sheep Camps, and Bread Ovens

1. This carving deteriorated considerably between the time Mr. and Mrs. Phil Earl traced it and the time I videotaped it in 1989. In 1998 it was not possible to identify the object the priest is holding in his hands.

2. In William A. Douglass's opinion, the carving could be the work of a pious man who depicted the priest exorcizing an American whore (personal communication, 1999). But the carver, J. M. was not particularly religious—on the contrary.

3. I have often wondered about the similarity of these Basque and English words.

4. Conversation with P. L., Reno, August 1992.

5. Story documented in two sources—one of them from Portugal—as early as 1454. See Lope Garcia Salazar, *Bienandanzas y fortunas*, Códice del siglo XV. Primera impresión tel texto completo con prólogo, notas e índices, por Angel Rodríguez Herrero; Introducción por el Excmo. Sr. Marqués de Arriluce de Ybarra, 2 vols. (Bilbao, 1967); Luis Michelena, *Textos arcaícos vascos* (Madrid: Minotauro, 1964), 45–46.

6. It could be a sunset.

7. The only other "Gora Euskadi" by a French Basque is in White Pine County, Nevada, carved by Marilux Lapeyrade. He added "France" to his name.

8. *Enciclopedia general ilustrada del País Vasco*, vol. 5 (San Sebastián: Editorial Auñamendi, 1974), 490–91.

9. C. Elizabeth Raymond, *George Wingfield: Owner and Operator of Nevada* (Reno: University of Nevada Press, 1992), 195.

10. *Reno Evening Gazette*, 5 July 1931, 10. See also Phillip I. Earl, *This Was Nevada*, introduction by Peter L. Bandurraga (Reno: Nevada Historical Society, 1986), 161–64.

11. Today the Borda clan in Eagle and Carson Valleys, Nevada, comprises about 180 members, who occasionally meet to celebrate their common roots. My thanks to John Borda, Dana Borda (Borda's Arco Service, Carson City), and his daughter Bonnie Hoffecker for the information.

12. Personal communication, Dolores Taylor, Reno, 1998.

13. Author's video, 1993.

14. Information provided by Mallia of Gardnerville, Nevada.

15. Roberta L. McGonagle, "Frank Rodriguez, Sheepherder and Artist," *Women in Natural Resources* 12, no. 1 (1990): 40–41.

16. Ibid.

17. Conversation with Severino Ibarra, March 1994.

18. Actually, after thirty more years sheep grazing ended on Bates Mountain. McGonagle, "Frank Rodriguez."

19. I found this note in my papers, but no other reference or detail.

20. According to *Land Status Map of Nevada*, 2d ed., comp. and cartography by Ira A. Lutsey and Susan A. Nichols (Nevada Bureau of Mines and Geology, University of Nevada, Reno, 1972).

21. Conversation with a Peruvian herder, Copper Basin, Elko County, Nevada, July 1997.

22. Personal communication with several Forest Service archaeologists, Elko, Nevada.

23. See Carrie E. Smith, "Sheep Camps, Ovens, and Aspen Trees: Discovering the Basque on the East Side of the Tahoe National Forest," *Society for California Archaeology Newsletter*, 27, no. 1 (1993): 11–13.

24. Severino and his brother Juan were born in Nafarroa and arrived in Reno in November 1952 with three other herders. They came without a contract. Juan worked three years with the sheep, and Severino six. None of them carved trees, but Juan told me that 95 percent of the herders did; personal communication, 1994. Bengoetxea was from Gizaburuaga, Bizkaia, according to Joe Bengoechea.

25. Conversation with Severino Ibarra, Reno, March 1994.

26. From M. O'Halloran's draft, 20 August 1992. Lavari acquired his skills working for sheep owners McPherrin and Zugadi, according to Severino Ibarra of Reno.

27. Whiskey Creek had been partially recorded on 3 October 1973 by Susan Lindstrom and Rollin Kehlet; further supplemented in July 1990 by Richard Markley, Susan Rose, and Carrie Smith; and completed in August 1992 by Barbara Southerland, according to USDA Tahoe National Forest IMACS Form, Permanent Trinomial CA-PLA-701-H.

28. Letter to Ms. Kathryn Gualtieri, state historic preservation officer in Sacramento, California, 22 July 1991. Ms. Gualtieri did not reply, so Skinner made the decision to preserve Whiskey Creek.

29. NABO also helped finance some of the materials used in the restoration.

30. The local media from Truckee, California, reported on the restoration effort. See Abby Hutchison, "Restoring Basque Culture in the Granite Chief Wilderness," *Sierra Sun*, 29 July 1999, 9A.

31. Personal communication, July 1999.

32. Personal conversation with Carrie Smith, Truckee, California, 1999.

33. I am grateful to Judy Mendeguia of Reno, 1999, for the information on the Russell Valley camp. She also said that, for example, in the old days the sheepmen left food at the camp for the hunters, who would replace whatever they used. But then, she added, collectors of antiques started taking things, among them a couple of woodstoves.

34. Calvert McPherrin, whose wife is Bizkaian, still runs sheep in Carmen Valley southeast of Portola, California. Conversations with Michael Baldrica, 1998 and 1999.

35. Jayo told me: "Amaikatxo suge urten eban" (The snakes just kept coming out); personal communication, 1993.

36. Information from Mikel Lopategui of Carson City and Bob Echeverria of Elko, January 1999. At this rate, the sheep camps in Nevada will soon be history. Some ovens fell victim to land developers, such as the one in Olympic/Squaw Valley, California. Fausto Lavari built the oven in 1954 on the Miller & Kuhn forest grazing allotment. The herders made bread in this oven until the 1970s. A man from Denmark named Jensen was the original sheepman of the area. Wayne Paulsen, the landowner and a nearby resident—not associated with sheepherding—baked bread in the Squaw Valley oven until 1981. When he sold the land to Squaw Creek Resort, the developer bulldozed the oven, probably in 1989 or 1990, before it could be videotaped or photographed. Valerio Zubiri the camptender was one of the last bakers in Squaw Valley. Interview with Paulsen, 1990.

37. In the original: "Frantses euskera inkau jat."

38. This project generated considerable local media coverage. See Lee Light, "Tahoe Clippings: Digs, Digs and a Basque Oven on the Tahoe National Forest," *Mountain Messenger* (Downieville, Calif.), 13 August 1992, 1; Barbara Barté, "Archaeology Highlighted at State Park," *Sierra Sun*, 6 May 1993, 10; Catherine Gibbs, "Remembering the Basques: Sierra History Often Overlooks Shepherds," *Sierra Sun*, 6 May 1993, 10.

39. Probably not the original oven.

40. *Kyburz Flat: 2000 Years of History* (USDA Forest Service brochure, 1994). Many people participated in the project. I would like to thank everyone for their help, especially Michael Baldrica and the Sierraville Forest Service District, California, and Carrie Smith and the archaeology department at the Truckee Ranger District, California, as well as my son, Erik Mallea.

41. "The Kyburz Flat Area," *Sierra Booster* (Loyalton, California) 44, no. 17, 14 August 1992; "Zen Bat Gara," *Sierra Booster* 45, no. 11, 21 May 1993.

42. Conversation with Abel and Judy Mendeguia, 10 October 1992.

43. Personal communication, M. I., Winnemucca, 1990. "The sheepherders, on their return trip to the valley areas in the fall, habitually set fires behind them"; W. S. Brown and S. B. Show, *California Rural Land Use and Management: A History of the Use and Occupancy of Rural Lands in California* (USDA Forest Service, California Region, 1944), 155.

Incidentally, with the fast decline of aspen population thoroughout the western United States, forest managers appear to be having second thoughts about such fires. Similarly, in some parts of the West the Forest Service is inviting the sheep back to help keep the grass around the pine seedlings in check. It is an experiment, an alternative to spraying the weeds with 2,4-D, which may not be safe for the environment, or hand cleaning with rake and hoe, which is costly. Pete Meyer of California's Plumas National Forest was pleased with the sheep's appetite for weeds, but not pines. See Jane Braxton, "Hungry Sheep Patrol Doing Its Part to Reforest Sierra," *Reno Gazette-Journal*, 5 October 1986, 3A. Similar experiments with brush-devouring goats in Laguna, California, proved very effective, but the laid-off workers sued the city government. In the spring of 1999 Carson City, Nevada, hired a sheepherder and his sheep to clear the underbrush from an area close to the population.

44. Conversation with A. M., 1993.

45. Personal communication, Tomas Zabala, Winnemucca, Nevada, 1988. However, most of the many stone piles found around Gerlach, Nevada, were not built by sheepherders.

46. William A. Douglass, "Lonely Lives under the Big Sky," *Natural History* 82, no. 3 (1973): 28–39.

47. José Miguel de Barandiaran, *Obras completas*, tomo V, Ikuska 3 (Bilbao: Editorial La Gran Enciclopedia Vasca, 1974), 381–85.

48. P. B. Villareal de Berriz, *Máquinas hidraúlicas de molinos y herrerías y gobierno de los montes y árboles de Vizcaya* (San Sebastián: Sociedad Guipuzcoana de Publicaciones, 1973), 123; see also Fermin Leizaola, *Euskalerriko artzaiak* (Donostia: Etor, 1977), 30.

49. The fishermen on the coast of Newfoundland in eastern Canada still build cairn-balises that look very much like the *harri mutilak*; see Azkarate Hernandez and J. Nuñez, *Balleneros vascos* (Gasteiz: Eusko Jaurlaritza, 1992), 173.

Chapter Six. Conclusions

1. As one condescending ex-herder reminded me, printed history is not true history because true history can be and usually is offensive to a number of people. Publishers do not want to offend people, especially those in power, so books are censored before they are printed. He continued: "When two countries are at war, you cannot believe every-

thing either one says or writes about the war. Believe 50 percent of what they say and then you may have a better picture of what really happened" (conversation with J. L., Elko, Nevada, 1990).

2. I have conducted little research in this area, but several people who took part in the 1997 and 1998 PIT projects organized by John Kaiser, archaeologist of Fremont National Forest, Lakeview, Oregon, told me they found relatively few carvings by the Irish herders, who were the majority until the 1930s.

3. This explains why so many Basques are gardeners, even in Nevada, where growers face a short season and unpredictable weather. They grow the vegetables popular in the Old Country. They plant fruit trees and brag about their peppers, leeks, and garlic as much as they once did about their lambs and sheepdogs.

4. In Bilbao, where fifty years ago use of Euskara was a rarity among young people, today you can ask any youngster a simple question in Basque and be understood. Most are capable of responding in that language.

5. Published in Paris by the Libraire Armand Colin, 1929–1938, 1939–42, 1945.

6. Those who resist using orally transmitted tales as a source of history should remember that the tales must conform to a set of time-honored rules no less rigid than the academic canons. Oral stories are tightly knit, well organized, and always deliver a punch line. Overall, I find oral stories commensurably more interesting than printed books.

7. According to Jared Diamond, New Guineans' brains are more active than and superior to brains of Westerners because their lifestyles require them to remember more; see *Guns, Germs, and Steel* (New York: W. W. Norton, 1997), introduction. Diamond was awarded the Pulitzer Prize for this work. I mention this not to prove the superiority of the Basque brain, but to illustrate the "active" nature of information in an oral culture.

8. "Culture. Kultura: Mugica or Muxika?" *Euskal Etxeak*, no. 39 (1998): 24–25.

9. Ibid.

10. For a fuller description of *auzolan*, see Luis Michelena, *Diccionario general Vasco-Orotariko Euskal Hiztegia*, vol. 3: *Ase-Bapuru* (Bilbao: Euskaltzaindia, 1989), 499–500. In rural Euskal Herria the individual was valued, but so was the *baserri*, life's center. In fact, some say that the *baserri* has more rights than the people who live in it. People come and go, but the building remains. All of which explains the many farmsteads carved on aspens.

11. This same observation was made to me by William Smallwood of Bhuel, Idaho, 1998. Smallwood, a non-Basque, is familiar with Idaho Basques and their culture and has written a novel that takes place in the Basque Country.

12. Russell, *History of Nevada*, 378, passim.

13. See Wilbur Shepperson, *Restless Strangers: Nevada Immigrants and Their Interpreters* (Reno: University of Nevada Press, 1970).

14. Today no one would deny that the Basques in California are assimilated into American culture. However, even third-generation Basques still speak some Euskara.

15. I think these pressures on immigrants of all kinds and ages, and the resultant changes through adaptation, can be viewed as an integral part of Frederick Jackson Turner's theory that the western frontier was a deciding agent and factor in U.S. history. He saw the frontier expanding under the push of easterners moving west; a lot of them were foreigners. Turner first enunciated the theory in 1893; see *Frontier in American History* (New York, 1920). But this "frontier" can also be seen as a universal phenomenon associated with the clash of cultures, that is, the immigrant groups. See Wilbur R. Jacobs, *On Turner's Trail: 100 Years of Writing Western History* (Lawrence: University Press of Kansas, 1994), 145.

16. Languages are the oldest recorded form of history. Every time a language dies in the world, a great deal more than sounds and a way of speaking are lost.

17. Samuel Noah Kramer, quoted by Elinor W. Gadon, *The Once and Future Goddess: A Symbol for Our Time* (San Francisco: Harper and Row, 1989), 127–28.

18. Homestead Sheep Camp, located north of Gerlach, Nevada.

19. Luis Pedro Peña-Santiago, *Arte popular vasco*, 5th ed. (San Sebastián: Editorial Txertoa, 1985), 19.

20. I am often ambivalent about which slides of arborglyphs are appropriate and which are not. The publishers of tree-carving art may be faced with the same dilemma that confronted the *Biblical Archaeological Review* in July 1991, when they deliberated whether or not to print the sex scenes found in the ruins of the ancient city of Ashkelon. The readers were divided, and in the end the magazine decided to print them, but in a removable format. See George W. Cornell, "Bible Archaeology Buffs Confront Lewd Art," *Glenwood (Colo.) Post*, 19 July 1991, 11.

21. For example, in Paradise Valley, Nevada, sheepherders were barred from attending community dances, according to Howard W. Marshall, *Paradise Valley: The People and Buildings of an American Place* (Tucson: University of Arizona Press, 1995), 9; for additional information, see William A. Douglass and Jon Bilbao, *Amerikanuak: Basques in the New World* (Reno: University of Nevada Press, 1975), 274–77.

22. It is also documented that one out of three cowboys in the American West was either Mexican, Indian, or African American—see *The Old West: The Cowboys*, by the editors of Time-Life Books, text by William H. Forbis (Alexandria, Va.: Time-Life Books, 1973), 18—but it is not likely that you will see it in the movies or read about it in many western novels.

23. Eugenio Sarratea's statement, interview by Curt Daniels, KNPB–Channel 5, Reno, 1989.

24. Sixteenth-century documents from Mexico and other Spanish colonies inevitably use the term *Spaniards* or *Christians* whenever the colonials were confronted with a hostile army of Native Americans, but when the danger disappeared and life reverted to normalcy, the "Spaniards" regressed to their Old Country mode and were identified as Old Castilians, New Castilians, Basques, Granadans, Galicians, etc.

25. Personal communication from Penny Rucks, Lake Tahoe Basin Management Unit archaeologist, 1997.

26. Goñi, a native of Mugaire, Nafarroa, herded sheep in several states before settling down in Reno. Goicoechea, from Gorriti, Nafarroa, after a short stint as a herder, managed a sheep company for eight years. When he settled in town, unlike the great majority of other ex-sheepherders, who went into construction or landscaping, he became a car salesman in Wyoming.

27. Those interested in this ancient craft of the Basques can read Gorka Aulestia's *Improvisational Poetry from the Basque Country* (Reno: University of Nevada Press, 1995).

28. *Martin Goikoetxea Bertsolaria Ameriketan*, San Francisco, California, 1995.

29. Audiocassette in the author's possession.

Appendix I. List of Herders by Year(s) and by Geographical Location

1. There was a sheep boss by that name based in Rocklin, California.

2. He and Jose Azcarraga saved the Paiute cutthroat trout from becoming extinct by moving some specimens to a new location.

3. According to the Frances R. Wallace and Hans Reiss Papers at UNR's Special Collections, Elcano is dated 1901 and he was working somewhere in this area. I have not seen such a carving. But I did find a J. Elcano dated 1901 near Sierraville, California.

4. Several names and dates for the Spooner area were recorded and made available to me by Dr. Susan Lindstrom, archaeologist from Truckee, California.

5. One John Inchauspe, nineteen years old, herder, is listed in the 1900 U.S. Census, Elko.

6. When a name appears twice within a decade, as here, it occurred in more than one year during that period.

7. The area surrounding the lake itself is Indian land, which I did not research.

8. *Pastor* means "sheepherder," and it is sometimes used as a first name. There is a possiblity that the arborglyph means "Uriarte, sheepherder."

9. It took him three years to travel from the Caribbean to Table Mountain!

Appendix II. Sheepherders on Peavine Mountain, Washoe County, Nevada

1. The name is not questionable but the date is. I know he was on Peavine during the 1960s, and if 1932 is accurate, then he was an old herder.

2. He forgot to carve the last digit.

3. This is Joaquin Urreaga's nephew, who worked for John Espil of Gerlach, Nevada, for a number of years.

4. Ojer is rather uncommon today. In the late Middle Ages it was used as a first name as well; for example, one of the sailors who discovered America with Columbus in 1492 was Ojer Berastegi.

5. Rather than herding, Arla may have been just visiting Peavine.

6. Arrosa may be his birthplace rather than his surname.

7. Pete Salvage is probably a nickname.

8. The names of nonherders—campers, hikers, and the like—are not included in this list.

9. Information supplied by Goyenetch's nephew, Jerry Etcheberry of Winnemucca, Nevada, personal communication, 1998.

Appendix III. Names in Copper Basin, Elko County, Nevada

1. I want to thank Dan Urriola and Oyvind Frock of Reno, members of the AM-ARCS group, for originally compiling this list from the data gathered in the field. *Eskerrik asko* (Thank you).

2. During my first visit to Copper Basin I recorded earlier carvings by Kopentity dated 1908.

Appendix V. Methodology for Recording Arborglyphs

1. I am indebted to the library and the audiovisual center of Reno's Truckee Meadows Community College for allowing me to use their equipment.

2. Computer experts at the University of Nevada, Reno, provided invaluable assistance, particularly Stephen Foster, who also helped with scanners and related technical problems.

3. Of course, the resolution of a printout is usually lower than that of a video.

4. This, however, will not totally eliminate the functionality of still photos and color slides.

5. I am grateful to the University of Nevada Graduate School for providing the necessary hardware and software.

6. Do not use zoom—it drains the battery fast; move the camcorder instead.

GLOSSARY

Many Basque toponyms have two or more versions, a reflection of the fact that the Basque Country is a meeting place of different cultures and political entities, namely Basque, French, and Spanish. Euskal Herria is the traditional name by which Basques call their country, although today the term *Euskadi* is often substituted. Lapurdi, Benafarroa, and Zuberoa form part of the Pyrénées Atlantiques. Nafarroa, or Navarra, is autonomous. Araba, Gipuzkoa, and Bizkaia constitute the Autonomous Government of Euskadi. Most of the population and economic power of the Basque Country are in Euskadi. País Vasco and Pays Basque are the Spanish and French names, respectively, for Euskal Herria.

Traditionally, Euskal Herria consists of seven historic regions (non-native forms—French or Spanish—appear in parentheses): Lapurdi (Labourd), Zuberoa (Soule), Nafarroa (Navarra/Navarre, later subdivided further into Benafarroa/Basse Navarre), Araba (Álava), Gipuzkoa (Guipúzcoa), and Bizkaia (Vizcaya, or Biscay, for the English). Today in the Basque Country the old forms Vizcaya and Guipúzcoa are no longer used.

Since Euskal Herria lacks political sovereignty, Basques in the United States have traditionally been differentiated according to their nationality, that is, Spanish or French. If there was some confusion in other people's minds regarding their identity, the Basques themselves never doubted that they were a distinct ethnic group. The majority of those who immigrated to the United States spoke Euskara (the Basque language) as their primary tongue. Although in Nevada and the West the herders were known as Basques or Bascos, they identified themselves by the Old Country marker *Euskaldun*. The reader is reminded that in Basque the ending *a* vowel attached to nouns usually indicates the article "the" and is not part of the word. Thus, *Euskaldun* = speaker of the Basque language; *Euskalduna* = the speaker of the Basque language. *Eskualdunak* is the plural form of *Euskaldun*. Northern Basques say *Eskualdun*. Everyone who is not *Euskaldun* is *Erdaldun*, or a speaker of foreign language (but in the United States the non-Basques, especially Anglos, are referred to as *Amerikanuak*).

In this work I use Basque native forms of place-names, which are now officially sanctioned, but whenever necessary I will also provide their Spanish or French transcriptions.

Basque renderings of English place-names and terms are marked with an asterisk.

Amerikanu Basque who lives or used to live in America. Also a non-Basque in the United States.

Ardi Sheep (ewe in particular).

Arno/Ardo Wine. Bizkaians say *ardau*.

Artzain Sheepherder; pl. *artzainak*. Also carved *arzain* and *arsain*.

Auzolan Cooperative work at the village or hamlet level.

**Bakiarda* Backyard.

**Balamont* Battle Mountain.

Baserri Basque farmstead, in southern dialects.

Baserritar Dweller of a *baserri*, peasant (sometimes used to mean farmer); pl., *baserritarrak*.

Batua Unified modern Basque language.

**Ben* Bend.

Bertso Improvised verse or poetry.

Bertsolari Improviser/singer of verses.

Biba Hurrah, hail. Also carved *viva* (Spanish).

**Bokarrantxa* Monkey wrench.

Borda Basque farmstead, in northern dialects.

**Burnos* Burns.

Borreguero Spanish for "sheepherder," often misspelled.

Cabrón Bastard (Spanish). Basques say *kabroi*.

**Dokabale* Duck Valley.

**Estet* Stead.

Fok (Basque rendition of "fuck"; used similarly by Spanish herders).

**Fokerraun* Fuck around.

Harri mutil A sheepherders' monument or marker made of a pile of superimposed stones, found in highly visible spots throughout the American West.

Hegoalde Southern Basque Country, comprising the regions of Nafarroa, Bizkaia, Araba, and Gipuzkoa, in Spain.

Ikurrin Basque flag.

Iparralde Northern Basque Country, part of Pyrénees Atlantiques in France, comprising the regions of Lapurdi, Benafarroa, and Zuberoa.

**Irurika* Eureka.

**Izpikerije/izpikeridse* Spaghetti.

Jani Basque rendition of "honey."

**Jobritx* Jarbidge.

**Jolarrantxu* Holland Ranch.

**Kacepik* Castle Peak.

Kaioti. Bizkaian rendition of "coyote," pronounced *káioti*.

Kaka Shit, also carved *caca*.

Kanpo handi Main summer sheep camp.

Kanpo ttipi Temporary sheep camp.

**Karson Siri* Carson City.

**Konputerra* Computer.

Labe Oven (for bread baking).

**Labeloks* Lovelock.

Lauburu Four heads; Basque swastika, symbol of the sun.

Lona Lawn.

Makila Walking stick.

**Meresbill* Marysville.

**Minemuka/Binamuka* Winnemucca.

**Munlai* Moonlight.

Mus A Basque card game.

**Mustan* Mustang.

Navarro One from Nafarroa; pl. *Navarros*. Sometimes refers to non-Basque speakers.

Ostatu Basque boardinghouse.

**Pikapa* Pickup.

**Pikinije/pikinidse/pikinika* Picnic.

**Polipar* (Humboldt County) Folly Farm.

Puta Prostitute (in Basque and Spanish).

**Reino* Reno.

Sanfermin July holidays in Nafarroa in honor of San Fermin, its patron saint.

**Saniskibi/Saniskibil* Thanksgiving.

**Siper* Sheepherder.

**Sispaka* Six pack.

**Sotuerra* Software.

**Susunbill* Susanville.

Tranpa Sneaky.

**Trela* Trail.

**Troka* Truck.

Txamizu Sagebrush.

Txapel Typical Basque beret.

**Txepena/xepena* Shipping (the lambs).

**Txirte* Chick grass.

**Uintxila* Windshield.

Zato, xahaku, or *bota* Wine skin.

ANNOTATED BIBLIOGRAPHY

Literature on Tree Carvings

Atxaga, M. "Jose Mallea Olaetxe. Nevadako zuhaitzak euskaraz mintzo dira" [Jose Mallea Olaetxe. The trees in Nevada contain Basque words]. *Deia*, 5 July 1992, 4SD.

Basque Country U.S.A. Reno, Nev.: Basque Studies Program brochure, 1992.

"Basque Sheepherder Studies." *PIT Traveler: Passport in Time* 8, no. 1 (1997): 26.

"Basque Tree Carvings (Arboglyphs) from Jarbidge Area." *Chippings* (AM-ARCS of Nevada, Reno) (August 1998): 1.

Beesley, David, and Michael Claytor. "Adios, California: Basque Tree Art of the Northern Sierra Nevada." *Basque Studies Program Newsletter* (University of Nevada, Reno), no. 19 (1978): 3–5.

———. "Aspen Art and the Sheep Industry in Nevada and Adjoining Counties." *Nevada County Historical Society Bulletin* 33, no. 4 (1979): 25–31.

———. "The Basque and Their Carvings." *Sierra Heritage* (June 1982): 18–21.

Bosarge, Paul. "Carving the Basque Experience." *Reno Magazine* 2, no. 4 (1980): 18.

Brunvand, Jan Harold, and John Abramson. "Aspen Tree Doodlings in the Wasatch Mountains: A Preliminary Survey of Traditional Tree Carvings." In *Forms upon the Frontier*, ed. Austin Fife, Alta Fife, and Henry H. Glassie, 89–102, Monograph Series 16, no. 2. Logan: Utah University Press, 1969.

Burgess, Jan. "The Basque Sheepherders and Their Carvings: Why?" Term paper, University of Nevada, Reno. No date. 16 pp.

Cendagorta, Christine. "The Sheepherders." *Nevada: The Magazine of the Real West* (May–June 1981): 36–37.

Cockle, Richard. "Basque Sheepherders Etched Loneliness on Trees." *Oregonian*, 19 December 1999. Reprinted, *Reno Gazette-Journal*, 9 January 2000.

Cortazar, Iban. "Historia demokratiko baten sustraiak." Aitzina. Herria. *Gara* (1999. irailaren 7a): 4–5.

Dekorne, James B. *Aspen Art in the New Mexico Highlands*. Santa Fe: Museum of New Mexico Press, 1970.

De Leon, Darcy. "Basque Carvings: Deep Feelings in Wood." *Reno Gazette-Journal*, 8 October 1990, 1, 14.

Douglass, William A. "Lonely Lives under the Big Sky." *Natural History* 82, no. 3 (1973): 28–39.

Earl, Philip I. "Basque Folk Art in Nevada: Aspen Tree Carvings Are Called 'Mountain Picassos.'" *Las Vegas Sun Sunday*, 14 November 1982.

———. "Carving the Basque Experience: The Etchings of Sheepherders Become Legitimate Culture." *Reno Magazine* 2, no. 4 (1980): 13.

———. *El Basco: Basque Aspen Art of the Sierra Nevada*. Reno: Earl Enterprises, n.d.

———. "History Takes Root through Nevada's Trees." *Reno Gazette-Journal*, 6 June 1999, 8B.

———. "Lonely Sheepherders Carved up Aspens." *Reno Gazette-Journal*, 19–25 October 1992, 3.

Etxarri, Joseba. "Emeki, emeki aurrera goaz." *Eguna. Katea Ez Da Eten*. II garaia, V urtea, no. 321 (22 November 1990): 1, 7.

———. *Euskaldunen Amerika: Bidaia bat EEBBetan zehar*. Donostia: Kronika-Elkar, 1994.

Ferrin, Lynn. "Table Mountain." *AAA Motor World* (March–April 1991): 20.

Fiscus, Chris. "Flagstaff's Telltale Trees." *Phoenix Gazette*, 23 October 1994, G1, G4.

Flannery, Kit. "A Dissertation on Arboreal Anaglyphs or Tree Graffiti." *Holiday* 55, no. 4 (1974): 49, 66.

"La historia que crece en los árboles." *Euskal Etxeak* (Eusko Jaurlaritza: Lehendakaritzaren Idazkaritza), no. 21 (1992): 4–7.

"History Fans Talk It Over." *Reno Gazette-Journal*, 30 August 1992, 11C.

Holliday, J. S. "The Lonely Sheepherder." *American West* 1, no. 2 (1964): 2, 36–45.

Hopkins, A. D. "Saving the Mountain Picassos." *Nevadan*, 9 November 1980, 6J, 23J.

Horne, Yvonne Michie. "Treeffiti." *Westways* 70, no. 11 (1978): 28–29.

Hutchison, Abby. "Restoring Basque Culture in the Granite Chief Wilderness." *Sierra Sun* (Truckee, Calif.), 29 July 1999, 9A.

Kanblong, Ramuntxo. "Nevadako zuhaitzak euskaraz mintzo dira." *Herria*, no. 2164 (22 July 1992).

Lane, Richard. "Basque Tree Carvings." *Northeastern Nevada Historical Society Quarterly* 1, no. 3 (1971): 1–7.

"Lertxun Marrak. The Nevada Experience." *Take 5* (Channel 5, Reno) 7, no. 3 (1990).

Licensing News. California State Board of Forestry and Fire Protection Newsletter 18, no. 1 (1999).

[Light, Lee]. "Messenger Reporter Receives Award at Forest Service Kyburz Gala." *Mountain Messenger* (Downieville, Calif.), 22 May 1997, 1, 5.

Macias, Sandra. "Recording Basque History on Trees." *Nevada State Journal*, Nevada Arts Section, 13 January 1980, 7A.

Mallea-Olaetxe, J. "Basque Aspen Carvings." *Basque Studies Program Newsletter* (University of Nevada, Reno), no. 43 (April 1991): 3–6.

———. "Basque Aspen Carvings, Harri Mutil, and Bread Ovens." *Neon: Artcetera from the Nevada State Council on the Arts* (Carson City, Nev.) (spring–summer 1996).

———. "History that Grows on Trees: Basque Aspen Carving in Nevada." *Nevada Historical Society Quarterly* 35, no. 1 (1992): 21–39.

———. "Indioek ala bakeroek hil zituzten Nevadako hiru euskaldunak? Erahiltza eta Lertxun Marrak." *Argia*, no. 1337 zenbakia (5 May 1991): 22–25.

———. "Lertxun Marrak. Gure artzainen arte bakana Estatu Batuetan." *Elhuyar* (Donostia), no. 29 (1989): 26–28.

———. [Photograph of tree carving with caption.] Program of the Western Society for French History Nineteenth Annual Conference, Reno, Nevada, 6–9 November 1991.

———. "Secrets in the Aspens." *Chippings* (AM-ARCS of Nevada, Reno) (July 1991): 3–4.

———. "The Secrets of the Aspens." *23nd Annual Reno Basque Festival* (August 1990): 23–29.

———. "Whiskey Creek Sheep Camp on the National Register of Historic Places." *28th Annual Reno Zazpiak Bat Basque Festival* (19–20 August 1995): 21–23.

McGonagle, Roberta L. "Frank Rodriguez, Sheepherder and Artist." *Women in Natural Resources* 1, no. 1 (1990): 40–41.

"Mileposts. Talking Trees." *Arizona Highways* (June 1995): 46.

Mitchell, Roger. *Western Nevada Jeep Trails*. Glendale, Calif.: La Siesta Press.

Muldoon, Jack. "The Art of Boredom." *International Native Arts and Crafts*, no. 1 (1976): 15–18.

Osa "Sakona," Andres. "Artzantza. Euskal Artzaien Aztarnak Estatu Batuetan. Mallearen Erakusketa." *Deia Igandea*, 1 January 1995, 36.

Perales, Josu. "Zuhaitza, artzainaren bakardadearen lekuko." HABE *Euskara ikasten ari garenon Aldizkaria* no. 161 (15 October 1989): 18–19.

Pike, Deidre. "The Forests Bear Stories." *Silver and Blue* (University of Nevada, Reno) (March–April 1999): 22–23.

"Quaken Aspen Carries Tale of Josh Jones, Early Trapper." *Idaho Recorder Herald* (Salmon, Ida.), 1 June 1950. Report on the aspen tree found by George E. Shoup.

San Sebastian, Koldo. "Las *marcas* de los pastores vascos en el oeste americano. La historia que crece en los arboles." *Deia*, 21 June 1992, S. 55.

Slater, John. *El Morro: Inscription Rock*. Los Angeles: Plantin Press, 1961.

Smith, Carrie E. "Sheep Camps, Ovens, and Aspen Trees: Discovering the Basque on the East Side of the Tahoe National Forest." *Society for California Archaeology Newsletter* 27, no. 1 (1993): 11–13.

Sneed, David. "Mallea to Discuss Basque Aspen Tree Carvings." *Union* (Grass Valley–Nevada City, Calif.), 28 March 1996, B4.

———. "Researchers Dig into Stage Stop History." *Union* (Grass Valley–Nevada City, Calif.), 8 August 1992, 3. On Kyburz Flat stage stop, oven, etc.

Txostena 1993 Memoria. Euskal Arkeologia, Etnografia eta Kondaira Museoa. Bilbao, 1995.

Vogel, Ed. "Lifestyle. Doodlings." *Nevada Appeal* (Carson City), 29 December 1989, C1.

Wallace, Frances, and Hans Reiss. "Basquos." *Basque Studies Program Newsletter* (University of Nevada, Reno), no. 5 (1971): 3–5.

Other Works

Abel, Ernest L., and Barbara E. Buckley. *The Handwriting on the Wall: Toward a Sociology and Psychology of Graffiti*. Westport, Conn.: Greenwood Press, 1977.

Abercrombie, Thomas J. "Europe's First Family: The Basques." *National Geographic* 188, no. 5 (1995): 74–97.

Abram, David. *The Spell of the Sensuous: Perception and Language in a More-Than-Human World*. New York: Pantheon Books, 1996.

Agirre, Domingo. *Garoa*. 1912. Oñati: Arantzazuko Frantziskotar Argitaldaria, 1966.

Alpers, Paul. *The Singer of the Eglogues: A Study of Virgilian Pastoral*. Berkeley: University of California Press, 1979.

Amorrortu, Estibaliz. "Retention and Accommodation in the Basque of Elko, Nevada." Master's thesis, University of Nevada, Reno, 1994.

Anderson, Tim. "Deal Protects Carson-Area Land from Development." *Reno Gazette-Journal*, 19 December 1997, D1, D6.

Arrizabalaga, Marie-Pierre. "A Statistical Study of Basque Immigration into California, Nevada, Idaho, and Wyoming." Master's thesis, University of Nevada, Reno, 1986.

Artzaintza cultura pastoril. Bilbao: Euskal Arkeolojia, Etnografia eta Kondaira Museoa, 1988.

Aulestia, Gorka. *Basque-English Dictionary*. Reno: University of Nevada Press, 1989.

———. *Improvisational Poetry from the Basque Country*. Trans. Lisa Corcostegui and Linda White. Reno: University of Nevada Press, 1995.

Axular, Pedro. *Gero*. Bordelen: G. Milanges erregeren inprimazaillea baithan, 1643.

Azkarate, Agustin, Jose Antonio Hernandez, and Julio Nuñez. *Balleneros vascos. Amerika Eta Euskaldunak. America y los vascos*. Gasteiz: Eusko Jaurlaritza, 1992.

Baca, Albert A. "El Pastor." *New Mexico Magazine* (November 1982): 42–49.

Bakeless, John. *Daniel Boone*. Harrisburg, Pa.: Stackpole, 1939.

Baldrica, Michael, and Carrie Smith. "Basqueing in the Sun: Research and Interpretation of Basque Sites on the East Side of the Tahoe National Forest." Paper Presented to the Society for California Archaeology, 7–11 April 1993.

Bannon, J. F., ed. *Bolton and the Spanish Borderlands*. Norman: University of Oklahoma Press, 1964.

Barandiaran, Jose Miguel de. *El hombre primitivo en el País Vasco*. Buenos Aires: Ekin, 1953.

———. *Obras completas*. Tomo II: *Eusko-Folklore*. Bilbao: Editorial La Gran Enciclopedia Vasca, 1973.

———. *Obras completas*. Tomo V, Ikuska 3. Bilbao: Editorial La Gran Enciclopedia Vasca, 1974.

Barandiaran'dar J. M. *Euskalerri'ko Leen Gizona*. Donostia: Benat Idaztiak, 1934.

Barté, Barbara. "Archaeology Highlighted at State Park." *Sierra Sun*, 6 May 1993, 10.

"Basque Sheepherders: Breed Apart." *Country Home* (February 1992): 96–102, 115–16. Photographs by Perry Struse.

Basque Sheepherders of the American West. Photographs by Richard H. Lane. Text by William A. Douglass. Reno: University of Nevada Press, 1985.

Bassett, Carol Ann. "Spring Sheep Drive on Arizona Trails. Taking the High Road in Basque Tradition." *American West* 20, no. 3 (1983): 22–28.

Beal, Merrill D. *History of Idaho: Personal and Family History*. Vol. 3. New York: Lewis Historical Publishing, 1959.

Beal, Merrill D., and Merle W. Wells. *History of Idaho*. New York: Lewis Historical Publishing, 1951.

Beltran Paris: *Sheepman of the West*. As told to William A. Douglass. Reno: University of Nevada Press, 1979.

Berenguer, Magín. *Prehistoric Man and His Art*. Trans. Michael Heron. Park Ridge, N.J.: Noyes Press, 1973.

Bergon, Frank. *Shoshone Mike*. Reno: University of Nevada Press, 1994.

Bilbao, Iban, and Chantal Eguiluz. *Lista de vascos arribados al puerto de Nueva York, 1897–1902*. Vitoria: Diputación Foral de Alava, 1981.

Biztanleriaren inkesta iharduearen arabera, I-1988. Gasteiz / Vitoria: Euskal Estatistika Erakundea, 1988.

"The Bounding Basque Awarded Decision." *Reno Evening Gazette*, 5 July 1931, 10.

Bowlby, John, and others. *Loneliness: The Experience of Emotional and Social Isolation*. Cambridge: MIT Press, 1973.

Boyd, Bob. *Amerikanuak: Basques in the High Desert*. Bend, Ore.: High Desert Museum, 1995.

Boyd-Bowman, Peter. *Indice geobiográfico de pobladores españoles de America en el siglo XVI*. Vol. 1: 1493–1519. Bogotá: Instituto Caro y Cuervo, 1964.

Braxton, Jane. "Hungry Sheep Patrol Doing Its Part to Reforest Sierra." *Reno Gazette-Journal*, 5 October 1986, 3A.

Breuil, Henry, and Raymond Lantier. *The Men of the Old Stone Age*. Trans. B. B. Rafter. London: George G. Harrap, 1965.

Brown, W. S., and S. B. Show. *California Rural Land Use and Management: A History of the Use and Occupancy of Rural Lands in California*. USDA Forest Service, California Region, 1944.

Burgess, Jan. "The Basque Sheepherders and Their Carvings: Why?" Term paper, University of Nevada, Reno. N.d. 16 pp.

Caesar, Julius. *The Conquest of Gaul*. Trans. S. A. Handford. New York: Penguin Books, 1981.

Campbell, Craig. "The Basque-American Ethnic Area: Geographical Perspectives on Migration, Population, and Settlement." *Journal of Basque Studies* 6 (1985): 83–89.

Cardaberaz, Agustin. *Eusqueraren berri onac, eta ondo escribitceco, ondo iracurteco, ta ondo itzegiteco Erreglac*. Iruñea: A. Castilla, 1761.

Caro Baroja, Julio. *El mito del carácter nacional: meditaciones a contrapelo*. Madrid: Seminarios y Ediciones, 1970.

Cassell's Spanish Dictionary. London: Macmillan, 1982.

Castellano, Rafael. *El comportamiento sexual de los vascos*. San Sebastián: Editorial Txertoa, 1981.

———. *Erotismo vasco*. San Sebastián: L. Haranburu, 1977.

Cavalli-Sforza, L. Luca, Paolo Menozzi, and Alberto Piazza. *The History and Geography of Human Genes*. Princeton, N.J.: Princeton University Press, 1994.

Census of the United States Taken in the Year 1910. Vol. 3: Agricul-

ture, 1909 and 1910. Washington, D.C.: Government Printing Office, 1913.

Clark, Robert P. *The Basques: The Franco Years and Beyond*. Reno: University of Nevada Press, 1980.

Cleland, Robert Glass. *The Cattle on a Thousand Hills: Southern California, 1850–1880*. San Marino, Calif.: Huntington Library, 1964.

Code of Federal Regulations. Parks, Forests, and Public Property, 36, pts. 1–1991. Rev. 1 July 1992. Washington, D.C.: Government Printing Office, 1992.

Cole, Sally J. *Legacy on Stone: Rock Art of the Colorado Plateau and Four Corners Region*. Boulder, Colo.: Johnson Publishing, 1992.

Collins, Roger. *The Basques*. New York: Blackwell, 1986.

Conlin, Joseph R. *Bacon, Beans, and Galantines: Food and Foodways on the Western Mining Frontier*. Reno: University of Nevada Press, 1986.

Corcuera, Ignacio. "Vascos en los Estados Unidos." *Imagen Vasca* 23 (October 1989): 7–39.

Cornell, George W. "Bible Archaeology Buffs Confront Lewd Art." *Glenwood (Colo.) Post*, 19 July 1991, 11.

Cunningham, Arnie Lynn. "Analysis of the Interrelationship of Archeological Surface Assemblages: An Approach to Petroglyphs." Master's thesis, University of Nevada, Las Vegas, 1978.

DeByle, Norbert V., and Robert P. Winokur, eds. *Aspen: Ecology and Management in the Western United States*. Report RM-119, USDA Forest Service, Fort Collins, Colo., n.d.

Dechaud, Siege. "Basque Sheepherder." *Range Magazine* (Reno) (spring 1991): 16–19.

Destinos. [A Spanish Language Course.] New York: McGraw-Hill, 1992.

Diamond, Jared. *Guns, Germs, and Steel*. New York: W. W. Norton, 1997.

Douglass, William A., and Jon Bilbao. *Amerikanuak: Basques in the New World*. Reno: University of Nevada Press, 1975.

Dryer, Carolyn. "Basque'ing in Glory." *Glendale Star*, 14 January 1998, A10.

———. "Basques Plan Winter Gathering in Glendale." *Glendale Star*, 10 December 1998, A3, A11.

Dundes, Alan. *Life Is a Chicken Coop Ladder: A Portrait of German Culture through Folklore*. New York: Columbia University Press, 1984.

Earl, Phillip I. *This Was Nevada*. Introduction by Peter L. Bandurraga. Reno: Nevada Historical Society, 1986.

Echeverría, Jerónima. "Basque Tramp Herders on Forbidden Ground: Early Grazing Controversies in California's National Reserves." *Locus* 4, no. 1 (1991): 41–57.

———. "Californiako ostatuak: A History of California's Basque Hotels." Ph.D. diss., North Texas State University, 1988.

Egan, Ferol. *Frémont: Explorer for a Restless Nation*. Reno: University of Nevada Press, 1985.

Elliott, Russell R., with William D. Rowley. *History of Nevada*. 2d rev. ed. Lincoln: University of Nebraska Press, 1987.

Enciclopedia general ilustrada del País Vasco. Vol. 5. San Sebastián: Editorial Auñamendi, 1974.

Etxabe, Baltasar. *Discursos de la antiguedad de la lengua cántabra*. Mexico, 1607.

Etxeberri, Joanes. *Eliçara erabiltçeco liburua*. Bordelen, 1636.

Etxezarreta, Ramon. *Hiztegi erotikoa*. Gasteiz: Hordago, 1983.

"Euskal artzain baten . . . bizitza ta asmoak." *Agur*. Hilerokoa. Markina, Bizkaia, 1967. Translation of an article that appeared in the *Hanford (Calif.) Sentinel*, no author, date, or page number.

Euskaltzaindia. *Euskararen liburu zuria*. Bilbao: Euskaltzaindia, 1978.

Euskararen jarraipena. La continuidad del Euskera. La continuité de la langue basque. Vitoria-Gasteiz: Eusko Jaurlaritza-Gobierno Vasco, 1995.

A First Book of Tree Identification. New York: Random House, 1951.

Gadon, Elinor W. *The Once and Future Goddess: A Symbol for Our Time*. San Francisco: Harper and Row, 1989.

Gallop, Rodney. *A Book of the Basques*. Reno: University of Nevada Press, 1970.

Garat, Margaret Brenner. "Family Archives: Garat and Indart, Brenner and Eastham." Unpublished manuscript, Oakdale, Calif., 1989.

Garibay, Esteban. *Compendio historial*. 1571. Reprint, Lejona: Editorial GerardoUña, 1988.

Georgetta, Clel. *Golden Fleece in Nevada*. Reno: Venture Publishing, 1972.

Gibbs, Catherine. "Remembering the Basques: Sierra History Often Overlooks Shepherds." *Sierra Sun*, 6 May 1993, 10.

Gimbutas, Marija. *The Civilization of the Goddess: The World of Old Europe*. Ed. Joan Marler. San Francisco: Harper, 1990.

———. *The Gods and Goddesses of Old Europe, 7000 to 3500 BC: Myths, Legends and Cult Images*. Berkeley: University of California Press, 1974.

Gómez-Tabanera, José M. *La caza en la prehistoria. Asturias, Cantabria, Euskal-Herria*. Madrid: Colegio Universitario de Ediciones Istmo, 1980.

Graham, Lanier. *Goddesses in Art*. New York: Artabras, a Division of Abbeville Publishing Group, 1997.

Grant, Campbell. *Rock Art of the American Indian.* New York: Thomas Y. Crowell, 1967.

Grant, Michael C. "The Trembling Giant." *Discover* (October 1993): 82–89.

Gudde, Erwin G. *California Place Names: The Origin and Etymology of Current Geographical Names.* Berkeley: University of California Press, 1969.

Heizer, Robert F., and Martin A. Baumhoff. *Prehistoric Rock Art of Nevada and Eastern California.* Berkeley: University of California Press, 1962.

Herrero Garcia, Miguel. "Ideología española del siglo XVII. Concepto de los vascos." *Revista Internacional de Estudios Vascos* 18 (1927): 549–69.

Holliday, J. S. *The World Rushed In: The California Gold Rush Experience.* New York: Simon and Schuster, 1981.

Holme, Charles, ed. *Peasant Art in Austria and Hungary.* London, Paris, New York: The Studio, 1911.

Humboldt, Alexander von. *Political Essay on the Kingdom of New Spain.* 4 vols. Trans. John Black. London: Longman, Hurst, Rees, Orme, and Brown, 1822.

Hyde, Dayton O. *The Last Free Man: The True Story behind the Massacre of Shoshone Mike and His Band of Indians in 1911.* New York: Dial Press, 1973.

"An Interview with Frank Yparraguirre." Conducted by R. T. King, 22 May 1984. University of Nevada, Reno, Oral History Program, 1984.

Intxausti, Joseba. *Euskara, Euskaldunon Hizkuntza.* Vitoria-Gasteiz: Eusko Jaurlaritza, Hizkuntza Politikarako Idazkaritza Nagusia, 1990.

Isaacs, Jennifer, comp. and ed. *Australian Dreaming: 40,000 Years of Aboriginal History.* 1980. Reprint, Sydney: Ure Smith Press, 1992.

Jacob, James E. *Hills of Conflict: Basque Nationalism in France.* Reno: University of Nevada Press, 1994.

Jacobs, Wilbur R. *On Turner's Trail: 100 Years of Writing Western History.* Lawrence: University Press of Kansas, 1994.

James, E. O. *The Cult of the Mother-Goddess.* New York: Barnes and Noble, 1959.

Jones, Oakah L. *Nueva Vizcaya Heartland of the Spanish Frontier.* Albuquerque: University of New Mexico Press, 1988.

Kinsey, Alfred C. *Sex in the Human Female.* Philadelphia: W. B. Saunders, 1953.

Kurlansky, Mark. *The Basque History of the World.* New York: Walker, 1999.

———. *Cod: A Biography of the Fish that Changed the World.* New York: Penguin U.S.A., 1998.

Kyburz Flat. 2000 Years of History. USDA Forest Service brochure, 1994.

Lamm, Robert C. *The Humanities in Western Culture: A Search for Human Values.* 4th ed. Madison: Brown and Benchmark, 1996.

Land Status Map of Nevada. 2d ed. Nevada Bureau of Mines and Geology. University of Nevada, Reno, 1972. Compilation and Cartography by Ira A. Lutsey and Susan A. Nichols.

Lane, Richard Harris. "The Cultural Ecology of Sheep Nomadism: Northeastern Nevada, 1870–1972." Ph.D. diss., Yale University, 1974.

Larousse Concise Spanish-English, English-Spanish Dictionary. 1993.

The Last Shepherds. Text by David Outerbridge. Photographs by Julie Thayer. New York: Viking Press, 1979.

Laxalt, Robert. "Basque Sheepherders, Lonely Sentinels of the American West." *National Geographic* (June 1966): 870–88.

———. "Land of the Ancient Basques." Photographs by William Albert Allard. *National Geographic* 134 (August 1968): 240–77.

———. *Nevada: A History.* Reno: University of Nevada Press, 1991.

———. *Sweet Promised Land.* New York: Harper and Row, 1957. (Reprinted numerous times by the University of Nevada Press.)

Legarda, Anselmo de. *Lo "vizcaíno" en la literatura castellana.* San Sebastián: Biblioteca Vascongada de los Amigos del País, 1953.

Lehmann, V. W. *Forgotten Legions: Sheep in the Rio Grande Plain of Texas.* El Paso: Texas Western Press, University of Texas Press, 1969.

Leizaola, Fermin. *Euskalerriko artzaiak.* Donostia: Etor, 1977.

Liber Sancti Jacobi. Codex Calixtinus. Ed. A. Moralejo, C. Torres, and J. Feo. Santiago de Compostela, 1951.

Liberty (New York), 30 August 1941. Weekly magazine with cover photograph of smiling Basque sheepherder Aniceto Vergara with a lamb on his shoulders and his sheepherder's crook. Photograph by Preston Duncan.

Louis, Adrian C. *Wild Indians and Other Creatures.* Reno: University of Nevada Press, 1996.

Lozano, Claudio. *La educación en los siglos XIX y XX.* Madrid: Editorial Síntesis, 1994.

Mack, Effie Mona. *The Indian Massacre of 1911 at Little High Rock Canyon of Nevada.* Sparks, Nev.: Western Printing and Publishing, 1968.

Madariaga, Salvador. *Anglais, Français, Espagnols.* 12th ed. Paris: Gallimard, 1930.

Mallea-Olaetxe, J. "Ameriketako artzainak azkenetan." *Zeruko Argia.* Kultura Gaiak (3 January 1971).

———. "Mexicoko lehen euskaldun meatzariak." In El

origen de la comunidad vasca de México, ed. Koldo San Sebastián, 81–98. Getxo-Gernika: Harriluze, 1993.
Mallory, J. P. *In Search of the Indo-Europeans: Language, Archaelogy and Myth*. London: Thames and Hudson, 1989.
Marshall, Howard W. *Paradise Valley: The People and Buildings of an American Place*. Tucson: University of Arizona Press, 1995.
Martínez Salazar, Angel. *Diego de Borica y Retegui 1742–1800. Gobernador de California*. Vitoria-Gasteiz: Diputación Foral de Alava/Arabako Foru Aldundia, 1992.
Martínez Salazar, Angel, and Koldo San Sebastián. *Los vascos en México, estudio biográfico, histórico y bibliográfico*. Vitoria-Gasteiz: Eusko Jaurlaritza-Gobierno Vasco, 1992.
Martynov, Anatoly I. *The Ancient Art of Northern Asia*. Trans. Dimitri B. Shimkin. Urbana: University of Illinois Press, 1991.
Mathers, Michael. *Sheepherders: Men Alone*. Boston: Houghton Mifflin, 1975.
McKay, John P., Bennett D. Hill, and John Buckler. *A History of Western Society: From Absolutism to the Present*. Vol. 2. 5th ed. Boston: Houghton Mifflin, 1995.
Meaker, Reginald. *Nevada Desert Sheepman*. As told to Sessions S. Wheeler and Gerald Meaker. Sparks, Nev.: Western Printing and Publishing, 1981.
Mecham, Lloyd J. *Francisco Ibarra and Nueva Vizcaya*. Durham, N.C.: Duke University Press, 1927.
Michelena, Luis. *Diccionario general vasco-Orotariko euskal hiztegia*. Vol. 3: *Ase-Bapuru*. Bilbao: Euskaltzaindia, 1989.
———. *Textos arcaícos vascos*. Madrid: Minotauro, 1964.
Moguel, Juan Antonio de. *Peru abarka*. 1881. Reprint. Bilbao: La Gran Enciclopedia Vasca, 1981.
Molina, Tirso de. *La prudencia en la mujer*. Colección Austral No. 369. Madrid: Espasa-Calpe, 1964.
El Morro Trails. 18th ed. El Morro National Monument, New Mexico. Tucson: Southwest Parks and Monuments Association, 1987.
Motten, Clement G. *Mexican Silver and Enlightenment*. New York: Octagon Books, 1972.
Nanclares, Sonia. "El batua se inventó en América antes que aquí, asegura Mallea. Investiga las huellas de los vascos en EE UU." *Correo Español–El Pueblo Vasco* (Bilbao) (5 July 1992): Cultura, 59.
Nevada Range Sheep Owners Association. Papers in the Basque Studies Program Archive, University of Nevada, Reno.
1910 Nevada Census Index. Comp. Bryan Lee Dilts. Salt Lake City, Utah: Index Publishing, 1984.
1992 United States Census of Agriculture. Vol. 1, pt. 28: *Nevada*. Washington, D.C.: Government Printing Office, 1994.
Nordbladh, Jarl. "Some Problems concerning the Relation between Rock Art, Religion and Society." In *Acts of the International Symposium on Rock Art*, ed. Sverre Marstrander, 185–210. Oslo: Universitets-forlaget, 1972.
The Norton Anthology of World Masterpieces. Exp. ed. Gen. ed. Maynard Mack. Vol. 1. New York: W. W. Norton, 1995.
Noticias de Euskadi. Boletín de Información sobre la Comunidad Autónoma del País Vasco. No. 16 (Septiembre 1998): n.p.
The Old West: The Cowboys. Alexandria, Va.: Time-Life Books, 1978.
O'Neal, Bill. *Cattlemen vs. Sheepherders: Five Decades of Violence in the West, 1880–1920*. Austin, Tex.: Eakin Press, 1989.
Ormaechea, Nicolas "Orixe." *Euskaldunak*. Donostia: Auñemendi, 1976.
Patterson, Edna B., Louise A. Ulph, and Victor Goodwin. *Nevada's Northeast Frontier*. Sparks, Nev.: Western Printing and Publishing, 1969.
Paul, Virginia. *This Was Sheep Ranching: Yesterday and Today*. Seattle: Superior Publishing, 1976.
"Pedro Altube, Spanish Ranch Founder, Honored." *Elko Free Press*, 13 February 1960, 3.
Peña-Santiago, Luis Pedro. *Arte popular vasco*. 5th ed. San Sebastián: Editorial Txertoa, 1985.
Pepper, Elizabeth, and John Wilcock. *A Guide to Magical and Mystical Places: Europe and the British Isles*. New York: Harper and Row, 1977.
Pipkin, George C. *Pete Aguereberry, Death Valley Prospector and Gold Miner*. Trona, Calif.: Murchison Publications, 1971.
"Precious Paiute Trout Perseveres." *Reno Gazette-Journal*, 15 October 1992, 3D.
Quintana, Xabier. "Literatura euskerica de los siglos XVI y XVII." Conferencia en el Ateneo de Madrid, 27 January 1976.
Recondo, José Ma. *San Francisco de Javier. Vida y obra*. Madrid: Biblioteca de Autores Cristianos, 1988.
Reisner, Robert. *Graffiti: Two Thousand Years of Wall Writing*. New York: Cowles Book Company, 1971.
Reynold, Reginald. *Cleanliness and Godliness*. London: George Allen and Unwin, 1943.
Ritter, Dale W., and Eric W. Ritter. "Medicine Men and Spirit Animals in Rock Art in Western North America." In *Acts of the International Symposium on Rock Art*, ed. Sverre Marstrander, 97–125. Oslo: Universitets-forlaget, 1972.
Rowley, William D. *U.S. Forest Service Grazing and Rangelands*. College Station: Texas A & M University Press, 1985.
Salazar, Lope Garcia de. *Bienandanzas y Fortunas*. Códice del siglo XV. Primera impresión tel texto completo con prólogo, notas e índices por Angel Rodríguez Herrero. Intro-

ducción por el Excmo. Sr. Marqués de Arriluce de Ybarra. 2 vols. Bilbao, 1967.

Santa Teresa, Aita Prai Bartolome. *Euscal-errijetaco olgueeta, ta dantzeen neurrizco-gatz-ozpinduba*. Iruñea: Joaquin Domingo Nausijaren, eta Gaztiaren Liburuguillaan, 1816. Facsimile edition. Donostia: Hordago, 1978.

Satustregui, Jose M. *El comportamiento sexual de los vascos*. San Sebastián: Editorial Txertoa, 1981.

Seagel, Jeanne. "Erotic/Fragment: Ariane Lopez-Huici's Tactile Photographs." *Arts* 4, no. 31 (1990): 98–100.

Shadduck, Louise. *Andy Little, Idaho Sheep King: An Anecdotal Biography of a Memorable Idaho Sheep Man*. Caldwell, Ida.: Caxton Printers, 1990.

Shaffer, Mark. "Arizona's Aspens: White Trees Disappearing from Majestic Landscapes." *Arizona Republic* (Flagstaff), 1 August 1999, A1, A18.

———. "Trees Are Living History." *Arizona Republic* (Flagstaff), 1 August 1999, A19.

Shepherd, William. *Prairie Experiences in Handling Cattle and Sheep*. London: Chapman and Hall, 1884.

Shepperson, Wilbur S. *Restless Strangers*. Reno: University of Nevada Press, 1977.

Shover, John L. *First Majority–Last Minority: The Transforming of Rural Life in America*. De Kalb: Northern Illinois University Press, 1976.

Sinclair Drago, Henry. *Whispering Sage*. New York: A. L. Burt, 1922.

Snyder, James, James B. Murphy, and Robert W. Barrett. *Wilderness Historic Resources Survey. 1988 Season Report*. Studies in Yosemite History, no. 1. Yosemite Research Library, Yosemite National Park, California.

———. *Wilderness Historic Resources Survey. 1989 Season Report*. Studies in Yosemite History, no. 2. Yosemite Research Library, Yosemite National Park, California.

Steidlmayer, Heldy. "Basque Sheepherders Came, Stayed." *Colusa County Sun-Herald*, 27 September 1991, 1, 7.

Strutin, Michele. "The Basques: Lords of the Range." *Rocky Mountain Magazine*, 28–35.

Takaki, Ronald. *A Different Mirror: A History of Multicultural America*. Boston: Little, Brown, 1993.

Tippets, David. "Acid Test for New Perspectives." *Forestry Research West* (USDA Forest Service, Fort Collins, Colo.) (November 1991): 1–11.

Turner, Frederick Jackson. *The Frontier in American History*. New York: Henry Holt, 1921.

Turner, Victor. *Dramas, Fields, and Metaphors: Symbolic Action in Human Society*. Ithaca: Cornell University Press, 1974.

Twelfth Census of the United States Taken in the Year 1900. Vol. 5: *Agriculture*. Washington, D.C.: U.S. Census Office, 1902.

Twitchell, Ralph Emerson. *The Leading Facts of New Mexican History*. Vol. 1. Albuquerque: Horn and Wallace, 1963.

Urza, Carmelo. *Solitude: Art and Symbolism in the National Basque Monument*. Reno: University of Nevada Press, 1993.

Van Gennep, Arnold. *The Rite of Passage*. Trans. Monika B. Vicedom and Gabrielle L. Caffee. Chicago: University of Chicago Press, 1960.

Varley, James F. *Lola Montez: The California Adventures of Europe's Notorious Courtesan*. Spokane, Wash.: Arthur H. Clark, 1996.

Los Vascos en México y su colegio de las vizcaínas. (Various authors.) Mexico, D.F.: Instituto de Investigaciones Históricas, UNAM, 1987.

Villareal de Berriz, Pedro Bernardo. *Máquinas hidraúlicas de molinos y herrerías y govierno de los árboles y montes de Vizcaya*. 1736. Edición Facsimile. San Sebastián: Sociedad Guipuzcoana de Publicaciones, 1973.

Wall, Sawyer Byrd. *Nevada Nomads: A Story of the Sheep Industry*. San Jose, Calif.: Harlan-Young Press, 1971.

Weaver, Donald E., Jr. *Images on Stone: The Prehistoric Rock Art of the Colorado Plateau*. Flagstaff, Ariz.: Museum of Northern Arizona, 1984.

Weber, Eugen. *Peasants into Frenchmen: The Modernization of Rural France, 1870–1914*. Stanford: Stanford University Press, 1976.

Weiss, Robert S., and others. *Loneliness: The Experience of Emotional and Social Isolation*. Cambridge: MIT Press, 1973.

Werner, Louis. "Contrato con la soledad. Pastores latinoamericanos recorren el terreno aspero de Nevada." *Américas* 43, no. 2 (1991): 14–22.

Zalbide, Mikel. "Euskal eskola, asmo zahar bide berri." *Euskal eskola publikoaren lehen kongresua. Primer congreso de la escuela vasca*. Vol. 1: 211–71. Vitoria-Gasteiz: Eusko Jaurlaritza-Gobierno Vasco, 1990.

Zavala, Vicente. *Francisco de Ibarra*. 2 vols. Bilbao: Ediciones Mensajero, 1988.

Zuberogoitia, Aitor. "Elkarrizketa: Bittor Kapanaga." *Argia: Euskal Herriko Astekaria*. 1.685 (27 September 1998): 26–29.

Zulaika, Joseba. *Basque Violence: Metaphor and Sacrament*. Reno: University of Nevada Press, 1988.

Zumalde, Inaki. "Un pionero vasco en el oeste americano." *Aranzazu* 51 (1971): 22–23.

Videos

Basque Tree Carvings: Legacy in Nevada. Instructional Media Services, University of Nevada, Reno, 1992.

"Lertxun Marrak." Prod. Curt Daniels. On *The Nevada Experience*. KNPB–Channel 5, Reno, 1989.

Urrearen Irrika/La fiebre del oro. Iñaki Bizkarra, producer. Euskal Telebista (Basque Television), 1993.

Internet and Radio

"J.K. Mallearekin elkarrizketa Lertxumarrari buruz." *Euskadi Irratia.* Donostia. 18 August 1998.
"This Was Nevada: Basque Aspen Carvings." http://www.clan.lib.nv.us/docs/MUSEUMS/HISTthiswas/basque.htm

Posters

George Milpurrurru, *Kangaroo.* Ochres on Bark. Sydney: Art Gallery of New South Wales, 1988. Poster 00094.

INDEX

References to illustrations are indicated in italics